SEVEN IDEAS THAT SHOOK THE UNIVERSE

Second Edition

NATHAN SPIELBERG
BRYON D. ANDERSON
Kent State University

JOHN WILEY & SONS, INC.
New York Chichester Brisbane Toronto Singapore

ACQUISITION EDITOR Cliff Mills / Erica Liu
MARKETING MANAGER Catherine Faduska
PRODUCTION EDITOR Erin Singletary
COVER DESIGNER Steve Jenkins
MANUFACTURING MANAGER Susan Stetzer
PHOTO RESEARCHER Mary Ann Price
ILLUSTRATION Gene Aiello

This book was set in 10/12 Times Roman by University Graphics, Inc. and
printed and bound by Courier Stoughton. The cover was printed by Phoenix Color.

Recognizing the importance of preserving what has been written, it is a
policy of John Wiley & Sons, Inc. to have books of enduring value published
in the United States printed on acid-free paper, and we exert our best.
efforts to that end.

The paper on this book was manufactured by a mill whose forest management programs include
sustained yield harvesting of its timberlands. Sustained yield harvesting principles ensure that
the number of trees cut each year does not exceed the amount of new growth.

Library of Congress Cataloging-in-Publication Data
Spielberg, Nathan.
Seven ideas that shook the universe / Nathan Spielberg, Bryon D.
Anderson. — 2nd ed.
p. cm.
Includes bibliographical references.
ISBN 0-471-30606-1 (cloth : alk. paper)
1. Physics. 2. Mechanics. 3. Astronomy. I. Anderson, Bryon D.
II. Title.
QC21.2.S65 1995
530—dc20 94-32986
CIP

Printed in the United States of America

10 9 8 7 6 5

SEVEN IDEAS THAT SHOOK THE UNIVERSE

Preface to Revised Edition

We have made a number of changes in this edition, other than correcting errors, in the hope of enhancing its usefulness. These fall into four categories: (1) updated material in view of recent developments (particularly Chapters 6 and 8); (2) expanded discussion of the impact of developments in physics on philosophy, literature, and the arts; (3) expanded discussion of electricity and magnetism in Chapter 6; and (4) optional sections, which can be skipped without loss of continuity, giving specific examples or more detailed, occasionally more quantitative, discussions of the development and implications of the basic ideas. Although some of these new sections can no longer be described as ''nonmathematical,'' the material is nevertheless suitable for students having some facility with numbers and a high school grasp of elementary algebra.

These optional sections appear in Chapters 2 through 8, and include material on satellite orbits and the effect of telescopic observations on astronomical distance measurements in Chapter 2; on work, energy, and power in Chapter 3 and 4; on the general gas and temperature scales in Chapter 5; on electrical circuits, power, and properties of materials in Chapter 6; on the calculation of the factor $\gamma = 1/\sqrt{1 - v^2/c^2}$ for special relativity and of the Schwarzschild radius for black holes in chapter 6; on energy levels and QED in Chapter 7; on nuclear reactions and radioactivity in Chapter 8. For the benefit of students studying the more quantitative optional material, we have added some problems to the study questions at the end of most chapters. We have also added an epilogue chapter.

We would like to acknowledge the very useful comments made by Professors Robert W. Cunningham, Kent State University; David H. Wolfe, Lycoming College; Ronald A. Brown, SUNY at Oswego; Jack A. Soules, Cleveland State University; and Michael Braunstein, Central Washington University, all of whom reviewed the previous edition of this book; and other comments by Professor Myra West of Kent State University. We appreciate the help and cooperation of Cliff Mills, physics editor, Erica S. Liu, assistant physics editor, and Erin Singletary, production editor at John Wiley & Sons, and their very capable staff. We are particularly grateful to Alice B. Spielberg and Joan Anderson for their encouragement and their arduous efforts in the preparation of the manuscript.

N. Spielberg
Kent, Ohio 1994 **B. D. Anderson**

Preface

This book is intended as a text for general education or cultural courses in science. A preliminary version of it has been in use for the past few years at Kent State University.

It is generally acknowledged that although science is one of the distinguishing characteristics of twentieth-century civilization, the majority of college students are inadequately informed about our scientific heritage. They have heard of such "heroes" of science as Galileo, Newton, or Einstein, for example, but they have only the vaguest notions of the contributions of these individuals to science. It is our hope that this text will help students to understand the great achievements of these and other scientists and their often tremendous historical significance.

We have adopted a descriptive approach to several major physical concepts that have developed over the past few centuries, relating them to the historical and philosophical context in which they are developed. The emphasis is on the origin, meaning, significance, and limitations of these concepts, in terms of our understanding of the physical nature of the universe in which we live. Relatively little is presented about applied science or technology, except as side or illustrative remarks; nor is there much discussion about the detailed implications of scientific developments for the benefit or detriment of society. We are not concerned in any major way with justifying science as a means of regulating society or of arriving at political decisions, or the scientific aspects of such problems as pollution, or other topical subjects arising out of the impact of science and technology on society. The historical and philosophical contexts are used as aids in teaching, so that students may relate the subject matter of the book to their other studies, and so that they may see how and why the major concepts of physics developed. The primary emphasis is on the major concepts and their meaning and interrelations. We believe that our experience over the past 12 years in teaching this material to large numbers of students demonstrates that the major concepts of physics are inherently intellectually interesting and challenging, when presented as part of a search for knowledge.

The major concepts dealt with are the Copernican approach to astronomy, the Newtonian approach to mechanics, the idea of energy, the entropy concept, the theory of relativity, quantum theory, and conservation principles and symmetries. Our style and approach are inspired by the success of magazines such as *Scientific American,* which are aimed at the lay public and present scientific and technological concepts and ideas in a ''nonmathematical'' yet authentic manner. In preparing this material, as well as in our own course, we have focused on a verbal rather than a mathematical discussion of the various concepts and their development and implications. The nature of the subject matter, however, requires a quantitative treatment; we have therefore resorted to graphs and diagrams as necessary. Close attention is paid to simple definitions. There are even a few equations and formulas. None of these are derived; they are simply stated, and occasionally justified on the basis of simplicity or plausibility. Analogies are freely used as aids to understanding. Review questions and references for further reading are supplied at the end of each chapter. In short, this book is intended to help students appreciate physics in the same way that students appreciate music or art. It is not necessary to know how to play an instrument to enjoy music; so, too, it is not necessary to engage in mathematics to enjoy physics.

We cover the entire text in a three-semester hour credit course. More material is presented than can be discussed in detail in lectures, but carefully paced lectures give adequate coverage. A coherent course covering less material and spending more time on individual topics can be taught by omitting most of the material in Chapter 1, the Introduction, all of Chapter 5, which deals with entropy, and all of Chapter 8, which deals with conservation principles. In addition, Parts A, B, and D of Chapter 2 on Copernican astronomy may be skipped.

As in all books of this type, it is difficult to acknowledge all the colleagues, sources, and textbooks we have consulted in addition to those specifically mentioned as references at the end of the various chapters or in the figure captions. We are grateful to Drs. David W. Allender, Wilbert N. Hubin, and John W. Watson for critical reading and useful comments on early versions of individual chapters; to Professors Robert Resnick, Rensselaer Polytechnic Institute; Dewey Dykstra, Jr., Boise State University; Ronald A. Brown SUNY Oswego; Jack A. Soules, Cleveland State University; Nelson M. Duller, Texas A&M University; Thad J. Englert, University of Wyoming; and Eugen Merzbacher, University of North Carolina, for reviewing the preliminary version of the text, and to Mr. Robert McConnin and the editorial, design, and illustration staffs of John Wiley & Sons, Inc., for seeing this text through the entire production process. Joan Anderson carried out the task of typing the entire manuscript, and we are most grateful. We are grateful also to Alice B. Spielberg, Joan Anderson, and William J. Ambrogio for their help in preparing the preliminary version of the textbook and, of course, to our students for giving us cause to prepare this book. Needless to say, errors of omission and commission are our responsibility.

N. Spielberg
Kent, Ohio **B. D. Anderson**

Contents

''The most incomprehensible thing about the universe is that it is comprehensible.''
—Albert Einstein. Nebulosity in *Monoceros.* (Photo courtesy Mount Wilson and Palomar Observatories.)

Chapter 1

INTRODUCTION

Matter and Motion:
The Continuing Scientific Revolution

A. Goals of This Book

1. The World's Greatest and Most Successful "Revolutionaries"

There is a certain glamour surrounding revolutions and revolutionary ideas. Participants in a political revolution often believe that by discarding the established order, they are heroically casting off shackles and bravely achieving some new freedom. Scientific revolutions carry an intellectual glamour with them, and represent the overthrow of a particular way of understanding the physical world in which we live. The cast-off shackles somehow interfered with a proper perspective of the material world. Scientific revolutions, perhaps even more than political revolutions, have profound, long-lasting, often unexpected, effects on the way in which we view and deal with our surroundings. In that sense, revolutionary scientific ideas can be looked upon as shaking the intellectual universe.

This book offers a "nonmathematical" introduction to seven of the most important and revolutionary ideas of physics. The individual participants in the development and elucidation of scientific thought over the past 500 years may well be regarded as being among the world's greatest and most successful "revolutionaries." The revolutions they brought about did not take place overnight, and in many cases required decades, if not centuries, for their completion. Nevertheless, their work has led to new ideas and new conceptions of the world and the universe, and the place of humans therein. Their successes have profoundly influenced modes of thought and reinforced belief in reason and rationality as operative tools for understanding the universe.

Virtually all social and political philosophies embrace or react to specific scientific concepts such as energy, relativism–absolutism, order–disorder, determinism–uncertainty. Science, justifiably or not, is often used to validate or to discredit various aspects of religious thought. The "scientific method" is considered by many as the way to deal with almost all human problems. This book describes the development of those physical ideas that have contributed to the modern view of the universe and to general acceptance of the scientific method.

2. Physics, the Most Fundamental Science

The oldest and most developed of the sciences, and a model for all the others, is physics. Physics addresses the most fundamental questions regarding the nature of our physical universe. It asks such questions as: What is the origin of the universe? How has the physical universe evolved and where is it headed? What are the basic building blocks of matter? What are the fundamental forces of nature? Because physics is the study of these and other basic questions, it provides the underpinnings for all of the physical sciences, and indeed for all of the biological sciences as well. The ultimate (i.e., microscopic) descriptions of all physical systems are based on the laws of the physical universe, usually referred to as the "Laws of Physics."

Because physics is the most fundamental science, no scientific conclusion that contradicts physical principles can be accepted as ultimately valid. For example, some of the scientific attacks that initially greeted Darwin's ideas of biological evolution were based on the branch of physics known as thermodynamics. Early calculations of the length of time it would have taken for the Earth to cool from a molten state to its present temperature indicated that there was not enough time to allow the necessary evolutionary processes to take place. It is said that Darwin was quite perturbed when informed of this. Only with the recognition in the late nineteenth century that radioactivity could supply additional internal heat was it possible to show that the Earth could actually have cooled slowly enough to allow time for evolutionary processes to take place. Radioactive dating techniques, derived from nuclear physics, now give more support to this idea.

Indeed, it has been said that all science consists of two parts: physics and butterfly chasing. This obvious overstatement emphasizes two aspects of scientific endeavor—the collection and classification of descriptive material and the understanding of the reasons for the various phenomena in terms of fundamental concepts. All science, including physics, necessarily includes both of these aspects. In physics perhaps more than any other science, however, real progress is considered to be made only when the understanding of the phenomena has been achieved.

The great scientific concepts to be discussed in this book have been selected because of their fundamental nature, and because of their inherent intellectual appeal. These concepts will be presented in broad and introductory terms, in the hope that the reader will acquire some understanding of their central themes, how they developed, and their implications for our understanding of our universe. It is also hoped that the reader will learn something about the terminology of physics. Specific discussions of the "scientific method" and the philosophy of science will be rather cursory, even though such concerns are important for establishing the validity of scientific concepts. Moreover, in this book, there is relatively little discussion of the practical fruits or products of the scientific enterprise, even though it is these fruits—ranging from video games and hand calculators to the preservation and extension of life, and to the horrors of nuclear warfare—that motivate very intensive cultivation of physical science throughout the modern world.

B. Dominant Themes in Physics

Two dominant themes run through the development of physics. These are (1) matter and motion, and (2) the search for order and pattern.

1. Matter and Motion, and Their Interplay

Physics deals with matter and motion. Sometimes greater emphasis is placed on matter than on motion, and vice versa, but it is also important to study the interplay between these two concepts. Indeed, in the modern theories of relativity and quantum mechanics, the distinction between matter and motion is blurred. In its ultimate fine structure, matter is considered to be always in motion. Even in the circumstances where causative motion is considered to cease, matter is still in a state of motion, called *zero-point motion* or *zero-point energy.*

Physics does not properly concern itself with such ideas as "mind over matter," that is, the use of the mind or of a Mind to control matter and motions, but rather with the *perception* of matter and motion. Physicists sometimes concern themselves with the "true nature of reality," but such considerations are more often left to philosophers. There are limits to scientific "understanding," and one of the goals of this text is to indicate what some of these limits are, at least at this time in the development of science.

2. Search for Order, Pattern, and Simplicity in the Universe

As a branch of human knowledge, science is concerned with the classification and categorization of objects and phenomena. There is a constant search for relationships among the objects and phenomena, and a constant effort to depict these relationships by means of geometrical representations such as diagrams, graphs, "trees," flowcharts, and so on. Often symmetries and recurrences are observed in these relationships, so that attention becomes focused on the relationships as much as on the objects or phenomena that are the subjects of the relationships. There is always a desire to find the simplest and most universal general relationships. This naturally leads to models and theories that simplify, exemplify, and expand various relationships.

For example, it is sometimes said that an atom is like a miniature solar system, in which the nucleus plays the part of the Sun and electrons play the part of the planets. In accordance with the solar system model, one can introduce the possibility that the nucleus and the electrons "spin" in the same way as their counterparts in the solar system. Thus it becomes possible to explore the extent to which the same relationships develop in the atom as in the solar system. The solar system serves as a model of the atom, albeit a flawed model, and thus contributes to our understanding of the atom.

C. Continuing Evolution of Scientific Knowledge

1. The Tentative Nature of Scientific Knowledge

Scientific knowledge is based on experiments by humans, and is not given in one awe-inspiring moment. There will always be further experiments that can be carried out to increase the amount of scientific knowledge available. The cumulative result of these experiments may be the perception of new patterns in the store of scientific knowledge, or the recognition that patterns that had once been thought to be all-embracing were not so universal after all. Previously established notions may be overthrown or greatly modified as new or different patterns of phenomena are recognized. It is this ongoing turmoil and upset, revision and updating, of scientific conceptions that led to the development of the seven major ideas of physics to be discussed in this book. A brief review of these ideas will indicate their major themes.

2. The Seven Ideas

1. ***The Earth is NOT the Center of the Universe—Copernican Astronomy.*** For approximately 2000 years, roughly from the time of Aristotle until somewhat after Columbus's voyages to the Americas, it was believed that the Earth was at the center of the universe, both literally and figuratively, in reality and in divine conception. The first of the major scientific upheavals to be discussed involved the revival, establishment, and extension of a contrary notion: The Earth is but a minor planet among many planets orbiting the Sun, which itself is but one star rather remote from the center of a typical (among multitudes of others) galaxy in an enormous, almost infinite (perhaps) universe. This first scientific revolution was possibly the most traumatic, not only because of the reception that greeted it, both within and without the intellectual world, but also because it contained some of the seeds and motivation for other revolutionary ideas to come. Not only did Copernicus revise astronomy, but he also made use of ideas of relative motion and of simplicity of scientific theories.

2. ***The Universe is a Mechanism that Operates According to Well-Established Rules—Newtonian Physics.*** All objects in the universe are subject to the laws of physics. Isaac Newton, in his presentation of his laws of motion and his law of universal gravitation, succeeded in showing the very solid physical foundation underlying the ideas of Copernican astronomy. He, and those who built on his ideas, showed further that these laws and other similar laws underlie the operation of the entire physical universe, both as a whole and in the minutest detail. Furthermore, these laws are comprehensive and comprehensible. As will be apparent from the detailed discussion in Chapter 3, this second idea, which associates cause and effect (causality)—aside from its governance of both natural and artificial phenomena, processes, and devices—has rather far-reaching implications for the two contrary doctrines of determinism (predestination) and free will.

3. ***Energy Drives the Mechanism—the Energy Concept.*** Despite its all-encompassing nature, Newtonian physics, the science of mechanics, does not provide a completely satisfactory description of the universe. There is a need to know what keeps the wonderful mechanism running. The Ancients at one time considered that there were "Intelligences," or even a "Prime Mover," which kept everything going. The third idea says that it is "Energy" that keeps the universe going. Energy appears in various forms that can be converted into each other. The recurring energy crises that cause so much alarm actually represent shortages of some particular one of these forms of energy, and problems resulting from the need to convert energy from one form to another.

We can get some idea of what energy "really" is by comparing energy to money. Money is a means of exchange and interaction between humans. So, too, energy is often exchanged in the interactions among various objects in the universe. Just as there is usually a limit on the amount of money available, there is also a limit on the amount of energy available. This limit is expressed as a *conservation principle,* which governs the dispensation of energy in its various forms. (Often, the principle of conservation of energy is stated as "energy is neither created nor destroyed, but merely changes from one form to another.") As will be discussed in detail in the seventh idea (Chapter 8), there are other quantities besides energy that are exchanged in interactions between physical objects, and that also obey conservation principles.

4. *The Mechanism Runs in a Specific Direction—Entropy and Probability.* Although energy is convertible from one form to another, without any loss of energy, there are limitations on the degree of convertibility. A consequence of these limitations on the convertibility of energy from one form to another is the establishment of an *overall time sequential order* of past events in the universe. The limitations on convertibility are governed by the same rules that govern the throwing of dice in an "honest" gambling game—that is, the laws of statistics. Such a statement brings forth the possibility that the "revolution" inherent in the second idea (the idea of determinism) may be at least modified by a "counter-revolution." The results of these considerations indicate that heat, one of the possible forms of energy, must be recognized as a "degraded" form of energy. Such a conclusion is especially important to remember in times of serious energy problems.

5. *The Facts Are Relative, but the Law Is Absolute—Electromagnetism and Relativity.* The concepts of the theory of relativity have their roots in some of the arguments that arose during the development of Copernican astronomy. Albert Einstein is almost universally associated with the theory of relativity, but he did not originate the idea that much of what we observe depends on our point of view (or, more accurately, our frame of reference). Late nineteenth-century advances in the study of electricity, magnetism, and light led not only to great technological advances, but also to a reopening of the question of how to find the "absolute" or "true" values of various physical quantities from a moving frame of reference. Indeed, Einstein originally developed his theory in order to find those things that are *invariant* (absolute and unchanging) rather than relative. Nevertheless, starting with the revolutionary idea (at that time) that the speed of light must be invariant, regardless of the reference frame of the observer, he was able to show that many things that were thought to be invariant or absolute, such as space and time, are instead relative. Einstein found it necessary to reexamine the basic concepts of space and time, and he showed that these concepts are closely intertwined with each other. The close examination of basic concepts and assumptions, which at that time was occurring in both physics and mathematics, showed physicists that they could not completely disregard the questions and demands imposed by metaphysical considerations.

6. *You Can't Predict or Know Everything—Quantum Theory and the Limits of Causality.* This idea, which rejects the complete and meticulous determinism proclaimed as a result of Newtonian physics, rose innocently enough from the attempt to achieve a sharper picture of the submicroscopic structure of the atom. It was recognized at the beginning of the twentieth century that atoms were built up from electrons and nuclei, and efforts were made to obtain more precise information about the motions of the electrons. The sharper picture is not attainable, however, and it is necessary to consider very carefully just what can be known physically, and what is the true nature of reality. Although we cannot obtain an extremely *sharp* picture of the substructure of the atom, we do have an extremely *accurate* picture. It is necessary to use new ways to describe atoms and nuclei: Systems can exist only in certain ''quantum states,'' and measured quantities are understood only in terms of probabilities. This new picture makes it possible to have a detailed understanding of chemistry, and to produce such marvels as transistors, lasers, microwave ovens, radar communications, super-strong alloys, antibiotics, and so on.

7. *Fundamentally, Things Never Change—Conservation Principles and Symmetries.* This seventh revolutionary idea, which contradicts the common notion that all things must change, is still under development, and the full range of its implications is not yet clear. It does say, however, that some quantities are ''conserved''—that is, they remain constant or unchanging. Despite the limitations imposed by the sixth idea, physics continues to delve into the ultimate structure of matter. Immense amounts of energy are involved in this ultimate structure. It may be that the ultimate building blocks of nature are now determined, and include particles such as *quarks,* from which protons and neutrons are made. The question naturally arises as to what rules govern the ultimate structure, and what these rules reveal about the nature of the physical universe. As already mentioned, there are other quantities besides energy that affect interactions among the ultimate constituents of matter, in accord with specific rules or conservation principles. Other questions arise: Is it possible to separate matter and energy from the space–time manifold in which they are observed? What is the nature of space–time, and how does its ''shape'' and ''symmetry'' affect the various conservation principles that seem to be universally operative? For that matter, is there an ultimate structure or basic unifying principle at work in the physical universe? All the recent progress in determining the basic building blocks of matter has followed directly from the recognition of the intimate relationship that exists between conservation laws and symmetries in our physical universe.

3. The Symphonic Analogy

The history of science indicates that there is no reason to suppose that these seven revolutionary ideas are the only ones that will ever develop for physics. It is possible to compare physics to an unfinished symphony, in that it should well be expected that new ''movements'' will appear. These new movements will invoke recurring themes as well as new themes, and themes to be developed out of previous movements. For example, the search for symmetries that goes on in the most advanced areas of elementary particle physics is not very different from the search for perfection in the physical world that characterized Greek science.

Because of both recurring and new themes, and their interplay, physics can be looked on as "beautiful," in the same way that symphonies or works of art can be looked on as being intellectually "beautiful." Frequently it is the very simplicity of physical ideas that is considered to be beautiful.

4. Order and Presentation of the Seven Ideas

There are several ways in which a symphonic composition can be presented to an audience. Often only one or two movements are presented, or the score may be somewhat rearranged from its original order. In this book, the movements of our symphony, the seven ideas, will be presented more or less in the order in which they were developed over a period of 2500 or 3000 years. Although from the standpoint of logical and straightforward development from basic principles, it is not necessarily the best order, this historical order of presentation does lead to easier understanding of the ideas and better recognition of the context in which they developed. It also leads to deeper insight into the continuing evolution of the science of physics.

As it turns out, the historical order of presentation also coincides roughly with the decreasing order of size of objects studied; that is, the first objects studied were the stars and the planets, and the most recent objects are the subatomic and subnuclear particles. Together with the decreasing size of objects under study, paradoxically enough, is increasing strength of forces acting on these objects, from the very weak gravitational force at work in the universe at large to the exceedingly strong force that binds together *quarks* (constituents of nucleons and mesons). Paradoxically also, as increasingly basic and fundamental aspects of matter and energy are studied, the discussion becomes increasingly abstract.

D. PHYSICS WITHOUT MATHEMATICS

1. The Nonmathematical Reader

As stated at the beginning of this chapter, this book is primarily nonmathematical. The reader with little training or ability or inclination for mathematics should be able to follow the conceptual development. Yet physics is a very quantitative science and owes its success to the applicability of mathematics to its subject matter. The success and failure of physical theories depend on the extent to which they are supported by detailed mathematical calculations, and it is not possible to ignore mathematics in an intelligent discussion of physical concepts. It is possible, however, to minimize the formal use of mathematics in discussions of science for the lay public, as has been demonstrated repeatedly in such successful publications as *Scientific American* and *Endeavor*. The basic concepts and methods can be communicated quite successfully with a very minimal amount of mathematics. Ultimately, nature appears to follow a few amazingly simple rules.

These rules, the laws of physics, can generally be described well with little formal mathematics. It is in the discovery, extension, and application of these laws that considerable mathematical ability is required.

2. Minimization of Formal Mathematics

It is possible to avoid the use of much mathematics by presenting quantitative or mathematical relationships by means of graphs. A well-chosen graph, like a picture, is worth a thousand words. (It must be added that a well-chosen mathematical formula is worth a thousand graphs!) Nevertheless, there are occasions when it is necessary to present formulas, but these are not many. In any case, this book contains very few derivations, brief or lengthy, of formulas.

There are some optional sections, which can be skipped without loss of continuity, for the benefit of those readers desiring a more quantitative or detailed discussion of some topics. Some of these sections contain minimal mathematics.

Moreover, although the material in this book is essentially nonmathematical, it is occasionally necessary to discuss the actual magnitudes of physical quantities in a comprehensible way. The size of the quantities involved in scientific phenomena range from the almost unbelievably small to the almost unbelievably large; from a very small fraction of the diameter of an atomic nucleus to the distance of the furthermost stars. In normal experience we deal with dimensions in meters or feet or inches, but when discussing atoms we deal with dimensions in terms of Angstroms (.025 of a millionth of an inch) and when discussing astronomical distances we deal with dimensions in terms of light years (the distance light travels in one year). In order to make calculations or to compare quantities, it is often desirable to express all dimensions in meters. Thus the diameter of a hydrogen atom is .000000000106 meters, that is, a decimal point followed by nine zeroes and then 106. On the other hand, the number of hydrogen atoms contained within a millionth of a millionth of a gram (28 grams per ounce) is approximately 600000000000, that is, 6 followed by 11 zeroes.

It is much easier to grasp the significance of these quantities, and even to write them, if **scientific notation** is used to express them. Basically scientific notation writes quantities as the product of two numbers, one giving a numerical value to the digits and the other saying where to place the decimal point. Thus the diameter of a hydrogen atom (in meters) is written as the product of 1.06 and 10^{-10}. The second number, 10^{-10}, is simply 1 divided by 10 ten times. The superscript -10 is called the exponent of ten and the combination of the two numbers means that the decimal point in 1.06 should be moved 10 places to the left and the intervening places filled with zeroes. The number of hydrogen atoms in a millionth of a millionth of a gram of hydrogen is the product of 6.02 and 10^{11}. In this case the second number, 10^{11}, represents 1 multiplied by 10 eleven times. In this case the superscript 11 is the exponent of 10 and the combination of the two numbers means that the decimal point should be moved 11 places to the right, and the intervening blank spaces should be filled with zeroes.

Scientific notation makes it easy to multiply two numbers having quite different magnitudes. All that is done is to multiply the digital parts of the two numbers to get a new digital part and add the exponents to get a new exponent of 10. If, for example, the number of hydrogen atoms in a millionth of a millionth of a gram were arranged in a straight line, we can calculate the length of that line very easily. The length of the line is the number of hydrogen atoms multiplied by the diameter of the hydrogen atom, that is, (6.02×10^{11}) times (1.06×10^{-10}). Now 6.02 times 1.06 = 6.36, and −10 plus 11 equals 1. Therefore the length of the line is $6.36 \times 10^{1} = 63.6$ meters.

Similarly to divide two numbers, divide the two digital parts and subtract the exponents. Thus the result of 1.06×10^{-10} divided by 6×10^{11} is found by dividing 1.06 by 6 to obtain .18 and subtracting 11 from −10 to obtain −21. So the result is $.18 \times 10^{-21}$ or 1.8×10^{-22}.

3. The Symphonic Analogy—A Reprise

It is generally accepted that a truly cultured individual should have some appreciation of music. Such an individual need not be a musician or even a student of music, but it is necessary to have some familiarity with, and appreciation for, the themes, rhythms, and sounds of music. Such an appreciation and familiarity can be acquired without having to attempt a single "toot" on a wind instrument. So, too, a truly educated individual needs to have some familiarity with, and appreciation for, the themes, rhythms, and facts of science. It is not necessary for such an individual to carry out calculations, simple or difficult. Of course, one's appreciation for the fine points and overall grandeur of science is greatly enhanced if one knows something about how calculations are carried out, or even more so if one is a scientist; but so, too, is one's appreciation for good music heightened if one is a musician. Nevertheless, music is not, and should not be, limited only to people with instrumental ability, and similarly science is not, and should not be, limited only to those with mathematical ability.

4. Michael Faraday

It is possible, although very difficult, to become an outstanding scientist, even a physical scientist, with very limited mathematical abilities. Michael Faraday (1791–1867), a celebrated British scientist of the nineteenth century, was such a person. A self-educated orphan, he made many contributions to both chemistry and physics. He read a great deal, but his mathematical abilities were rather limited in comparison with those of his contemporaries. Thus he found analogies extremely useful. He originated the concept of *lines of force* to represent electric and magnetic fields, using an analogy with rubber bands. He is generally credited with using this analogy to develop the idea that light is an electromagnetic wave phenomenon. Maxwell, who subsequently developed the formal mathematical theory of the electromagnetic nature of light, acknowledged that without Faraday's physical intuition, Maxwell himself could not have developed his more rigorous theory.

In this book, therefore, reliance is placed on reading ability rather than on mathematical ability. (It should be understood, however, that nowadays practitioners of science cannot succeed without strong mathematical preparation.)

E. Science and Other Areas of Human Endeavor—Distinctions and Similarities

It is worthwhile to close this chapter with a comparison of science and other activities, and with a brief discussion of the *scientific method.*

1. Science and Culture

Physical science now plays a significant role in the literary and dramatic arts (including theater, motion pictures, television, and radio as well as written literature), largely superseding, in some cases subsuming, the role of magic (comparison of science with magic is discussed further below). Science fiction (''sci-fi''), a major component of popular culture, often has a wondrous mixture of science, technology, and magic, along with a rollicking adventure story. Among the earliest authors to systematically employ science in their writing were Jules Verne and Samuel Langhorne Clemens (Mark Twain). Verne described powered submarines capable of extended submersion some 25 years before they were built and space travel a century before it was accomplished. Clemens employed the idea of bidirectional time travel in *A Connecticut Yankee in King Arthur's Court,* and this same idea has been used often in comic strips, motion pictures, and television (''Alley Oop,'' *Back to the Future,* ''Dr. Who'').

In many of these works the authors have given free rein to their imaginations and included as yet unrealized, often physically impossible, advances in science. In some cases the authors have been remarkably prescient in their imaginations, particularly Jules Verne. The physically impossible idea of bidirectional time travel by means of some sort of time machine (which has taken the place of a ''magic potion'') has been explored at some length. Space travel and intergalactic missions featured in different media (''Dr. Who,'' *Star Wars,* ''Star Trek,'' *Back to the Future, A Wrinkle in Time*) have taken place at impossible speeds of multiples of the speed of light, or by some sort of ''twist'' of the fabric of space–time. A very large ancillary literature and clubs and conventions have developed around the television series ''Star Trek,'' with meticulously detailed manuals and technical descriptions of the scientifically and technologically advanced devices employed in the course of the story.

On a more serious level, the natural human desire to unify various areas of knowledge results in efforts to apply successful scientific concepts to other areas of thought, sometimes literally, but often by analogy. This has been particularly true for some of the concepts discussed in this book, as will be noted in subsequent chapters.

2. Science and National Policy

The twentieth century has seen an increasing awareness among national policy makers of the importance of science and science-based technology as elements of military, economic, and social power. National treasuries have been used to fund major research centers and to ensure an adequate cadre of scientifically competent professional personnel.

Among the earliest examples of this are the German Max Planck Institutes (originally called Kaiser Wilhelm Institutes) for research in various areas of science. Currently the United States supports major national laboratories and science and technology centers. Many other nations do likewise.

Since the end of World War II in 1945, many countries have mounted major efforts to achieve nuclear weapons capability or more benignly to develop and deploy nuclear power sources. The U.S. federal government has expended billions of dollars annually for more than 30 years on its space program. The Japanese government has strongly encouraged the development of scientifically based industries as engines of economic progress. Many countries and international consortia of countries have invested very heavily in research at the most fundamental levels in the belief that in the long run the fruits of this research can be given practical application. Because these efforts involve very substantial amounts of money, there are continuing debates about values and priorities. A typical case is that of the SSC, the Superconducting Super Collider, which was intended as a site for very complex experiments probing the ultimate limits of the structure of matter. There was a nationwide competition among 25 states of the Union to have the SSC site within their boundaries, because the project would represent a huge boost to local economies, with the creation of thousands of jobs both during construction and during actual operation of the facility. The projected budgets for this facility were so large as to possibly impinge on funding for other national needs, both scientific and nonscientific. As a result questions as to priority of need and prospective of "payoff" were raised. As of 1994, the completion of this project was deferred for budgetary reasons. Possibly it may be reactivated at a later time, perhaps as an international cooperative project, following the precedent set by the very successful European Center for Nuclear Research (CERN) in Geneva, Switzerland.

Sometimes ethical and moral questions are raised as to whether particular scientific projects or techniques should be (or should have been) pursued because they may have unfortunate consequences. Although these are perfectly valid questions, they do not have simple scientific answers because they involve values to which individuals and societies subscribe, rather than scientific principles.

3. Science and Technology

It is important to distinguish between science and technology. A dictionary definition of technology describes it as the "totality of the means employed to provide objects necessary for human sustenance and comfort." Included in these means is applied science. Science, on the other hand, is a systematized body of knowledge, the details of which can be deduced from a relatively small number of general principles by the application of rational thinking. One may think of toothpaste, television, integrated circuits, computers, medicines, automobiles, airplanes, weapons, clothing, and so on, and the devices and techniques for their production, as products or components of technology. The "products" or components of science, on the other hand, are Newton's laws of motion, the photoelectric effect, the invariance of the speed of light in vacuum, semiconductor physics, and so on, and the techniques for their study.

The distinctions between science and technology are not always sharp and clear. Lasers and their study, for example, can be regarded as belonging either to the domain of science or the domain of technology. Although the emphasis in science is on understanding, and in technology on application, understanding and application clearly enhance and facilitate each other.

4. Science and Magic

Science and magic are usually regarded as opposites, particularly because the knowledge of magic is necessarily limited to a few initiated practitioners and is at least partially irrational, whereas the knowledge and practice of science is available to all who have sufficient amounts of the human faculty of reasoning. Yet both applied science and magic are motivated by some similar desires, for example, the desire to understand, and to exert control over, nature and to bend it somehow to human will. There is always a desire to explain apparently magical or miraculous phenomena in terms of scientific principles. Some branches of science have been derived at least partially from magic, for example, chemistry from alchemy. There are even certain laws and principles in magic that have parallels in science, and which can be traced into science.

Voodoo practice, for example, makes use of a law of similarity. A witch doctor makes a doll to represent a person, and inserts needles into the doll, or smears it with filth, in order to cause pain or illness for the person represented by the doll. In science or technology, models, mathematical or actual, are constructed to represent an object or phenomenon under study, and "experiments" are done with the model to see what would happen to the object or phenomenon under certain conditions. Of course, it is understood that the model is not the actual object, but valuable information is obtained from its use.

5. Science and Religion

Both science and religion may be regarded as avenues to truth. In medieval Europe, at least, science was considered to be the "handmaiden" to theology, and many people still wrestle with the problem of integrating their scientific and religious beliefs. Often major new scientific discoveries or advances are hailed as supplying support for or against a particular theological or atheistic viewpoint.

In such discussions or concerns, it is important to bear in mind some necessary characteristics of science. Science and scientific conclusions are always tentative and subject to revisions when new evidence is discovered. Therefore, basing religious beliefs on scientific foundations supplies very shaky support for religion. The goal of science is to explain phenomena in terms that humans can understand, without invoking any divine inspiration or interference. Scientific knowledge must be verifiable or testable, preferably in quantitative terms. Miracles, on the other hand, cannot be explained scientifically; otherwise they would not be miracles. In general, miracles are not repeatable, whereas scientific phenomena are. The reason that science has been so successful in its endeavors is that it has limited itself to those phenomena or concepts that can be tested repeatedly, if desired. Accepting such limitations, however, means that science cannot serve as a guide to human behavior, morality, and will, other than to specify what physical actions are rationally possible. In addition, science cannot logically address the question of the ultimate reason for the existence of the universe at all, although it is astounding how much of the history and development of the universe have been deduced by science.

6. Science and Philosophy

Science in general, and physics in particular, was once considered to be a branch of philosophy, known as *natural philosophy.* Metaphysics is a branch of philosophy that deals with the nature of reality, particularly in a secular rather than a theological sense. Such questions as to whether there are unifying principles in nature, whether reality is material or ideal or dependent only on sensory perceptions, belong to the field of metaphysics. Even such concepts as space and time, which play a great role in formulations of physics, lie within the province of metaphysics. In this sense, philosophy has much to say to physics. (Nevertheless, individual philosophers, as philosophers, have not contributed much to physics.) The experimental results of physics have had, in turn, much to say to metaphysics.

The results of physics also have some influence (sometimes based on misinterpretation) on ethics, primarily through their influence on metaphysics. Proponents of the idea of moral relativism, for example, have borrowed viewpoints from the theory of relativity, much to the chagrin of Einstein.

7. Scientific Method

Almost every textbook, particularly at an introductory or survey level, dealing with physical, natural, or social science, contains a description of the "scientific method," which is supposed to be *the* valid technique for discovering or verifying scientific information or theories, and for solving all sorts of scientific and human problems. The prescription for the scientific method may be summarized as follows: (1) Obtain the facts or data. (2) Analyze the facts and data in light of known and applicable principles. (3) Form hypotheses that will explain the facts. The hypotheses must be consistent with established principles as much as possible. (4) Use the hypotheses to predict additional facts or consequences, which may be used to further test the hypotheses by enlarging the store of facts in step (1). The sequence of steps is repeated systematically as often as necessary to establish a well-verified hypothesis. The scientific method is open-ended and self-correcting, in that hypotheses are continually subject to modification in the light of newly discovered facts. If the hypotheses are correct in every detail, then no modification is necessary.

Although this is the model procedure for scientific investigation, it is also true that even in physics, perhaps the most successful science, practitioners of the "art" of scientific investigation may not actually follow such a procedure. Often hunch or intuition plays a significant role. It is very important to be able to ask the right questions at the right time (i.e., the significant questions) as well as to be able to find the answers to the questions. Sometimes it is necessary to ignore alleged facts, either because they are not really facts or because they are irrelevant or inconsistent (sometimes with a preconceived notion) or are masking other more important facts or are complicating a situation.[1] Often dumb luck, sometimes called serendipity, plays a role either in revealing a key piece of information, or in revealing a particularly simple solution. The discoveries of the magnetic effects of electric currents and of x-rays came about in just this manner.

Ultimately, however, despite the vagaries of *private science,* as practiced by individual scientists, the demands of the scientific method, as enumerated above, must be satisfied. A scientific theory that does not agree with experiment must eventually either be modified satisfactorily or be discarded. So, too, every theory, however inspired its creator, must satisfy the prescribed requirements of the scientific method.

6. Enjoying the Symphony

A thorough study of science yields power, both in a large sense and in a very real, detailed, and practical sense. The resulting power is so great that there is concern that it not be used for self-destruction, rather than for the blessings it can bestow. Science also gives a deep insight into, and understanding of, the workings of nature. There is marvelous symmetry and rationality in the physical universe. The material to be presented in the succeeding chapters of this book, unlike the material in most physics textbooks, will not reveal very much about the power of physics; rather it is hoped that from this material the reader will get an idea of the beauty, simplicity, harmony, and grandeur of some of the basic laws that

[1]It is said that Einstein was once asked what he would have done if the famous Michelson-Morley experiment had not established the invariance of the speed of light, as required for his theory of relativity. His reply was that he would have disregarded such an experimental result, because he had already concluded that the speed of light must be regarded as invariant.

govern the universe. It is hoped that the reader's imagination will be both intrigued and satisfied. It is time to end the overture and to enjoy the first movement of the scientific symphony.

STUDY QUESTIONS

1. On what basis did many nineteenth-century scientists criticize Darwin's ideas of biological evolution?
2. What is implied by the statement that all science consists of two parts: physics and butterfly chasing?
3. Name two dominant themes that run through the development of physics.
4. Why is scientific knowledge tentative?
5. What is meant by saying that a quantity is conserved?
6. What does gambling have in common with physics?
7. What basic concepts did Einstein reexamine?
8. What is meant by determinism?
9. What is a quark?
10. Who was Michael Faraday?
11. What are the differences between (a) science and technology and (b) science and magic?
12. What are the elements of the scientific method?
13. Are there limitations on science?

REFERENCES

Charles C. Gillispie, *The Edge of Objectivity,* Princeton University Press, Princeton, N.J., 1960.
A standard book on the history and philosophy of science.

Gerald Holton and Stephen G. Brush, *Introduction to Concepts and Theories in Physical Science,* 2nd ed., Addison-Wesley, Reading, Mass., 1973.
A more mathematical textbook than this one, but with extensive discussions on the nature of scientific thought, etc., and with extensive bibliographies. Chapters 3, 12, 13, and 14 are pertinent to this chapter.

Stanley L. Jaki, *The Relevance of Physics,* University of Chicago Press, Chicago, 1966.
History and philosophy of science, and the effect of science on other areas of thought.

Thomas S. Kuhn, *The Structure of Scientific Revolutions,* 2nd ed., University of Chicago Press, Chicago, 1970.

This book has had a great impact on the way in which people think about "scientific revolutions."

Cecil J. Schneer, *The Evolution of Physical Science,* Grove Press, New York, 1964 (a reprint of *The Search for Order,* Harper & Row, New York, 1960).
An interesting historical perspective of the development of physical science. Especially good discussions of the early Greek contributions.

Nicolaus Copernicus. (Photo courtesy Harvard College Observatory.)

Chapter 2

COPERNICAN ASTRONOMY

The Earth is not the center of the universe

Almost everyone believes that the Earth is a planet that travels in a nearly circular path about the Sun, and that the Moon travels in a similar path about the Earth. In fact, in the present "space age" we look on the Moon as a natural satellite of the Earth, a satellite that in its motion is no different from the large number of artificial satellites that have been launched from the Earth since 1958. Travel to the Moon is a demonstrated fact, and space travel to and from distant planets—even galaxies—is commonplace in the popular imagination and on the motion picture and television screen. Yet 400 or 500 years ago, anyone who dared express such ideas was considered irrational, or even heretical. Indeed, the evidence of our five senses tells us that it is not the Earth that is moving, but rather the Sun and all the objects in the sky that travel in circular paths about the stationary Earth.

Every 24 hours the Sun, Moon, planets, and stars "rise" in the east, traverse the heavens, "set" in the west, and disappear from view for several hours before reappearing again. Until just a few centuries ago, the entire body of knowledge available to Western civilization seemed to point to the view that the Earth is both stationary and centrally located in the universe. This chapter is devoted to a review of the development of this *geocentric* (Earth-centered) model of the universe, and its overthrow and replacement by our modern concepts. Such a review reveals much about the development and nature of scientific thought.

A. Early Scientific Stirrings in the Mediterranean Area

1. Origins of Western Science in Eastern Mediterranean Cultures

The origins of Western science are difficult to trace. For now, it is sufficient to mention a few sources that seem to be generally recognized. Perhaps foremost among these was the Greek penchant for abstraction and generalization. The Greeks, however, were greatly influenced by their commercial and military contacts with the Mesopotamian and Egyptian civilizations. The early Indian (Hindu) civilizations also speculated on the nature of the physical universe, but the extent of their interactions and influence on the early Greeks is not clear. These cultures had accumulated large bodies of extremely accurate astronomical data, and had developed mathematical techniques to employ these data in commerce, surveying, civil engineering, navigation, and the determination of civil and religious calendars. With a knowledge of the calendar, it is possible to determine when to plant crops, the favorable seasons for commerce and war, and when various festivals and rites should be held. But the calendar itself is determined by the configuration of the heavenly bodies. For example, in the northern hemisphere, the Sun is high and northerly in the summer, low and southerly in the winter. Ancient cultures (and some not so ancient) found great portents for the future in various "signs" or omens seen in the heavens. Sailors on the sea, when out of sight of land, are able to navigate by measurements of the location of various heavenly objects.

The ancient civilizations did not make the sharp distinctions between secular and religious affairs that are common in modern society, so they naturally found connection between all aspects of human endeavor and knowledge, mythology and religion, and astronomy, astrology, and cosmology. For Western civilization, the development of ethical monotheism by the ancient Israelites, coupled with the Greek search for a rational basis for human behavior, probably helped motivate philosophers to attempt to unify all branches of knowledge.

The Greek philosophers—Socrates, Plato, and Aristotle—stressed that civilizations and nations needed to be governed wisely and according to the highest moral principles. This required an understanding and knowledge of the Good. A necessary prerequisite for understanding of the Good was an understanding of science—arithmetic, geometry, astronomy, and solid geometry. It was not enough to be skilled in these subjects, however, but rather it was also necessary to understand the essential nature of them, as related to the Good. This understanding could only be achieved after long and arduous study. Plato, for example, felt that one could not comprehend the essential nature of such subjects without a thorough mastery of their details and their uses.

2. Greek Science and Numerology

According to tradition, the recognition of mathematics as a subject worthy of study in its own right, regardless of its utility for practical matters, was achieved some 2600 years ago by Thales, who lived on the Asiatic coast of the Aegean Sea. Some time later the followers of Pythagoras (who lived in one of the Greek colonies that had been established in Italy) proclaimed that the entire universe was governed by numbers. The numbers they referred to were integers (whole numbers) or ratios of integers. They considered that all things were built up from individual blocks, which they called atoms. Because atoms are distinct units, they are countable. This meant to the Pythagoreans that geometry could be looked on as a branch of arithmetic.

It was soon realized, however, that there are numbers, such as π and $\sqrt{2}$, that could not be expressed as the ratio of integers. These numbers were therefore called irrational numbers. This caused quite a problem because it meant that many triangles could not be built up of atoms. An isosceles right triangle, for example, might have an integer number of atoms along each of its sides. Its hypotenuse would have $\sqrt{2}$ as many atoms, but this is impossible because $\sqrt{2}$ is irrational and any integer multiplied by $\sqrt{2}$ is also irrational and therefore cannot represent a whole number of atoms. According to legend, the Pythagoreans found the existence of irrational numbers rather disturbing and attempted to suppress the knowledge of their existence.[1]

[1]If $\sqrt{2}$ were exactly equal to 1.4, then $\sqrt{2}$ would equal the ratio of two whole numbers, 14 and 10, that is 14/10. If so, then an isosceles right triangle with 10 atoms on two sides would have exactly 14 atoms on its hypotenuse. But actually $\sqrt{2}$ is slightly greater than 14/10, and so the triangle should have more than 14 (but fewer than 15) atoms on the hypotenuse. It is impossible to find two whole numbers, the ratio of which is exactly equal to $\sqrt{2}$. Of course, this problem disappears if the assumption that atoms are the basis of geometry is discarded. This may be one reason that the Greeks never developed the concept of atoms significantly. The concept lay essentially dormant and undeveloped for about 2000 years.

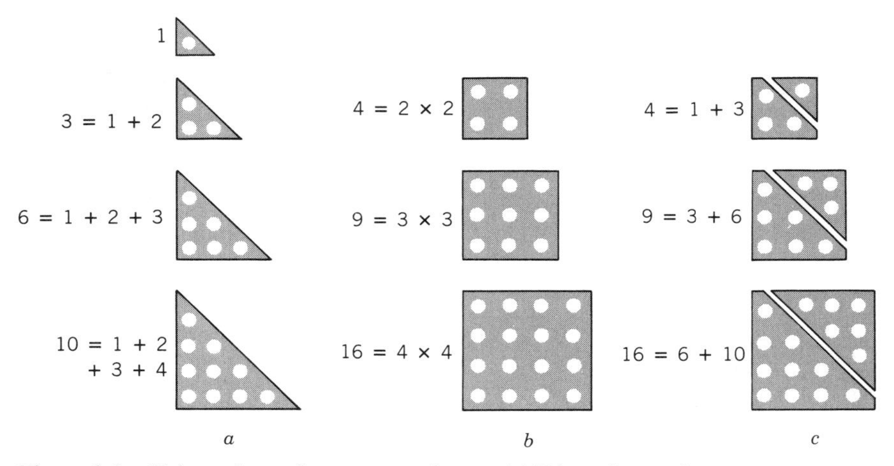

Figure 2.1. Triangular and square numbers. (a) Triangular numbers as arrangements of individual elements. (b) Square numbers as arrangements of individual elements. (c) Square numbers as combinations of two successive triangular numbers in which the first triangle is turned over and fitted against the second triangle.

Nevertheless, it was possible to establish connections between arithmetic and geometry through such formulas as the Pythagorean theorem relating the three sides of a right triangle ($A^2 + B^2 = C^2$), or the relationship between the circumference and radius of a circle ($C = 2\pi r$), or the relationship between area and radius of a circle ($A = \pi r^2$), and so on. It was found that some numbers could be arranged in geometric patterns, and relationships were discovered between these patterns. For example, ''triangular'' numbers are shown in Fig. 2.1a and ''square'' numbers in Fig. 2.1b. The combination of any two successive triangular numbers is a square number, as shown in Fig. 2.1c. The Pythagoreans were said to believe so strongly in the significance of mathematics that they established a religious cult based on numbers.

The assignment of significance to numbers was not, and is not, limited to the Greeks. According to the Jewish Cabala, the deeper meaning of some words can be found by adding up the numbers associated with their constituent letters. The resultant sum would be significant of a particularly subtle connotation of the word. Nowadays, there are lucky and unlucky numbers, such as 7 or 13. Card games and gambling games are based on numbers. Some buildings do not have a thirteenth floor, even though they may have a twelfth and a fourteenth floor. In nuclear physics, reference is made to *magic numbers,* in that nuclei having certain numbers of protons or neutrons are particularly stable, and their numbers are deemed *magic.* In atomic physics, at one time some effort was expended trying to find significance in the fact that the so-called *fine structure constant* seemed to have a value exactly equal to 1/137.

The ancient Greeks were also fascinated by the shapes of various regular figures and developed a hierarchy for ranking these shapes. For example, a square

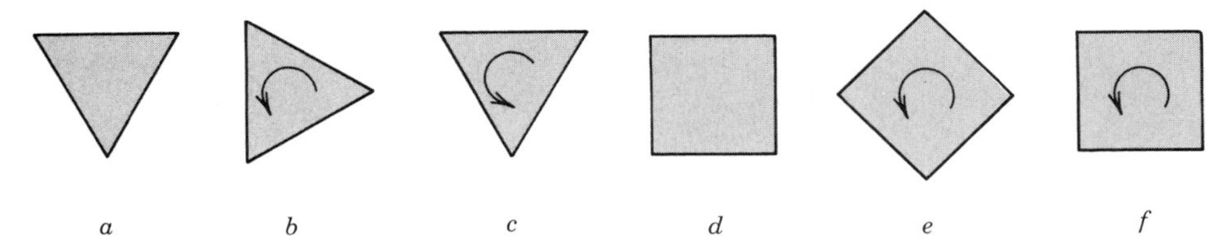

Figure 2.2. Geometrical shapes and symmetry. (a) Equilateral triangle. (b) The triangle rotated by 90°. (c) Rotation of the triangle by 120° from its original position to a new position, which is indistinguishable from its original position. (d) Square. (e) Rotation of the square by 45°. (f) Rotation of the square by 90° from its original position to a new position, which is indistinguishable from the original position.

may be considered to show a higher order of ''perfection'' than an equilateral triangle. If a square is rotated by 90° about its center, its appearance is unchanged (see Fig. 2.2). An equilateral triangle, on the other hand, must be rotated by 120° before it returns to its original appearance. A hexagon needs to be rotated by only 60° to preserve its appearance. An octagon needs only a 45° rotation; a dodecagon (12 sided figure) needs only a 30° rotation.

The more sides a regular figure has, the smaller the amount by which it must be rotated in order to restore its original appearance. In this sense, increasing the number of sides of a regular figure increases its perfection. As the number of sides of such a figure increases, it comes closer and closer in appearance to a circle, and so it is natural to look upon a circle as being the most perfect flat (or two-dimensional) figure that can be drawn. No matter how much or how little a perfect circle is rotated about its center, it maintains its original appearance. Its appearance is *constant.* It is important to note that perfection is identified with constancy—something that is perfect cannot be improved, so it must remain constant.[2]

The Greek concern with perfection and geometry set the tone for their approach to science. This is particularly illustrated in Plato's Allegory of the Cave. In this allegory, Plato (427–347 B.C.) envisioned humans as being like slaves chained together in a deep and dark cave, which is dimly illuminated by a fire burning some distance behind and above them. There is a wall behind them and a wall in front of them. Their chains and shackles are such that they cannot turn around to see what is going on behind them. There are people moving back and forth on the other side of the wall behind them, holding various objects over their heads and making unintelligible (to the slaves) sounds and noises. The slaves can see only the shadows of these objects, as cast by the firelight on the wall in front of them, and can hear only the muffled sounds reflected from the wall in front of

[2] E. A. Abbott's charming little book, *Flatland,* describes a fictional two-dimensional world, peopled by two-dimensional figures, whose characters are determined by their shape. Women, for example, the deadlier of the species, are straight lines, and are therefore able to inflict mortal wounds, just like a thin sharp rapier. The wisest male, in accordance with the Greek ideal of perfection is, of course, a perfect circle.

Figure 2.3. Time exposure of the night sky. (Fritz Goro/Life Magazine © Time, Inc.)

them. The slaves have seen these shadows all their lives, and thus know nothing else.

One day, one of the slaves (who later becomes a philosopher) is released from the shackles and brought out of the cave into the real world, which is bright and beautiful, with green grass, trees, blue sky, and so on. He is initially unable to comprehend what he sees because his eyes are unaccustomed to the bright light. Indeed, it is painful at first to contemplate the real world, but ultimately the former slave becomes accustomed to this new freedom. He has no desire to return to his former wretched state, but duty forces him to return to try to enlighten his fellow human beings. This is no easy task because he must become accustomed to the dark again, and must explain the shadows in terms of things the slaves have never seen. They reject his efforts, and those of anyone else like him, threatening such people with death if they persist (as indeed was decreed for Socrates, Plato's teacher). Plato insisted, however, that duty requires that the philosopher persist, despite any threats or consequences.

The first task of the philosopher is to determine the Reality, or the Truth, behind the way things appear. As a particular example, Plato pointed to the appearance of the heavens, which is the subject matter of astronomy. The Sun rises and sets daily, as does the Moon; and the Moon goes through phases on approximately a monthly schedule. The Sun is higher in the summer than in the winter; morning stars and evening stars appear and disappear. These are the "Appearances" of the heavens, but the philosopher (nowadays the scientist) must discover the true reality underlying these appearances—what is it that accounts for the "courses of the stars in the sky"? According to Plato, the true reality must be Perfect or Ideal, and the philosopher must look to mathematics, in particular, to geometry, to find the true reality of astronomy.

Most objects in the sky, such as the stars, appear to move in circular paths about the Earth as a center; and it is tempting to conclude that the truly essential nature of the motion of the heavenly objects must be circular, because the circle is the perfect geometrical figure. It does not matter whether all heavenly objects seem to move in circular paths or not; after all, human beings can perceive only the shadows of the true reality. The task of the philosopher (or scientist) is to show how the truly perfect nature of heavenly motion is distorted by human perceptions. Plato set forth as the task for astronomy the discovery of the way in which the motions of heavenly objects could be described in terms of circular motion. This task was called "saving the appearances." As will be seen, the goal of discovering the true reality is one of the major goals of science, even though the definition of true reality is now somewhat different from Plato's definition. This goal has strongly affected the way in which scientific problems are approached. In some cases it has led to major insights; in other cases, when too literally sought, it has been a severe handicap to the progress of science.

B. Geocentric Theory of the Universe

1. Appearance of the Heavens

If a time exposure photograph of the night sky is taken over a period of a few hours, the result will be as shown in Fig. 2.3. During the time that the camera shutter is open, the positions of the various stars in the sky change, the paths of their apparent motions being traced by arc-like streaks, just as photographs of a city street at night show streaks of light from the headlights of passing automobiles. If the camera shutter is left open for 24 hours (and if somehow the Sun could be "turned off"), many of the streaks, in particular those near the pole star, would become complete circles. Because the Sun cannot be turned off, the picture can only be taken during the hours of darkness, and thus only a fraction of the complete circles can be obtained, equal to the fraction of 24 hours during which the camera shutter is open. If the photograph is taken night after night, it will be almost the same, with a few exceptions to be discussed below.

It appears that the Earth is surrounded by a giant spherical canopy or dome, called the **stellar sphere** or the **celestial sphere,** and the stars are little pinpoints of light mounted on this dome (Fig 2.4a). The dome rotates about us once in 24 hours in a direction from east to west (rising to setting). The Sun, Moon, and planets are also mounted on this sphere. Of course, when the Sun is in view we cannot see the stars because the bright light of the Sun, as scattered by the Earth's

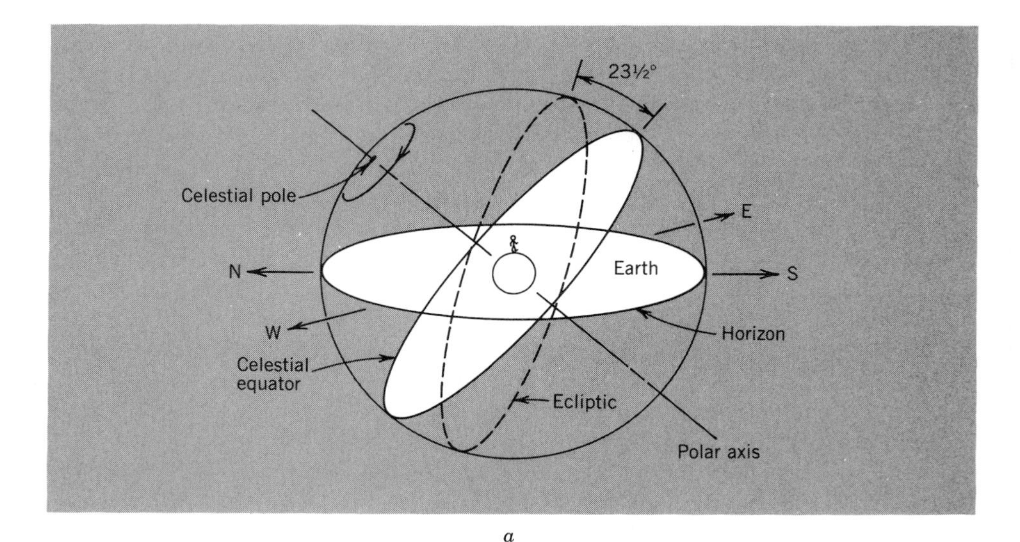

a

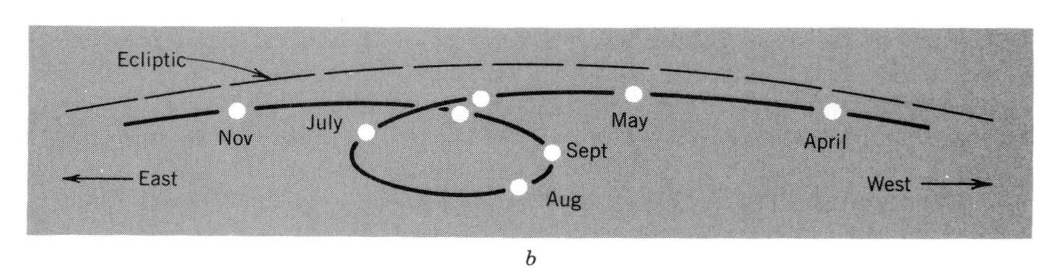

b

Figure 2.4. Celestial sphere. (a) Schematic of sphere showing Earth at center and the ecliptic. (b) Greatly enlarged view of part of the path of a planet along the ecliptic, showing retrograde motion.

atmosphere, makes it impossible to see the relatively weak starlight. (When there is an eclipse of the Sun, it is possible to see the stars quite well.)

By patient observation and measurement, the ancient astronomers and astrologers, who did not have photographic equipment, were able to observe and keep track of this daily rotation, called the **diurnal rotation** or **diurnal motion** of the celestial sphere. They were able to determine that all the heavenly objects, with a few exceptions, are fixed in position on the rotating celestial sphere. In the course of time it was recognized that the Earth itself is also spherical and appears to be located at the center of the celestial sphere. The part of the celestial sphere that can be seen depends on the location of the observation point on the Earth. Standing at the North Pole of the Earth, the centers of the circles will be directly overhead. At a latitude of 45° the celestial pole (the centers of the circles) will be 45° up from the north horizon (in the northern hemisphere), whereas at the equator, the celestial pole will be down at the north horizon.

Fairly early it was recognized that certain objects on the celestial sphere are not fixed; that is, these objects appear to move as compared with the general background stars. The position of the Sun on the sphere, for example, changes over the course of a year. It follows a path, called the **ecliptic,** shown in Fig. 2.4a as a dashed circle tilted at an angle of $23\frac{1}{2}°$ from the celestial equator. The direction of motion of the Sun is from west to east (opposite in direction to the daily rotation.)[3] The position of the Moon also changes, and it travels from west to east along the ecliptic making a complete circuit in $27\frac{1}{3}$ days, on the average. Similarly, the planets that are visible to the unaided eye—Mercury, Venus, Mars, Jupiter, and Saturn—travel roughly along the ecliptic, from west to east also, taking 90 days to 30 years to make the complete circuit.

The motion of the Sun is called **annual motion** and is fairly, but not completely, smooth and even. The motion of the Moon is also somewhat smooth, but less so than the Sun's motion. The motion of the planets, on the other hand, is variable in speed, and sometimes in direction as well. Indeed, this is the origin of the word *planet,* which means *wanderer.* The planets also normally move from west to east, relative to the fixed stars. When a planet occasionally varies its direction of motion, it moves from east to west relative to the fixed stars, rather than follow its overall west to east path. This motion, called **retrograde motion,** is illustrated in Fig 2.4b, which shows the planet not only changing direction, but wandering slightly off the ecliptic as well. Accompanying the retrograde motions are variations in the apparent brightness of the planet.

At times a planet is just ahead of the Sun on the ecliptic, and at other times it is just behind the Sun. This is called **alternate motion.** If the planet is just west of the Sun, the daily rotation will bring it into view as the *morning star;* if it is just east of the Sun, it will be seen as the *evening star* just after sunset.[4] Because of their varying speeds, there will be times when the two planets appear very close to each other. Such an event is sometimes called a *conjunction.* Ancient astrologers and soothsayers who searched for "signs" in the heavens attached great significance to conjunctions as portending or reflecting momentous events for humans. Even more rare (and portentous) are double conjunctions, when three planets appear very close to each other. (In current astronomical usage, conjunction has a different meaning, the discussion of which is beyond the scope of this book.)

As already mentioned, over the centuries the various great civilizations attached much importance to the appearance of the heavens. These appearances were used to establish the calendars necessary for the governance of extensive empires and recording important events in human affairs. They were also important for travel to distant places, because navigators could determine where they

[3]This does not contradict the common observation that the daily motion of the Sun is from east to west. It means rather that if one could observe the position of the Sun against the background of "fixed" stars (such as during an eclipse), one would see that the Sun slowly moves from west to east on the celestial sphere.

[4]The Moon and Sun were also considered to be planets, but they do not exhibit retrograde motion or the same degree of variation of speed as the other planets. In this book, the word *planet* does not include the Moon and Sun.

were by the appearance of the heavens in their particular location. In fact, the calculation and use of accurate navigation tables played a very important part in the exploration of the world by various seafaring nations, and in their ability to carry out commercial and military expeditions. Until the last few centuries, astrological predictions, based on the appearance of the heavens in general, or some unusual event such as a conjunction, or a comet, or a nova, have had widespread influence on political decision making. Even today horoscopes are surprisingly popular.

The Babylonians developed one of the great Mesopotamian civilizations. Their astronomers carried out rather accurate measurements and calculations of the heavenly appearances. They were more concerned with the accuracy and reliability of their astronomical measurements and predictions than with the development of a general understanding of the whys and wherefores of the workings of the universe.

The Greeks, on the other hand, although interested in accurate and reliable knowledge of the heavens, did not improve the Babylonian measurements significantly. The Greek contribution to astronomy came more from their concern with understanding on a philosophical level, as already indicated. They regarded the heavens as being the regions where perfection could be found, because the true nature of the heavens was perfection. The appearances of the heavens, they felt, had to be interpreted in terms of their inherent perfection and constancy. The Sun and Moon seem to have the "perfect" shape—namely, circular. The fact that the diurnal motion of the stars resulted in circular motion was appropriate because a circle represented the perfect figure, as discussed in Section A2 above.

Although the detailed motion of the planets seemed to depart from smooth and circular motion, on the average the motion was circular. Therefore it was asserted that the essential nature of their motion was circular, and the departures from circular motion that were observed represented but the shadows of the true reality, as understood from Plato's Allegory of the Cave. The Greek philosophers took seriously Plato's command to "save the appearances," that is, to explain all motions of heavenly bodies in terms of circular motions. In modern terms, they undertook to develop a *model* of the universe that would explain how it "really works."

2. Competing Models of the Universe

Many ancient civilizations developed models of the universe. The Egyptians, for example, believed the universe to be like a long rectangular box, with the Earth being a slightly concave floor at the bottom of the box and the sky being a slightly arched iron ceiling from which lamps were hung. There were four mountain peaks supporting the ceiling, and they were connected by ranges of mountains, behind which flowed a great river. The Sun was a god who traveled in a boat along this river and, of course, could only be seen during the daylight hours.

The models considered by the Greeks at a much later time were a great deal more advanced than this, and reflected considerably more knowledge. They considered two types of models, the **geocentric** or Earth-centered type and the **heliocentric** or Sun-centered type. In the geocentric models, the Earth in addition to being at the center of the universe was usually stationary. In the heliocentric models, the Earth circled about the Sun or about a central fire, like all the other planets, and usually rotated on its own axis. By and large, the Greeks preferred the geocentric models and rejected the heliocentric models.

In one of the simplest geocentric models, the Earth is a motionless small sphere surrounded by eight other rotating concentric spheres that carry the Moon, the Sun, Venus, Mercury, Mars, Jupiter, Saturn, and the fixed stars, respectively. Because the spheres all have the same center, this is called a **homocentric** model. Associated with each of these eight spheres are a number of other auxiliary spheres. (This model is credited to Eudoxus, one of Plato's pupils who took up the challenge to determine the true reality.) The Moon and Sun each have two auxiliary spheres, and the planets have three auxiliary spheres each. Counting the stellar sphere, this made a total of 27 spheres.

The purpose of the auxiliary spheres is to help generate the observed motions. For example, the sphere carrying the Moon rotates about an axis whose ends are mounted on the next larger sphere, so that the axis of the rotation of the Moon's sphere can itself rotate with the auxiliary sphere. In turn, the axis of this auxiliary sphere is mounted on the next auxiliary sphere, which can also rotate. By placing these axes at various angles with each other, and adjusting the rotation speeds of the spheres, it is possible to generate a motion for a particular heavenly object such that an observer on Earth will perceive the object, against the background of the fixed stars, following the appropriate path along the ecliptic.

In the early versions of this model the question as to how these spheres obtained their motion was ignored. It was simply assumed that the very nature of a perfect sphere in the heavens is such that it would rotate. The purpose of the model was simply to explain the nonuniform motion of heavenly objects as combinations of uniform circular motion, thereby "saving the appearances." To the extent that anyone might wonder about the causes of the motion, it was said to be due to various "Intelligences." By this time the Greek philosophers had more or less put aside such mythical explanations as gods driving fiery chariots across the sky and so on.

3. Aristotle's Conception of the Universe

The great philosopher Aristotle (384–322 B.C.) adopted the homocentric model of the universe and thoroughly integrated it into his philosophical system, showing its relationship to physics and metaphysics. (Metaphysics is that branch of philosophy that deals with the nature of reality.) He considered the universe as being spherical in shape and divided into two "worlds," the astronomical (or heavenly) world and the sublunar world. The sublunar world was made up of four **prime substances**—*earth, water, air, and fire*—whereas the astronomical world contained a fifth substance, *ether.* These substances had certain inherent characteristics. The laws of physics were determined by the inherent nature of these substances, and therefore the laws of physics in the sublunar world differed from the laws in the astronomical world. Aristotle's understanding of physics was quite different from our understanding of physics today, but it was tightly interlocked with his entire philosophical system. Aristotle's concepts of physics will be discussed in more detail in the next chapter.

Aristotle was not content with the simple description of planetary motions introduced by Eudoxus, because in Eudoxus's scheme the motions of the various sets of spheres were independent of each other. Aristotle felt a need to link the motions of the heavenly objects together into one comprehensive system, and so he introduced additional spheres between the sets assigned to specific planets. For example, the outermost auxiliary sphere of the set belonging to Saturn had its axis mounted on the sphere of the fixed stars. On the innermost sphere belonging to Saturn was mounted a set of additional spheres, called counteracting spheres, consisting of three spheres that rotated in such a way as to cancel out the fine details of Saturn's motion. On the innermost of these counteracting spheres was mounted the outermost auxiliary sphere belonging to Jupiter. Similarly, groups of counteracting spheres were introduced between Jupiter and Mars, and so on.

All told, in Aristotle's scheme there were 56 spheres. The outermost sphere of the 56 was called the **Prime Mover,** because all the other spheres were linked to it and derived their motion from it. Aristotle did not specify in detail how the motion was actually transmitted from one sphere to another. Later writers speculated that somehow each sphere would drag along the next inner sphere with some slippage. Many centuries later working models were built to illustrate this motion, with gears introduced as necessary to transmit the motions.

It was soon recognized that there were some weaknesses in the simple geocentric theory discussed by Aristotle. The measured size of the Moon, for example, may vary by about 8 to 10 percent at different times of observation. Similarly, the planets vary in brightness, particularly being brighter when undergoing retrograde motion. These phenomena imply that the distance from the Earth to the planet or to the Moon changes. But this is not possible in the homocentric system. Even more important, as a result of Alexander the Great's conquests, the Greeks became particularly aware of the large amount of astronomical data that had been accumulated by the Babylonians. They found that Aristotle's model did not fit the data. It thus became necessary to modify Aristotle's model, and this was done over the course of several hundred years.

4. Ptolemy's Refined Geocentric Model

These modifications culminated some 1800 years ago, about the year 150, when the Hellenistic astronomer Claudius Ptolemy, in Alexandria, Egypt, published an extensive treatise, *The Great Syntaxis,* on the calculations of the motions of the Sun, Moon, and planets. With the decline of secular learning accompanying the collapse of the Roman Empire, this work was temporarily lost to the Western world. It was translated into Arabic, however, and eventually reintroduced into Europe under the title of the *Almagest,* which means, ''the Majestic'' or ''the Great.''

Ptolemy abandoned Aristotle's attempt to link all the motions of the Sun, Moon, and planets to each other, dismissing physics as being too speculative. He felt it was sufficient to develop the mathematical schemes to calculate accurately the celestial motions, without being concerned about the reasons for such motions. The only criterion for judging the quality or validity of a particular scheme of calculation was that it should give correct results and "save the appearances." It was necessary to do this because the available data indicated that the appearances of the heavens were gradually changing. In order to be able to establish calendars and navigational tables, it was important to have accurate calculations, regardless of how the calculations were justified.

To this end, he made use of several devices (as the modifications were called), which had been suggested by other astronomers, as well as some of his own devising. Nevertheless, he preserved the concept of circular motion. Figure 2.5a shows one such device, the **eccentric.** This simply means that the sphere carrying the planet is no longer centered at the center of the Earth—it is somewhat off center. Therefore in part of its motion, the planet is closer to the Earth than in the rest of its motion. Then when the planet is closer to the Earth, it will seem to be moving faster. The center of the sphere is called the eccentric. In some of his calculations, Ptolemy even had the eccentric itself moving, although slowly.

Figure 2.5b shows an **epicycle,** which is a sphere whose center moves around or is carried by another sphere, called the **deferent.** The planet itself is carried by the epicycle. The epicycle rotates about its center, whereas the center is carried along by its deferent. The center of the deferent itself might be at the center of the Earth, or more likely, it too is eccentric.[5]

Depending on the relative sizes and speeds of the epicycle and deferent motion, almost any kind of path can be traced out by the planet. As shown in Fig. 2.5c, retrograde motion can be generated, and the planet will be closer to the Earth while undergoing retrograde motion. Combinations of epicycle and deferent motions can be used to generate an eccentric circle or an oval approximating an ellipse, or even an almost rectangular path.

For astronomical purposes, however, it is not enough to generate a particular shape path. It is also necessary that the heavenly object travel along the path at the proper speed; that is, the calculation must show that the planet gets to a particular point in the sky at the correct time. In fact, it became necessary to cause either the planet on the epicycle, or the center of the epicycle on the deferent, to speed up or slow down. Thus it was necessary to introduce yet another device, called the **equant,** as shown in Fig. 2.5d. Note that the Earth is on one side of the eccentric and the equant an equal distance on the other side of the eccentric. As seen from the equant, the center of the epicycle moves with uniform angular motion, thereby saving the appearance of uniform motion. Uniform angular motion means that a line joining the equant to the center of the epicycle rotates at a constant number of degrees per hour (e.g., like a clock hand). However,

[5]One can visualize the motion as being like that of certain amusement park rides, which have their seats mounted on a rotating platform that is itself mounted at the end of a long rotating arm. The other end of the arm is also pivoted and can move (see Fig. 2.6).

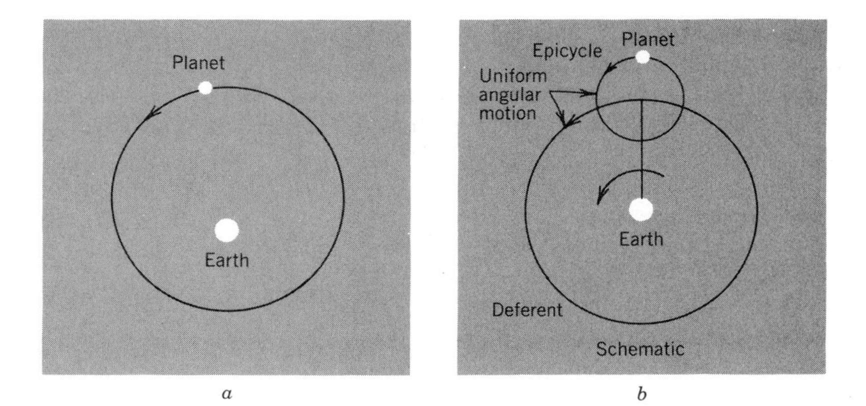

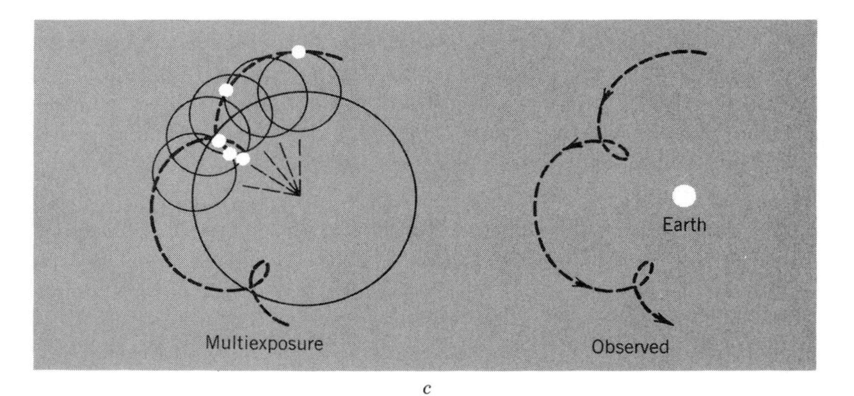

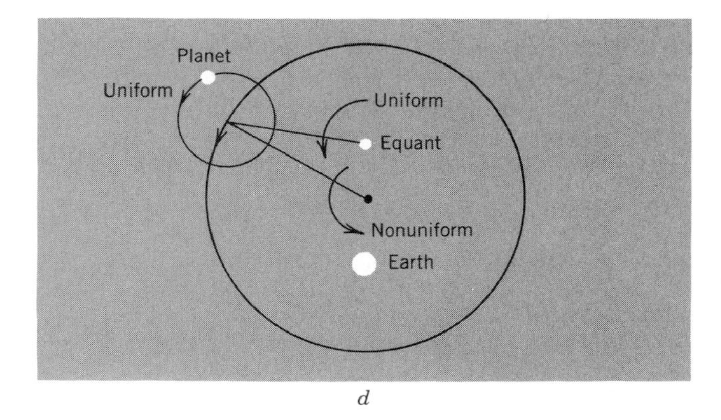

Figure 2.5. Devices used in Ptolemy's geocentric model. (a) Eccentric. (b) Schematic of epicycle on deferent. (c) Generation of retrograde motion by epicycle. (d) Equant.

Figure 2.6. Amusement park ride involving epicycles. (Charles Gupton/Stock Boston.)

uniform angular motion with respect to the equant is not uniform motion around the *circumference* of the circle. Because the length of the line from the equant to the planet changes, the circumferential speed (say, in miles per hour) changes proportionally to the length of the line.

Ptolemy made use of these various devices, individually or in combination, as needed, to calculate the positions of the planets. He sometimes used one device to calculate the speed of the Moon, for example, and another to calculate the change of its distance from the Earth. He was not concerned about using a consistent set of devices simultaneously for all aspects of the motion of a particular heavenly object. His primary concern was to calculate the correct position and times of appearances of the various heavenly bodies. In this respect, he was like a student who knows the answer to a problem, and therefore searches for a formula that will give him the correct answer without worrying about whether it makes any sense to use that formula.

Some later astronomers and commentators on Ptolemy's system did try to make the calculations agree with some kind of reasonable physical system. Thus they imagined that the epicycles actually rolled around their deferents. This meant, of course, that the various spheres had to be transparent, so that it would be possible to see the planets from the Earth. Thus the spheres had to be of some crystal-like material, perhaps a thickened ether. The space between the spheres was also thought to be filled with ether because it was believed that there was no such thing as a vacuum or empty space.

To the modern mind, Ptolemy's system seems quite far-fetched; however, in his time and for 1400 years afterward, it was the only system that had been worked out and that was capable of producing astronomical tables of the required accuracy for calendar making and navigation. Ptolemy himself, as already mentioned, had one primary goal in mind: the development of a system and mathematical technique that would permit accurate calculations. Philosophical considerations—and physics was considered a branch of philosophy—were irrelevant for his purposes. In short, Ptolemy's scheme was widely used for the very pragmatic and convincing reason that it worked!

C. The Heliocentric Theory—Revival by Copernicus

1. Historical Background

According to the **heliocentric** (or Sun-centered) theory, the Sun is the center of planetary motion, and the Earth is a planet, like Mars or Venus, that travels in a path around the Sun, and also rotates or spins on its axis. As early as the time of Aristotle, and perhaps even earlier, the possibility that the Earth spins on its axis had been considered. There was also an even older suggestion that the Earth, as well as the Sun, moves about a "central fire." These ideas were rejected, and some of the arguments against them are discussed below. The Greeks by and large thought the idea of a moving and spinning Earth was untenable, and it was considered impious and dangerous, as well as ridiculous, to entertain such ideas.

Nevertheless, about 600 years ago, in the latter part of the fourteenth century, a few individuals suggested that there were some logical difficulties with the concept of the Earth-centered universe. For one thing, if the daily rotation of the stars was due to the rotation of the celestial sphere, then because of its huge size, the many stars on the surface of the celestial sphere would be moving at speeds that seemed impossibly high. It seemed just as reasonable to assume that only the small Earth was spinning, and the daily rotation was simply an optical illusion reflecting the relative motion. Some time later there was a suggestion that an Infinite Creator of the universe would have made it infinite in both time and space, and therefore *any* place could be chosen as its center. These were only speculations, however, and no one undertook the necessary detailed calculations to support such suggestions, until the sixteenth century, when a minor Church official at the Cathedral of Frauenberg (now called Frombork, Poland) took up the problem.

This person, Nicolaus Copernicus, studied at universities in Cracow, Poland, and Bologna and Padua, in Italy, where he acquired a fairly substantial knowledge of mathematics, astronomy, theology, medicine, canon law, and Greek philosophy. His duties at the Cathedral did not require much time, and so he was able to undertake a prolonged and extensive study of astronomy.

Copernicus ultimately concluded that the Ptolemaic system was too complicated. It seems clear that Copernicus was influenced by Neoplatonism, a philosophical outlook that had strong roots in classical Greek thought. Consistent with this outlook was a rule called **Ockham's razor,** which in essence says that simple explanations are preferable to complicated explanations. (In this connection, King Alfonso X of Castille, who in 1252 sponsored a new calculation of astronomical tables using Ptolemy's techniques and additional data, is reported to have complained that God should have consulted him before setting up such a complicated system.) Copernicus was dissatisfied with the inconsistent way in which Ptolemy had applied the various devices, as discussed above. He regarded the use of the equant as being particularly contradictory to the very idea of perfect circular motion. He therefore reexamined the heliocentric model, which had been proposed over 1500 years earlier by Aristarchus, a Greek astronomer who lived between the times of Aristotle and Ptolemy.

2. Simple Copernican Theory

Copernicus, following Aristarchus's suggestions, proposed that the Sun and the fixed stars should be regarded as stationary, and that the Earth be regarded as a planet that circled about the Sun in the same manner and direction as the other five planets. The Moon alone should be regarded as circling about the Earth. The length of time for any planet to complete one round trip around the Sun should be greater the further away the planet from the Sun. Retrograde motion and all the slowing down and speeding up of the planets is simply an ''optical illusion,'' which occurs because the direction in which one of the other planets is seen from the Earth depends on the relative positions of the Earth and the planet in their respective orbits. These relative positions will change over time in the same way as the relative positions of racing cars in a long-distance race around an oval track. A fast planet will increase its lead over a slow planet until it gains a whole ''lap'' and will pass the slower planet again and again. Therefore as viewed from the Earth the other planets will sometimes seem to be ahead and sometimes behind the Earth. A conjunction is nothing more special than the observation that one planet is about to ''lap'' another planet.

Figure 2.7a shows schematically how retrograde motion is seen from the Earth. Figure 2.7b shows the relative positions of the Sun, Earth, and Mars at seven different times. The arrows show the directions (compared to the direction of a distant star) in which an observer on Earth would have to look to see Mars at those times. In Fig. 2.7c, the corresponding directions are superimposed on each other, as they would appear to astronomers (or anyone else) looking from Earth toward Mars. In addition, the lengths of the arrows are adjusted to correspond to the changing distances from Earth to Mars in the seven different positions. The result is a very simple and ''natural'' explanation of retrograde motion.

To explain the daily rotation (diurnal motion), Copernicus also proposed that the Earth is spinning on its own axis once in 23 hours and 56 minutes.[6] The spin

[6]Note the time of rotation is not 24 hours. The other 4 minutes to make up the time for an apparent complete rotation of the heavens is needed because the Earth has moved to a different position in its orbit about the Sun.

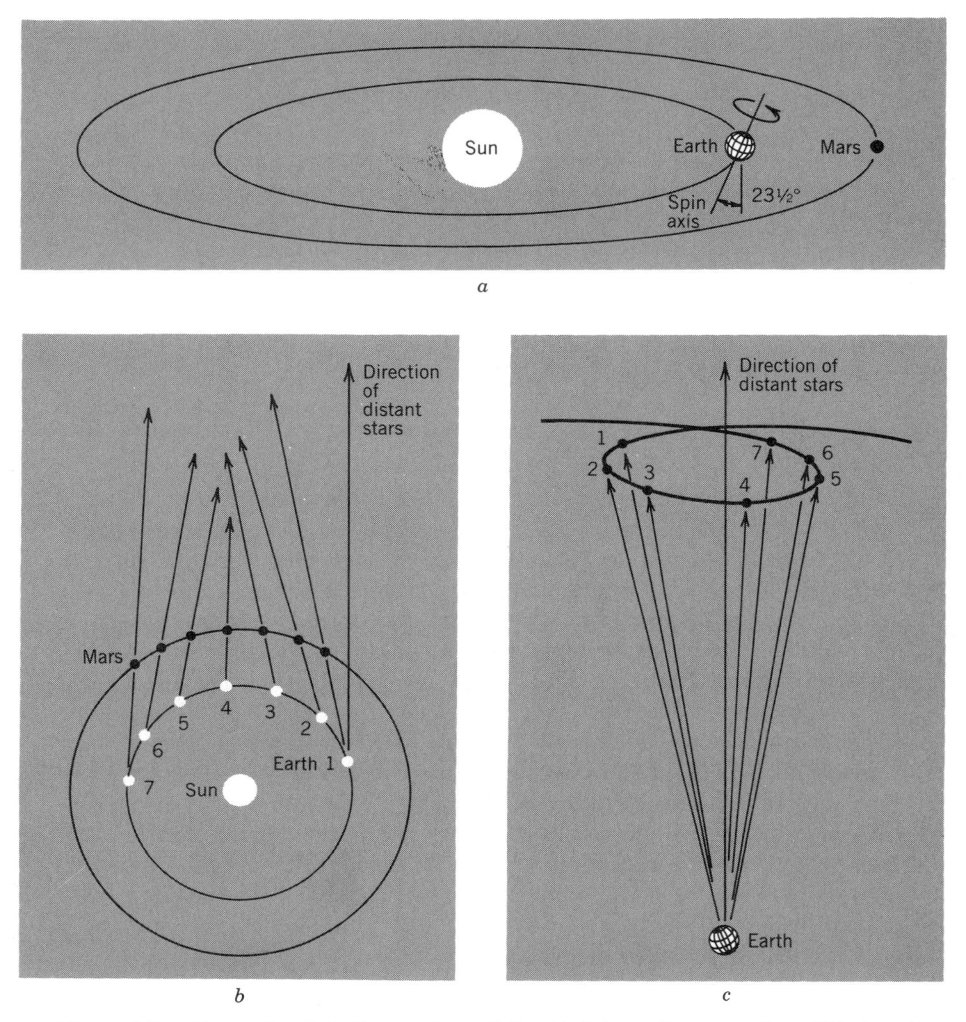

Figure 2.7. Copernicus's heliocentric model. (a) Schematic perspective of Earth and Mars orbiting the Sun, showing the Earth spinning on its axis. (b) Relative positions of Earth and Mars showing different angles of sight from Earth to Mars. (c) Superposition of angles of sight as seen from the Earth itself.

axis is tilted with respect to the axis of the orbit by an amount equal to the angle (about $23\frac{1}{2}°$) that the ecliptic makes with the celestial equator. (Copernicus also asserted that this tilt axis is rotating very slowly about the orbital axis, but maintaining the same tilt angle, thereby explaining a very slow change in the overall appearances of the fixed stars, which had been observed since the time of Ptolemy.)

In effect, Copernicus found that by shifting the point of view of the planetary motions from the Earth to the Sun, motions that had theretofore seemed very complicated became very simple. If one could stand on the Sun, then it would appear that all the planets are simply traveling in circular paths, with the planets closest to the Sun traveling the fastest. He then deduced from the astronomical measurements the correct order of the planets from the Sun, starting with the closest: Mercury, Venus, Earth, Mars, Jupiter, Saturn. He also calculated the ratios of their distances from the Sun fairly accurately, although his values for the actual distances were quite wrong. The Earth, for example, is actually 93 million miles from the Sun, whereas Copernicus took this distance to be about 4.8 million miles.

Having set up his general scheme, Copernicus then proceeded to make calculations as to exactly how the heavens would appear to astronomers on Earth. In particular, he calculated the appearances of the planets, time and place, for comparison with the astronomical measurements. He found, however, that his calculated positions gave much poorer results than the Ptolemaic theory! In other words, his beautiful and elegant conception, which worked so well qualitatively, had failed quantitatively.

3. Modified Copernican Theory

Undaunted by this outcome, Copernicus reintroduced some of the devices that Ptolemy had used, that is, the eccentric, deferent, and epicycle, but not the equant, which he despised. Moreover, in introducing these devices, he applied them in a much more consistent manner than was possible using Ptolemy's methods. He also used a much smaller number of spheres of various types, 46 compared with more than 70 used in some of the later Ptolemaic-type calculations.

The result of all this effort was that his calculated astronomical tables were overall no better than astronomical tables calculated using the geocentric theory and Ptolemy's methods. There was one advantage to the Copernican approach, so far as practical astronomers were concerned: The calculations were somewhat simpler to carry out.

Copernicus completed his work in 1530, but delayed publishing it in full detail for a number of years, possibly fearing the reaction it might generate. He knew that in the past proponents of the idea that the Earth is in motion had been severely ridiculed. Finally, in 1543, on the day that he died, his book, *On the Revolutions of the Heavenly Spheres,* was published. It was dedicated to Pope Paul III, and its preface (written by someone else) stated that the Copernican scheme should not be taken as representing reality, but rather should be regarded as simply a useful way to calculate the positions of various heavenly objects. Indeed the Copernican system and calculations were used in determining the Gregorian calendar, which was first introduced in 1582 and is still in use today.

4. Reception of the Copernican Theory

The initial reaction to the Copernican scheme was largely to regard it in the way suggested by the preface to Copernicus's book—useful for calculations but not necessarily connected with reality. Copernicus was aware of many of the objections to the heliocentric theory, or any theory that involved motion of the Earth, because they had been raised during the time of the ancient Greeks. He attempted in his book to refute these objections. They can be divided into two categories: (1) scientific objections, and (2) philosophic and religious objections. In the minds of most of his contemporaries, the two categories were equally weighty.

Some of the scientific objections were based on astronomical observations. For example, if the Earth actually moved in a circular orbit about the Sun, then the distance from the Earth to a particular portion of the equator of the celestial sphere should vary by as much as 9.6 million miles (Copernicus's value for the diameter of the Earth's orbit). The angular separation between two adjacent stars (which were thought to be about 80 million miles away) should therefore change noticeably, in the same way that the angular separation between two trees in a forest changes, depending upon how close one is to the trees. This effect, called **star parallax,** had never been observed. Copernicus asserted that the stars were much more than 80 million miles distant, so that even a change of 9.6 million miles was such a small fractional change in the total distance to the stars that the parallax would be too small to be seen by the human eye. His critics, however, rejected this contention because they believed that the celestial sphere was only slightly larger than the sphere of Saturn. In fact, the nearest stars are so far away that star parallax was not observed until about 200 years later, when sufficiently powerful telescopes were available.

A related objection involved the apparent size of the stars. As seen with the unaided eye, all stars seem to have the same apparent diameter because they are so far away. If their true diameters are calculated using the distance from the Earth to the stars, as asserted by Copernicus, they would then turn out to be several hundred million miles in diameter and therefore much, much larger than the Sun. This was also too incredible to believe. As it turns out, the apparent size of the stars as seen with the unaided eye is misleadingly large because of the wave nature of light and the relatively small aperture of the eye; but this was not understood until about 300 years later, when the principles of diffraction optics were worked out. (In fact, when seen through a telescope, the stars seem smaller than when seen with the naked eye!)

There were also objections based on physics, as the subject was understood at that time. This understanding was based primarily on the writings of Aristotle. For example, it was asserted that if the Earth is truly in motion, then its occupants should feel a breeze proportional to the speed of the Earth, just as a person riding in an open vehicle feels a breeze. The speed of the Earth as it travels around its orbit is actually about 67,000 miles per hour. (According to Copernicus's calculation the speed would be about 3500 miles per hour.) The resulting "breeze" would surely destroy anything on the surface of the Earth, including houses, trees, and so on (a storm with wind velocities greater than about 75 miles per hour is considered to be a hurricane). Obviously this does not happen.

OPTIONAL SECTION 2.1 *Calculation of Earth's Orbital Speed*

The speed of the Earth in its orbit can be calculated very easily once the distance from the Earth to the Sun is known. Assuming that the Earth's orbit is circular (a very good approximation), then the Earth–Sun distance is the radius of the orbit, and the circumference of a circle of that radius is $2\pi r = 2\pi \times 93$ million miles. It takes the Earth one year to complete an orbit. Thus the speed of Earth in orbit is 580 million miles per year. There are 365 days in a year; thus dividing 580 million miles by 365 gives 1.6 million miles per day. If this result is divided by 24, the number of hours per day, then the speed of the Earth in orbit is about 67,000 miles per hour.

Similarly, it was claimed that if the Earth is moving, an apple falling from a tree would not simply fall straight down to the ground, but would land several miles away because the tree and the ground beneath it would have moved that distance in the time it took the apple to reach the ground. It takes one second for an apple to fall to the ground after it breaks off from a tree branch that is 16 feet above the ground. During that time the Earth traveling at a speed of 67,000 miles per hour will have moved a distance of 18.6 miles.

For that matter, it has to be asked, how many horses (in modern terms, how big an engine) are needed to pull something as massive as the Earth at such a stupendous speed? There is no evidence of any agency to keep the Earth moving. Similarly, a spinning Earth is like a giant merry-go-round, and the resulting centrifugal force should be so large that all the inhabitants would have to hang on for dear life.[7] Because none of these things occur, then it was hard to believe that the Earth is actually moving. The proponents of the Copernican system did eventually overcome these objections (as will be discussed in the next chapter), but it took roughly a century to do so.

The philosophical and theological objections were equally weighty. It must be borne in mind that physics at that time was not accorded the overwhelming respect it has today. Indeed, theology was considered to be the Queen of the Sciences. Aristotelian philosophy had been integrated into the theological dogma of western Europe in the sixteenth century. Interestingly enough, 1000 years earlier, the Aristotelian outlook had been roundly condemned as pagan, and the very idea of a round Earth was considered heretical. In Copernicus's time, however, the Aristotelian outlook, which was essentially rational, was dominant, and any flaws in it could be taken as indicative of flaws in the entire accepted order of things. The Earth was considered to be unique and at the center of the universe, and so too was man.

As indicated earlier (Section B3), the Earth and the surrounding sublunar region up to the sphere of the Moon were considered to be imperfect. The Moon, Sun, planets, and stars were perfect because they were in the heavens—indeed

[7]The idea that the Earth could exert a gravitational attractive force was not developed until about a century later, as was the distinction between centrifugal and centripetal force.

that was why their motion was circular—perfect objects carrying out the perfect motion that was inherent in their nature. The Earth, being in the sublunar region, was inherently different from the heavenly bodies and could not be a planet like the other planets. Therefore the Earth could not undergo the same kind of motion as the planets. The planets were also considered to be different because they exerted special influences on individual behavior—individual personalities are sometimes described as mercurial or jovial or saturnine, for example, and these traits were attributed to the planets Mercury, Jupiter, or Saturn.[8] Similarly, a heliocentric theory also weakens the mystic appeal of astrology.

The heliocentric theory was attacked by the German leaders of the Protestant Reformation because it ran counter to a literal reading of the Bible. Certainly the central role of the Sun in the theory could be considered as leading to a return to pagan practices of Sun worship. The reaction of the ecclesiastical authorities of the Catholic Church was initially rather mild, but ultimately they too condemned heliocentrism as heretical. Some historians have suggested that if Copernicus had developed his ideas 50 years earlier, well before the onset of the Reformation, they might have received a more sympathetic audience. But coming as they did in the midst of the theological struggles of the sixteenth and seventeenth centuries, they were seen as a challenge to the competent authority of the Roman Church (or the Bible, in the Protestant view), which had long proclaimed a different cosmology, and had to be combated.[9]

The challenge perhaps seemed to be even more severe because it led to wild speculations such as those of Giordano Bruno, a monk who proclaimed some 50 years after Copernicus's death that if the Earth were like the planets, then the planets must be like the Earth. They were surely inhabited by people who had undergone historical and religious experiences similar to those that had transpired on Earth, including all the events recorded in the Bible. Bruno asserted that the universe was infinite in extent and the Sun was but a minor star, and that there were other planetary systems. The uniqueness of mankind, of God's concern for mankind and of various sacred practices and institutions was thereby lost. (Bruno was ultimately burned at the stake because he refused to recant and repudiate this and other heresies.)

All things considered, in view of the various objections raised against it and the fact that even after refinement it gave no better results than the Ptolemaic theory, there seemed to be no compelling reason to prefer the Copernican theory. Its chief virtue seemed to be that it was intellectually more appealing to some astronomers and lent itself to easier calculations. It did demonstrate, however, that there was more than one possible scheme for "saving the appearances," and

[8]It may be argued that such beliefs are inconsistent with the rationality that should have been imposed by Aristotelian philosophy, but all cultures have inconsistencies. Moreover, medieval Europe had not acquired its knowledge of Aristotelian philosophy from the original Greek texts, but from Latin translations of Arabic translations. The various translators inevitably added their own ideas to the original text.

[9]Copernicus, and others who supported the heliocentric view, felt that it could be reconciled with the Bible. They felt that the Sun simply represented God's glory and power.

it crystallized the idea that the Ptolemaic theory was no longer satisfactory. The time was at hand for new and better astronomical theories.

D. Compromise Theory

1. New Data

One of the weaknesses of Copernicus's work was that he made use of large amounts of very ancient, possibly inaccurate, data. He himself made only a limited number of astronomical observations. In 1576, the Danish astronomer Tycho Brahe built an astronomical observatory on an island near Copenhagen, under the patronage of the King of Denmark. Brahe was a nobleman of fearsome appearance (he had a silver nose as a result of a duel), but he was an unusually keen and meticulous observer and was able to measure the position of a planet or star quite accurately. He spent the following 20 years making a large number of extremely accurate measurements of the positions of the Sun, Moon, and planets. Where previous measurements were reliable to an angular accuracy of about 10 minutes of arc, Brahe's data were accurate to about 4 minutes of arc (there are 60 minutes in a degree). To give some idea of what this means, if a yard-long pointer is used to sight a planet, a shift of the end of the pointer by $\frac{1}{25}$ inch would give an angular error of 4 minutes of arc. While most previous observers were content to observe the position of planets at certain key points of the orbit, Brahe tracked them throughout their entire orbit. His observations were not done with telescopes, which were not invented until 1608, but were done with various "sighting tubes" and pointer instruments.

2. Failure of Both Copernican and Ptolemaic Theories

Brahe found that neither the Copernican theory nor the Ptolemaic theory agreed with his new data. In fact, he knew even 13 years before he built his observatory that both theories had serious errors. He observed a conjunction of Jupiter and Saturn, and, on checking the astronomical tables, found that the Ptolemaic theory had predicted this conjunction would occur a month later, whereas the Copernican theory predicted it would occur several days later. Tycho Brahe was also impressed by the various arguments against the motion of the Earth, and he realized that it was possible to construct yet another theory based on perfect circular motion.

3. Tycho Brahe's Theory

In his new theory, Brahe proposed that the Earth is at the center of the sphere of the fixed stars and at rest, and that the Sun and Moon travel in orbits around the Earth. The other five planets travel in orbits centered about the Sun. A theory of this type, in which the Earth is still at the center of the universe, but in which the other planets revolve about the Sun, is called a Tychonic theory. In Tycho Brahe's version of the theory, the outer sphere of the stars rotates once in 24 hours, giving rise to the diurnal motion. In other Tychonic theories, the stellar sphere is fixed, and the Earth rotates daily on its axis, while remaining at the center of the sphere.

The Tychonic model came to be generally accepted by most astronomers of the time because it preserved something of the intellectual unity of the heliocentric theory and the relative ease of calculations of the Copernican scheme. At the same time Brahe's theory avoided all the difficulties associated with the movement of the Earth. Although this theory was subsequently superseded, it is indeed notable because it represented the recognition of the failure of the Ptolemaic model. Brahe did not work out his theory completely; on his deathbed, he charged his assistant and ultimate successor, Johannes Kepler, with the task of carrying out the detailed calculations necessary for the theory.

E. New Discoveries and Arguments

Both Copernicus and Brahe were part of a long-standing and continuing tradition in astronomy—the observation and calculation of the positions of heavenly bodies. Starting in the late sixteenth century, however, new categories of information began receiving considerable attention.

1. Transient Phenomena in the Heavens

In 1572, a new and very bright star was seen in western Europe. This star flared up and became very bright, brighter than any other star and brighter even than the planets; and then it faded away from visibility over a period of several months. Careful observations by Brahe and several other astronomers showed that the star was as far away and as fixed in location as any other star. Another similar star was seen in 1604. Such a star is called a *nova,* and is not an uncommon occurrence. In 1576 a comet was observed moving across the heavens beyond the lunar sphere, and in such a direction that it must have penetrated through several of the planetary spheres.

Such occurrences had been reported in the records of the Greek astronomers, but the quality of the observations had been such that they had been thought to take place in the sublunar region. In 1572, 1576, and 1604 the observations were specially significant, however, because they clearly showed that the phenomena took place in the astronomical or heavenly world. This meant that even the heavens changed. If the heavens change, then they cannot be perfect—why else would they change? Yet according to Aristotelian physics, circular motion was reserved to the heavens because of their perfection, whereas the Earth could not be in circular motion because of its imperfection. But the new observations suggested that the heavens were not perfect, and so there was no reason why the motion of the Earth should be any different than the motion of a planet.

2. The Telescope

The observations of novas and comets were just the beginning of a large mass of new observations of a qualitatively different kind. In 1608 the telescope was invented. Initially it was intended for military purposes, but in 1609 an Englishman, Thomas Harriot, used a telescope to examine the surface of the Moon. Late the same year, Galileo Galilei (1564–1642), Professor of Mathematics at the University of Padua in Italy, constructed a greatly improved telescope, which he turned to the heavens. In his hands, the telescope became the source of many significant discoveries. Early in 1610 he published a slim book, the *Sidereus Nuncius* (or *Starry Messenger*), describing his observations.

He found that the surface of the Moon was very much like that of the Earth, with mountains and craters whose heights and depths he could estimate, and that the surface of the Sun had spots that developed and disappeared over time. Both of these discoveries indicated that, contrary to Aristotelian philosophy, not all heavenly objects are "perfect." He also found that there were other centers of rotation in the universe than the Earth—the planet Jupiter has moons and the Sun spins on its own axis. He was able to see that the planet Venus went through phases like the Moon; that is, it could be fully illuminated like a full moon or partially illuminated like a crescent moon. When fully illuminated it was much further away from the Earth than when it appeared as a crescent. This could only be explained if Venus circled about the Sun.

Galileo was also able to see that there were many more stars in the sky than had been previously realized—stars that were too far away to be seen by the naked eye. Moreover, even though his telescope in effect made objects seem as if they were 30 times closer than when seen by the naked eye, the stars remained mere pinpoints of light, thereby indicating that they were indeed at least as far away as Copernicus had claimed. In fact, as already mentioned in the discussion of the Copernican theory (Section C4), when studied with a telescope, the stars are actually smaller in apparent diameter than when seen with the naked eye, because of the greater aperture of the telescope. All this represented an experimental refutation of some of the objections raised against a heliocentric theory.

3. Beginnings of a New Physics

Earlier Galileo had begun the development of new principles of physics in opposition to those propounded by Aristotle. With these experiments Galileo was able to counter the scientific opposition to the Copernican system. This part of Galileo's work will be described in the next chapter. He was also a skilled publicist and polemicist, and participated in the scientific debates of the day. Indeed he was so devastating in written and oral argument that he made a number of enemies as well as friends.

4. Dialogues Concerning the Two Chief World Systems

By 1616, the religious schism arising from the Reformation and the Counter-Reformation had become so great that tolerance of deviations from established doctrines decreased significantly. In that year Galileo was warned not to advocate publicly the Copernican system. In 1623, however, Cardinal Barberini, a notable patron of the arts and sciences, became Pope. Thinking that a more favorable climate was at hand, Galileo felt that he could embark on a more vigorous advocacy of the Copernican system. He ultimately wrote a book, entitled *Dialogues Concerning the Two Chief World Systems,* which was directed to the educated lay public. The style of the book was that of Socratic dialogue, with three characters: an advocate of the Ptolemaic system, an advocate of the Copernican system, and an intelligent moderator. At the end of the book the Copernican advocate suddenly conceded the validity of the Ptolemaic system, although it was quite obvious to the reader that the Copernican system had been shown to be the correct one. In addition, the Ptolemaic advocate was depicted as somewhat of a simpleton.

Somehow Galileo managed to get the manuscript of the book approved by the ecclesiastical censors, and it was printed in 1632. Unfortunately, he had many enemies in high places, and they were quick to point out the true intent and implications of the book. In addition, it was asserted that Galileo intended to ridicule the highest officials of the Church. A few months later, the book was banned and all remaining copies were ordered destroyed. Galileo was called to trial before the Inquisition and, under threat of torture, compelled to repudiate his advocacy of Copernican doctrine. Perhaps because of his advanced age, poor health, and submission to the will of the Inquisition, his punishment was relatively light: house arrest and forced retirement. He died nine years later in 1642, shortly before his seventy-eighth birthday. He no longer advocated the Copernican system, but devoted his time to writing another book, *Discourses on Two New Sciences,* in which he detailed the results of his researches in mechanics and optics. This book was not published in Italy, but in the Netherlands, where the intellectual climate was more hospitable.

F. KEPLER'S HELIOCENTRIC THEORY

1. Pythagorean Mysticism

Both Tycho Brahe and Galileo Galilei used a relatively modern approach to scientific problems. Brahe insisted on the importance of systematic, extensive, and accurate collection of data. Galileo was not content simply to make observations, but recognized the necessity to alter experimental conditions in order to eliminate extraneous factors. Both men were very rational and logical in their approach to their work. Copernicus, on the other hand, and Kepler (of whom more will be said later) even more so, were concerned with the philosophical implications of their work. Kepler was well aware that the ancient Pythagoreans had set great store on finding simple numerical relationships among phenomena. It was the Pythagoreans who recognized that musical chords are much more pleasing if the pitches of their component notes are harmonics, that is, simple multiples of each other. The Pythagoreans also discovered that the pitch of notes from a lyre string, for example, are simply related to the length of the string, the

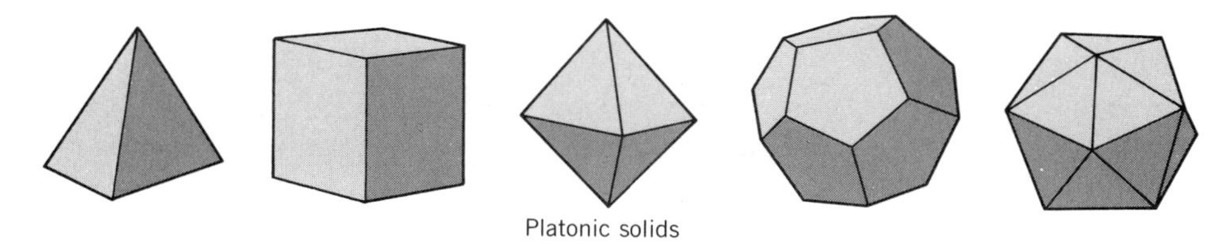

Figure 2.8. The Platonic solids. From left to right: tetrahedron, cube, octahedron, dodecahedron, icosahedron.

higher harmonics being obtained by changing the length of the string in simple fractions. Turning to the heavens, they asserted that each planetary sphere, as well as the sphere of the stars, emits its own characteristic musical sound. Most humans never notice such sounds, if only because they have heard them continuously from birth. The spatial intervals between the spheres were also supposed to be proportional to musical intervals, according to some of the legends about the Pythagoreans.

Johannes Kepler was born some 28 years after Copernicus's great work was published. He loved to indulge in neo-Pythagorean speculations on the nature of the heavens. His personality was undoubtedly affected by an unhappy childhood, relatively poor health, and a lack of friends in his youth. He showed mathematical genius as an adolescent, and as a young man succeeded in obtaining positions as a teacher of mathematics. His mathematical ability was soon recognized, and as a result his deficiencies as a teacher and his personality problems were often overlooked. In fact, he was such a poor teacher that he did not have many students, and this allowed him more time for astronomical studies and speculations.

Kepler was convinced of the essential validity of the heliocentric theory, and looked upon the Earth as one of the six planets that orbited the Sun. Because of his fascination with numbers, he wondered why there were six planets rather than some other number, why they were spaced as they were, and why they moved at their particular speeds. Seeking a connection to geometry, he recalled that the Greek mathematicians had proved there were only five "perfect" solids, known as the "Pythagorean" or "Platonic" solids, aside from a sphere.

A perfect solid is a multifaced figure, each of whose faces is identical to the other faces. Each of the faces is itself a perfect figure (see Fig. 2.8). A cube has six faces, all squares; a tetrahedron has four faces, all equilateral triangles; the eight faces of an octahedron are also equilateral triangles; the 12 faces of a dodecahedron are pentagons; and the 20 faces of the icosahedron are all equilateral triangles. Kepler proposed that the six spheres carrying the planets were part of a nested set of hollow solids, in which the spheres alternated with the Platonic solids. If the innermost and outermost solids are spheres, then there could be only six such spheres, corresponding to the number of known planets. The size of the five spaces between the spheres might be determined by carefully choosing the particular order of the Platonic solids, as shown in Fig. 2.9.

The largest sphere is the sphere of Saturn. Fitted to the inner surface of that

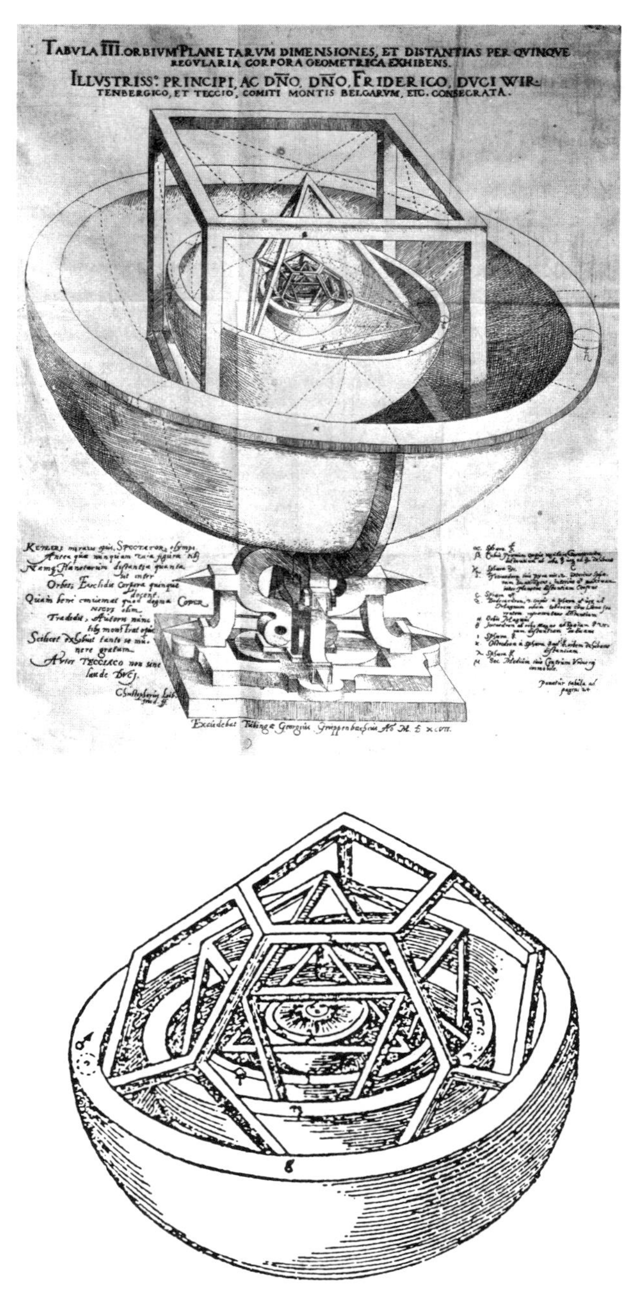

Figure 2.9. Kepler's nesting of spheres and Platonic solids to explain spacings and order of the planters. The lower drawing shows the spheres of Mars, Earth, Venus, and Mercury in more detail. Note the Sun at the center. (Top: Burndy Library; bottom: From illustrations in Kepler's *Mysterium Cosmographicum.*)

sphere is a cube with its corners touching the sphere. Inside the cube is the sphere of Jupiter, just large enough to be touching the cube at the middle of its faces. Inside the sphere of Jupiter is a tetrahedron, and inside that the sphere of Mars. Inside the sphere of Mars is the dodecahedron, then the sphere of the Earth, then the icosahedron, the sphere of Venus, then the octahedron, and finally the smallest sphere, the sphere of Mercury. With this order of nesting, the ratios of the diameters of the spheres are fairly close to the ratios of the average distances of the planets from the Sun.[10] Of course, Kepler knew that the spheres of the planets were not homocentric on the Sun; Copernicus had found it necessary to introduce eccentrics in his calculations. Kepler therefore made the spheres into spherical shells thick enough to accommodate the meanderings of the planets from perfect circular motion about the Sun. The results did not quite fit the data, but came close enough that Kepler wanted to pursue the idea further. He published this scheme in a book called the *Mysterium Cosmographicum* (the *Cosmic Mystery*), which became very well known and established his reputation as an able and creative mathematician and astronomer.[11]

2. Kepler's Mathematical Analysis of Tycho Brahe's Data

Kepler was not content to propose esoteric schemes for the order of planets in the heavens and to be satisfied if they worked approximately. He believed that any theory should agree with the available data quantitatively. He was aware, as were all the astronomers in Europe, of the great amount of accurate data that had been obtained by Tycho Brahe. Brahe, on the other hand, was aware of Kepler's mathematical ability. In 1600, Kepler became an assistant to Brahe, who had moved to the outskirts of Prague (now in the Czech Republic) and had become Court Mathematician to Rudolph II, Emperor of the Holy Roman Empire (the fanciful name by which the Austro-Hungarian Empire was once known). Kepler and Brahe did not always get along well, but Kepler found Brahe's data a very powerful attraction. Brahe died 18 months later in 1601, and Kepler immediately took custody of all the data. Shortly thereafter, in 1602, Kepler was appointed Brahe's successor as Court Mathematician to Rudolph II.

As mentioned earlier (Section D3), Kepler had been urged by Brahe to try to refine Brahe's compromise theory to fit the data, but he was not successful. He then returned to the heliocentric theory, but did not share Copernicus's objection to the equant (Section C3), so he included that in the calculations. He also made some improvements on some of Copernicus's detailed assumptions, making them

[10]The Earth even retains some uniqueness in this scheme. Its sphere rests between the two solids having the greatest number of faces, the dodecahedron and the icosahedron.

[11]Because we now know that there are at least nine planets, this scheme is of no value. The questions may still be asked about why the planets have the spacings they have, and if there is any rule that determines how many planets there are. Another scheme that was once thought valid (but no longer) is described by Bode's law or the Titius-Bode rule. A discussion of this rule is beyond the scope of this book, but a fascinating account of its history and the criteria determining whether such a rule is useful is given on pages 156–160 of the book by Holton and Brush, listed as a reference at the end of this chapter.

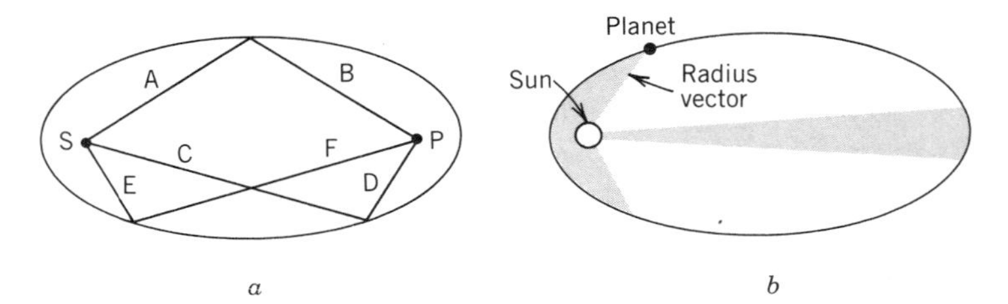

Figure 2.10. Elliptical orbits. (a) An ellipse: *S* and *P* are the two foci. The sum of the distances to any point on the ellipse from the foci is always the same. (b) Kepler's second law: The planet takes the same amount of time to traverse the elliptical arcs bounding the two shaded areas if the shaded areas are of equal size.

more consistent with a truly heliocentric theory. He concentrated particularly on the calculations for the orbit of Mars, because this seemed to be the most difficult one to adjust to the data. The best agreement he could get was on the average within 10 minutes of arc. Yet he felt that Brahe's data were so good that a satisfactory theory had to agree with the data within 2 minutes of arc. He felt that the key to the entire problem for all the planets lay in that 8-minute discrepancy. Gradually and painstakingly, over a period of two decades, punctuated by a variety of interruptions, including changes of employment, the death of his first wife, and a court battle to defend his mother against charges of witchcraft, he worked out a new theory.

3. Kepler's Laws of Planetary Motion

After spending five years studying the data for the planet Mars, Kepler despaired of finding the right combination of eccentric, deferent, epicycle, and equant that would make it possible to describe the orbit in terms of circular motion. He finally decided to abandon the circular shape, saying that if God did not want to make a circular orbit then such an orbit was not mandatory. Kepler then realized that his task was to determine the true shape of the orbit. Five years later, in 1609, he announced as a partial solution to the problem his first two laws of planetary motion. Still later, in 1619, he announced the rest of the solution, his third law of planetary motion.

Kepler's laws of planetary motion, as expressed today, are:

I. The orbit of each planet about the Sun is an ellipse with the Sun at one "focus" of the ellipse.

An ellipse can be contrasted with a circle. A circle surrounds a special point called the center, and the distance from the center to any point on the circle is always the same (constant). An ellipse surrounds *two* special points, called foci, and the *sum* of the distances from each focus to any point on the ellipse is always

Table 2.1. Data Illustrating Kepler's Third Law of Planetary Motion for the Solar System

Planet	T	D	T^2	D^3
Mercury	0.24	0.39	0.058	0.059
Venus	0.62	0.72	0.38	0.37
Earth	1.00	1.00	1.00	1.00
Mars	1.88	1.53	3.53	3.58
Jupiter	11.9	5.21	142	141
Saturn	29.5	9.55	870	871

the same (constant). This is illustrated in Fig. 2.10a, where the sum of the lengths of the pairs of lines *A* and *B*, or *C* and *D*, or *E* and *F*, is always the same.

II. The line joining the Sun and the planet (called the radius vector) sweeps over equal areas in equal times as the planet travels around the orbit.

In Fig. 2.10b, the planet takes the same length of time to travel along the elliptical arcs bounding the two shaded areas because the areas are the same size.

III. The square of the period of revolution (time for one complete orbit) of a planet about the Sun is proportional to the cube of the average distance of the planet from the Sun.

Mathematically, this means that the ratio T^2 to D^3, where T stands for the period and D the average distance, is the same for all planets.

The third law is illustrated in Table 2.1 above, which gives the period T (in years) of the planets; their average distance, D (in astronomical units, with one astronomical unit equal to 93 million miles); the period squared, T^2; and the average distance cubed, D^3. With these units, the ratio is equal to one, that is $T^2 = D^3$.

With these three laws, Kepler completed the program set forth by Copernicus. Copernicus made the Sun the center of the planetary system; Kepler discarded the idea that circular motion was required for heavenly bodies. By using elliptical orbits, the true beauty and simplicity of the heliocentric theory was preserved, in excellent agreement with the data, and with no need for such devices as the eccentric, epicycle, or equant. The fact that the focus of an ellipse is not in the center supplies eccentricity. Kepler's second law performs the same function as an equant, in that it accounts for the varying speed of the planet as it travels around the orbit.

Kepler went further and pointed to the Sun as being the prime mover or causative agent, which he called the **anima motrix.** He put the cause of the motion at the center of the action, rather than at the periphery as Aristotle had done. Kepler had a primitive idea about gravitational forces exerted by the Sun but failed to connect gravity to planetary orbits. Rather he asserted that the Sun exerted its influence through a combination of rays emitted by the Sun and the natural magnetism of the planet. He was influenced in this belief by some clever demonstrations of magnetic effects by William Gilbert, an English physician and physicist.

Interestingly enough, Kepler discovered his second law, dealing with the speed of the planet as it travels around the orbit, before he discovered the exact shape of the orbit, as given by the first law. The complete translated title of the book he published in 1609 on the first two laws is *A New Astronomy Based on Causation, or a Physics of the Sky Derived from Investigations of the Motions of the Star Mars, Founded on Observations of the Noble Tycho Brahe.* The book is usually referred to as the *New Astronomy.* Kepler had actually completed an outline of the book in 1605, but it took four years to get it printed because of a dispute with Tycho Brahe's heirs over the ownership of the data.

Kepler's third law, called the Harmonic Law, was described in a book called *Harmony of the World,* published while he held a new and lesser position as Provincial Mathematician in the town of Linz, Austria (his former patron, the Emperor Rudolph II, had been forced to abdicate his throne, and so Kepler lost his job). In this book Kepler returned to the Pythagorean mysticism of his early days, searching for harmonic relationships among the distances of the planets from the Sun (Section F1). Instead of finding a relationship between the "musical notes" of the heavenly spheres and their size, he found the relationship between the speed of the planets (i.e., their periods), which he somehow associated with musical notes and harmony, and the size of their orbits. He supported the connections with harmony by some surprisingly accurate calculations.

OPTIONAL SECTION 2.2 *Kepler's Laws and Satellite Orbits*

As will be discussed in Chapter 3, Kepler's Laws of Planetary Motion apply equally well to satellites orbiting planets. For example, the time it takes an artificial satellite to make a complete orbit about the Earth depends upon its altitude above the Earth, and can be easily calculated from a knowledge of the distance to the Moon and the period of the Moon. Suppose it were desired to place a satellite in a 3-hour circumpolar orbit, that is, passing over the North and South Poles. Kepler's Third Law says that $T_1^2 = k\ D_1^3$ for the satellite and $T_2^2 = k\ D_2^3$ for the Moon, with k the constant of proportionality. If the first relation is divided by the second, k cancels out and the result is $(T_1/T_2)^2 = (D_1/D_2)^3$ or $D_1^3 = D_2^3\ (T_1/T_2)^2$ where T_1 and T_2 are the times for a complete orbit of the satellite and the Moon, respectively, and D_1 and D_2 are the distances of the satellite and the Moon from the center of the Earth. We want T_1 to equal 3 hours, and we know that $T_2 = 28\frac{1}{4}$ days $= 678$ hours. We also know the Earth–Moon distance is 250 thousand miles. Putting these numbers into the formula above, $D_1^3 = (250)^3\ (3/678)^2$. At this point, it is best to take the cube root and therefore $D_1 = 250\ (3/678)^{2/3} = 6.7$ thousand miles from the center of the Earth after rounding off the answer. Since the radius of the Earth is about 4 thousand miles, the satellite should be placed in an orbit which is $6.7 - 4 = 2.7$ thousand miles above the surface of the Earth.

Note how close the satellite is to the surface of the Earth as compared to the Moon. This illustrates the general rule of Kepler's Laws, the closer the satellite to the Earth (or planet to the Sun), the faster it travels.

It is clear that Kepler was the right person at the right time to discover the laws of planetary motion. First and perhaps most significantly, Kepler was an intense Pythagorean mystic. He was an extremely capable mathematician who believed that the universe was full of mathematical harmonies. Because he was so interested in figures and shapes, it was relatively easy for him to consider elliptical orbits—once he finally decided that circular constructions must be wrong. Finally, Kepler was Brahe's assistant, so that he had access to Brahe's measurements and knew how accurate they were. No one else would have had the ability or the interest to perform all the mathematical calculations required to discover the laws of planetary motion, and no one else would have regarded an 8-minute angular discrepancy as significant.

OPTIONAL SECTION 2.3. *Accepted Astronomical Dimensions Prior to the Telescope*

At the time that Copernicus revived the heliocentric theory, there was a generally accepted set of data on the sizes of the various planets and their distances from the Earth. Although individual investigators used somewhat different values for these quantities, with very few exceptions the differences were not very large. This was true of the supporters of both the heliocentric and the geocentric systems.

In the geocentric systems these distances were expressed in terms of the Earth's radius. In the Ptolemaic system, the distance from the Earth to the Moon varied from about 33 to about 64 times the Earth's radius, with an average of about 48. This was determined by the use of "eclipse" diagrams, which noted that a total eclipse of the Sun by the Moon, as seen from one location on the surface of the Earth would be seen as a partial eclipse from some other location on the Earth's surface. Knowing the distance between the two locations on the Earth, the distance from the Earth to the Moon can be calculated from geometry. Using this distance, it is possible to return to the eclipse diagram and calculate the distance to the Sun as ranging between about 1100 and 1200 times the Earth's radius, corresponding to 4.4 to 4.8 million miles. (The correct value is about 20 times larger. The method requires accurate measurements of the apparent size of the Sun as seen in different positions. The calculations are of such a nature that the results are very sensitive to small errors in these measurements and to correction factors that are necessary because of different thicknesses of the atmosphere traversed by the line of sight to the Sun.)

Taking into account the dimensions of the deferents and epicycles for the various other planets, it is possible to calculate the range of distances of each planet from the Earth. These distances were used to specify the dimensions of a spherical shell belonging to each planet. It was also required that there be no empty or unused space in the heavens (Aristotle had asserted that Nature abhors a vacuum). This meant that each shell had to fit tightly inside the next larger shell, so that the universe could be imagined as being like a giant onion with distinct layers. (The layers were made up of Aristotle's fifth element, the ether, which was essential for celestial motion.) For example, the shell for Mars had an inner radius of 1260 and an outer radius of 8820 earth radii, while Jupiter's shell had radii of 8820 and 14,187 Earth radii and Saturn's shell had radii of 14,187 and 19,865 Earth radii. The outermost layer was the sphere of the "fixed stars," considered to have essentially zero thickness, with a radius about 19,000 or 20,000 Earth radii, thus determining the size of the universe. The radius of the Earth had quite early been determined to be about 4000 miles, and thus it was generally accepted that the radius of the universe should be about 80 million miles.

Copernicus chose the distance from the Earth to the Sun to be about the same as that chosen by Ptolemy, that is, about 1200 Earth radii. After Copernicus adapted eccentrics and epicycles to his system, there was a substantial range of distances of each planet from the Sun, but the resulting shell was still thin enough as to leave large volumes of "empty space" that were not traversed by a planet. The range of distances for Saturn, the most distant planet from the Sun, was 9881 to 11,073 earth radii. If the sphere of the fixed stars had a radius of 11,073 earth radii, then the Copernican universe would have been even smaller than the Ptolemaic universe, and star parallax as a result of the Earth's movement along its orbit should have been very easily observed. It was for this reason that Copernicus suggested that the sphere of the fixed stars should be very large, because no parallax was observed. Tycho Brahe calculated that in the Copernican universe the fixed stars should be at a distance of at least 8 million earth radii, that is, 32,000 million miles; otherwise he would have detected star parallax. This meant even more unused or empty space, which was simply unacceptable to him and most other learned people, who were thoroughly imbued with the Aristotelian concepts of nature. (Remarkably enough, there was one medieval astronomer, Levi ben Gerson, who set the distance to the fixed stars as being even 5000 times larger than the value that Tycho had rejected.)

One objection to a universe containing empty space was that there seemed to be no way to explain the sizes and positions of the void regions and the planetary shells. Thus one of the goals of Kepler's *Mysterium Cosmographicum* was to give a reason for the sizes of the shells for each planet.

Within the 30 or 40 years after Galileo first began using the telescope to study the heavens, the measurements of the apparent size of the Sun and of the planets became sufficiently refined that it was recognized that the previously accepted values for the distance from the Earth to the Sun were far too small. By the year 1675 a new type of telescope had come into widespread use and new measuring techniques were developed. As a result the accuracy of the measurements was significantly improved. By the year 1700 it was generally accepted that the distance from the Earth to the Sun was over 90 million miles. This compares well with the modern value of 93 million miles. Once this distance is known, it is possible to scale the distances to all the other planets by use of Kepler's third law of planetary motion.

4. Slow Acceptance of Kepler's Theory

Kepler sent copies of his work to many of the best known astronomers of his day and corresponded copiously with them. In general, although he was greatly respected, the significance and implications of his results remained unrecognized for a long time, except by a few English astronomers and philosophers. Most astronomers had become so accustomed to the relatively poor agreement between theory and data, and so imbued with the idea of circular motion, that they were not impressed with Kepler's results.

Galileo, for example, totally ignored Kepler's work even though they had corresponded and Galileo was quick to seek Kepler's approval of his own telescopic observations. It is not clear whether Galileo was put off by Kepler's Pythagorean mysticism, or whether he was so convinced of circular motion as the natural motion for the Earth and other planets that he simply could not accept any other explanation. The fact is, however, that Galileo did not embark on his telescope work until after Kepler published the *New Astronomy,* nor did he publish his *Dialogues* until 1632, two years after Kepler's death.

Only after Kepler calculated a new set of astronomical tables, called the *Rudolphine Tables* (which were published in 1627), was it possible to use his ideas and the data of Tycho Brahe to calculate calendars that were much more accurate than any previously calculated. For example, discrepancies in tables based on the Copernican theory had been found fairly quickly by Tycho Brahe within about 10 years of their publication; the *Rudolphine Tables* were used for about 100 years. During that time Kepler's work gradually became more and more accepted. It was not until some 70 years later, however, when Isaac Newton published his laws of motion and his law of universal gravitation in order to explain how Kepler's laws had come about, that they became completely accepted. The work of Newton is discussed in the next chapter.

5. The Course of Scientific Revolutions

The replacement of the geocentric theory of the universe by the heliocentric theory is often referred to as the *Copernican revolution.* The Copernican "revolution" is the archetype of scientific revolution, against which other scientific revolutions are compared. In fact, no other scientific revolution can match it in terms of its effect on modes of thought, at least in Western civilization. The only one that comes close to it in that respect is the Darwinian revolution, which is associated with the development of biological evolutionary theory.

The Copernican revolution is often pointed to as illustrating the "conflict between Science and Religion"; yet its leaders, Copernicus, Galileo, and Kepler, were devoutly religious. On the other side, within the religious establishment, there were many who did not follow the narrowly literal interpretation of religious dogma. Initially, at least, the biggest fear of the revolutionaries was not censure by religious authorities, but rather ridicule by their scientific peers. How could they challenge the established body of scientific knowledge and belief?

The revolution, moreover, did not take place overnight. The time span from the serious studies of Copernicus to Newton's publication of his gravitational studies was about 150 years. The seeds of the revolution went back a long time, about 2000 years before Copernicus, to Philolaus, a Pythagorean. As the older theories began to accumulate more and more anomalous and unexplained facts, the burden of necessary modifications simply became overwhelming, and the older theories broke down. Nevertheless, the older theories had been phenomenally successful. In fact, there were only a few heavenly bodies out of the whole host of the heavens (i.e., the planets) that caused major difficulties for the established theory. The revolution came about because of quantitative improvements in the data and the recognition of subtle, rather than gross, discrepancies, not only in the data, but in the way in which the data were assimilated. The development of new technologies and instruments that yielded entire new classes and qualities of data also played a large role. At the same time, the new concepts themselves initially could not account for all the data, and their advocates were required to be almost unreasonably persistent in the face of severe criticism. A new theory is expected, quite properly, to overcome a large number of objections, and even then some of the advocates of the old theory are never really convinced.

Not all revolutionary ideas in science are successful, or even deserve to be. If they fail to explain the existing facts or future facts to be developed, or if they are incapable of modification, development, and evolution in the face of more facts, they are of little value. The heliocentric theory was successful because it was adaptable. Kepler, Copernicus, and others were able to incorporate into the scheme such things as the eccentrics and epicycles, and finally a different shape orbital figure, the ellipse, without destroying the essential idea, so that it could be stretched to accommodate a wide variety of experimental facts.

Study Questions

1. What is meant by irrational numbers? Give some examples.
2. What were the connections perceived by the Pythagoreans between irrational numbers, geometry, and the concept of atoms?
3. Why did the ancient Greeks consider a circle to be a perfect figure?
4. What does Plato's Allegory of the Cave have to do with science?
5. Why did the ancient Greeks consider that heavenly objects had to move in circular paths?
6. What is the stellar or celestial sphere? Describe its appearance.
7. What is the celestial equator, celestial pole, celestial horizon, polar axis?
8. What is the ecliptic?
9. What is diurnal rotation?
10. Describe the annual motion of the Sun.

11. What is retrograde motion? alternate motion?
12. What is meant by a geocentric model of the universe?
13. What is meant by a homocentric model of the universe?
14. Name the prime substances described by Aristotle.
15. Name and describe the devices used by Ptolemy in his calculations.
16. What particular aspects of Ptolemy's approach did Copernicus not like?
17. How does the heliocentric theory account for retrograde motion?
18. Which of Ptolemy's devices did Copernicus have to use and why?
19. List and describe the scientific objections to the Copernican theory.
20. What were the philosophical and religious objections to the Copernican theory?
21. What are the common characteristics of Tychonic theories of the universe?
22. What was Tycho Brahe's real contribution to the science of astronomy?
23. What were the new discoveries and arguments that tipped the balance toward the heliocentric theory?
24. State Kepler's laws of planetary motion.
25. Why were Kepler's laws of planetary motion not immediately grasped by Galileo? (Part of the answer to this question is in the next chapter.)

QUESTIONS FOR FURTHER THOUGHT AND DISCUSSION

1. Discuss the similarities and differences between the reception of the heliocentric description of the universe and the reception of Darwin's ideas of evolution.
2. In light of what you understand about the history of scientific revolutions, what should be your attitude toward claims by proponents of extra-sensory perception (ESP), dianetics, precognition?
3. Some people contend that with the aid of modern computers, it is possible to construct a geocentric theory that would agree with astronomical observations—all that is necessary is to introduce enough eccentrics, epicycles, equants, and even elliptical orbits. Would such a theory be acceptable? Why or why not?

PROBLEMS

1. The calculated speed of the Earth in the orbit proposed by Copernicus is about 3500 miles per hour. Show how to calculate how far it will travel in 1 sec. (*Hint:* Convert miles per hour to miles per second.)

2. Calculate the length of a "pointer" that would give an accuracy of 1 sec of arc for a $\frac{1}{25}$-inch error in positioning its far end. (*Hint:* Tycho Brahe claimed an accuracy of 4 minutes of arc and this required a 3-ft-long pointer.) Does your answer seem to be a reasonable length, i.e., one that could be easily handled?

3. Satellites that are used for worldwide telecommunications are apparently stationary over one spot on the Earth. They are not really stationary but rather orbit the Earth with the same period as the Earth's rotation about the axis. Calculate the altitude of these satellites above the Earth's surface. (*Hint:* See the discussion of circumpolar orbits at the end of Section F3 above.)

References

E. A. Abbott, *Flatland,* Dover Publications, New York, 1952.
A very entertaining book, originally written a century ago, about a two-dimensional world.

Cyril Bailey, *The Greek Atomists and Epicurus,* Russell and Russell, New York, 1964.
This was originally written in 1928. Pages 64–65 give a brief discussion of the Indian philosopher Kanada and his atomic theory, and a few references. More recent material is contained in an article by Alok Kumar to be published in the forthcoming Encyclopedia of the History of Science, Technology and Medicine in non-Western Cultures (1995).

Arthur Berry, *A Short History of Astronomy: From Earliest Times Through the Nineteenth Century,* Dover Publications, New York, 1961.
Chapter II discusses the contributions of the Greeks.

J. Bronowski, *The Ascent of Man,* Little, Brown and Company, Boston, 1973.
Based on a television series of the same name. Chapters 5, 6, and 7 are pertinent to this chapter.

Herbert Butterfield, *The Origins of Modern Science,* rev. ed., Free Press, New York, 1965.
A brief book written by a historian, which discusses scientific developments up to the time of Isaac Newton.

F. M. Cornford, *The Republic of Plato,* Oxford University Press, London, 1941.
An account of Plato's Allegory of the Cave is found here.

J. L. E. Dreyer, *A History of Astronomy from Thales to Kepler,* 2nd ed., Dover Publications, New York, 1953.
Provides a fairly detailed account of the subject matter of this chapter.

Gerald Holton and Stephen G. Brush, *Introduction to Concepts and Theories in Physical Science,* 2nd ed., Addison-Wesley, Reading, Mass., 1973.
Chapters 1 through 5 and pp. 154–160 are pertinent, with many references to other sources.

Hugh Kearney, *Science and Change, 1500–1700,* McGraw-Hill, New York, 1971.
A small book that discusses the general scientific ferment in Europe during the period 1500–1700.

Arthur Koestler, *The Watershed* (Anchor Books), Doubleday, Garden City, New York, 1960.
A thin biographical discussion of Kepler's life and work.

Alexander Koyre, *Discovering Plato,* Columbia University Press, New York, 1945.
This book also discusses Plato's views about science and philosophy.

Thomas S. Kuhn, *The Copernican Revolution,* Random House, New York, 1957.
A detailed discussion of the subject of this chapter.

Thomas S. Kuhn, *The Structure of Scientific Revolutions,* 2nd ed., University of Chicago Press, Chicago, 1970.
See comment in references for Chapter 1.

Cecil J. Schneer, *The Evolution of Physical Science,* Grove Press, New York, 1964 (a reprint of *The Search for Order,* Harper & Row, New York, 1960).
See comment in references for Chapter 1.

Stephen Toulmin and June Goodfield, *The Fabric of the Heavens,* Harper & Row, New York, 1961.

Albert Van Helden, *Measuring the Universe: Cosmic Dimensions from Aristarchus to Halley,* University of Chicago Press, Chicago, 1985.
A thorough discussion by a historian of the various methods used by astronomers to determine distances and apparent sizes of heavenly objects, based on examination and evaluation of original documents.

Isaac Newton. (Photo courtesy Yerkes Observatory, University of Chicago.)

Chapter 3

NEWTONIAN MECHANICS AND CAUSALITY

The universe is a mechanism run by rules

It is now appropriate to discuss the development of that branch of physics called mechanics. This subject is concerned with the proper description and the causes of the motion of material objects. It was the development of mechanics, particularly the work of Newton, which finally "explained" the laws of planetary motion discussed in Chapter 2. The science of mechanics not only explains the motions of celestial bodies, but also the motions of earthly objects, including falling bodies and projectiles. In fact, following the work of Newton, it appeared that the laws of mechanics could explain nearly all the basic physical phenomena of the universe. This success was widely acclaimed and contributed significantly to the development of the "Age of Reason," a time when many scholars believed that all of the sciences, including the social sciences and economics, could be explained in terms of a few basic underlying principles.

A. Aristotelian Physics

1. Aristotle's Methods

Among the earliest contributors to the science of mechanics were the ancient Greek philosophers, particularly Aristotle. The Greeks inherited various strands of knowledge and ideas concerning natural phenomena from previous civilizations, especially from the Egyptian and Mesopotamian cultures. Although a large amount of knowledge had been accumulated (considerably more than most people realize), very little had been done to *understand* this information. Explanations were generally provided by the various religions, but not by a series of rational arguments starting from a few basic assumptions. Thus, for example, the Sun was said to travel across the sky because it rode in the sun god's chariot. Such an explanation clearly does not need a logical argument to support the conclusion.

It was the early Greeks, more than any of the earlier peoples, who began to try to replace the "religious" explanations with logical arguments based on simple, underlying theories. Their theories were not always correct, but their important contribution was the development of a *method* for dealing with physical phenomena. That method is now called the scientific method, and it is the basic mode of inquiry for science today. The essential objective of this scientific method was clearly stated by Plato in his "Allegory of the Cave" as discussed in Chapter 2, Section A2. The goal of science (or natural philosophy as it was previously called) is to "explain" the appearances, that is, to explain what is observed in terms of the basic underlying principles.

2. Prime Substances

So far as is known, Aristotle (384–322 B.C.) was the first philosopher to set down a comprehensive system on a set of simple assumptions, from which he could rationally explain all the known physical phenomena. Chapter 2, Section B3, gives a brief introduction to his system in connection with the movements of the heavenly bodies, but it needs to be discussed more fully. Aristotle wished to explain the basic properties of all known material bodies; these properties included their weights, hardnesses, and motions. His descriptions of the physical universe were *matter dominated.* He began from only a few basic assumptions, and then proceeded to try to explain more complicated observations in terms of these simple assumptions.

Aristotle believed that all matter was composed of varying amounts of four prime substances. These were earth, water, air, and fire. He believed that objects in the heavens were composed of a heavenly substance called ether, but Earthly objects did not contain this heavenly substance. Thus most of the heaviest objects were regarded as earth, whereas lighter objects included significant amounts of water or air. The different characteristics of all known substances were explained in terms of different combinations of the four primary substances. Furthermore, and more importantly for his development of the subject of mechanics, Aristotle believed that motions of objects could be explained as due to the basic natures of the prime substances.

3. Motion According to Aristotle

Aristotle enumerated four basic kinds of motion that could be observed in the physical universe. These were (1) **alteration,** (2) **natural local motion,** (3) **horizontal or violent motion,** and (4) **celestial motion.** The first kind of motion discussed by Aristotle, alteration (or change), is not regarded as motion at all in terms of modern definitions. For Aristotle, the basic characteristic of motion is that the physical system in question changes. Thus he regarded rusting of iron, or leaves turning color, or colors fading, as a kind of motion. Such changes are now considered to be part of the science of chemistry. We now consider motion to mean the physical displacements of an object, and rusting or fading involves such displacements only at the submicroscopic level. The understanding of these kinds of "motions" is beyond the scope and purpose of this discussion. Some aspects of molecular and atomic motions will be discussed later (but not the actual description of the particular motions present in chemical processes).

The second of Aristotle's kinds of motion, natural local motion, is at the center of his basic ideas concerning the nature of motion. For him, natural motion was either "up" or "down" motion. (The Greeks believed that the Earth is a sphere and that "down" meant toward the center of the Earth and "up" meant directly away from the center of the Earth.) Aristotle knew that most objects, if simply released with no constraints on their motion, will drop downward, but that some things (such as fire, smoke, and hot gases) will rise upward instead. Aristotle considered these downward and upward motions to be natural and due to the dominant natures of the objects, because the objects did not need to be pushed or pulled. According to Aristotle, the "natural" order was for earth to be at the bottom, next water, then air, and finally fire at the top. Thus earth (i.e., rocks, sand, etc.) was said to want to move naturally toward the center of the Earth, the natural resting place of all earthly material. Water would then naturally move to be on top of the earth.

Objects dominated by either of the prime substances fire or air, however, would naturally rise, Aristotle believed, in order to move to be on top of earth and water. Fire was "naturally" the one at the very top. Aristotle viewed all "natural" motion as being due to objects striving to be more perfect. He believed that earth and water became more perfect as they moved downward toward their "natural" resting place, and that fire and air became more perfect as they moved upward toward their "natural" resting place.

Aristotle did not stop with his explanation of why objects fall or rise when released, but he went further and tried to understand *how* different sorts of objects fall, relative to each other. He studied how heavy and light objects fall and how they fall in different media, such as air and water. He concluded that in denser media, objects fall more slowly, and that heavy objects fall relatively faster than light objects, especially in dense media. He realized that in less dense media (less resistive we say), such as air, heavier and lighter objects fall more nearly at the same speed. He even correctly predicted that all objects would fall with the same speed in a vacuum. However, he incorrectly predicted that this speed would be infinitely large. Infinite speeds would mean that the object could be in two places at once (because it would take no time at all to move from one place to another), he argued, which is absurd! Therefore he concluded that a complete vacuum must be impossible ("nature abhors a vacuum"). As will be discussed below, this last conclusion was later the center of a long controversy and caused some difficulties in the progress of scientific thought.

Although not all of Aristotle's conclusions were correct regarding the nature of falling objects, it is important to understand that the methods he employed constituted at least as much of a contribution as the results themselves. Aristotle relied heavily on careful observation. He then made theories or hypotheses to explain what he saw. He refined his observations and theories until he felt he understood what happened, and why.

Aristotle's third kind of motion concerned horizontal motions, which he divided into two basic types: objects that are continually pushed or pulled, and thrown or struck objects, that is, projectiles. He considered horizontal motion to be unnatural, that is, it did not arise because of the nature of the object and would not occur spontaneously when an object was released. The first type, which includes such examples as a wagon being pulled, a block being pushed, or a person walking, did not present any difficulty. The cause of the motion appears to reside in the person or animal doing the pulling, pushing, or walking.

The second type of horizontal motion, projectile motion, was harder for Aristotle to understand and represents one of the few areas of physical phenomena in which he seemed somewhat unsure of his explanations. The difficulty for him was not what caused the projectile to start its motion, that was as obvious as the source of pushed or pulled motion, but rather what made the projectile keep going after it had been thrown or struck. He asked the simple question: "Why does a projectile keep moving?" Although this seems to be a very obvious and important question, we will see below that there is a better and more fruitful question to ask about projectile motion.

Aristotle finally arrived at an explanation for projectile motion, although one senses that he was not completely convinced by his solution. He suggested that the air pushed aside by the front of a projectile comes around and rushes in to fill the temporary void (empty space) created as the projectile moves forward, and that this inrushing air pushed the projectile along. The act of throwing or striking the object starts this process, which continues by itself once begun. This entire process came to be known as **antiperistasis.** This particular part of Aristotle's physics was the first to be seriously challenged. Aristotle himself wondered if this were really the correct analysis, and therefore also suggested that perhaps the projectile motion continued because a column of air was set into motion in front of and alongside the object during the throwing process. This moving column of air, he suggested, might then ''drag'' the projectile along with it. This bit of equivocation by Aristotle is very unusual and emphasizes his own doubts about his understanding of projectile motion.

Aristotle's final kind of motion was celestial or planetary motion. He believed that objects in the heavens were completely different from earthly objects. He believed heavenly objects to be massless, perfect objects made up of the celestial ether. He accepted the idea of the Pythagoreans that the only truly perfect figure is a circle (or a sphere, in three dimensions), and thus argued that these perfect objects were all spheres moving in perfect circular orbits. This celestial system has been described in more detail in Chapter 2, Section B. It is important to note that Aristotle considered heavenly objects and motions to obey different laws than earthly, imperfect objects. He regarded the Earth as characterized by imperfection in contrast to the heavens. In fact, he believed that the Earth was the center of imperfection because things were not completely in their natural resting places. The earthly substances earth and water had to move downward to be in their natural resting place, and air and fire had to move upward to be in their natural resting place. The Earth was imperfect because some of these substances were mixed together and not in their natural places.

4. Significance of Aristotelian Physics

Because Aristotelian physics has been largely discarded, there is a tendency to underestimate the importance of Aristotle's contributions to science. His work represents the first ''successful'' description of the physical universe in terms of logical arguments based on a few simple and plausible assumptions. That his system was reasonable, and nicely self-consistent, is testified to by the fact that for nearly 2000 years it was generally regarded as the correct description of the universe by the Western civilized world. His physics was accepted by the Catholic Church as dogma. (Aristotle's physics was still being taught in certain Catholic schools and even in some Islamic universities in the twentieth century.)

But much more important than the actual system he devised was the method he introduced for describing the physical universe. We still accept (or hope) that the correct description can start from a few simple assumptions and then proceed through logical arguments to describe even rather complicated situations. Important also were the questions he asked: ''What are objects made of?'' ''Why do objects fall?'' ''Why do the Sun, Moon, and stars move?'' ''Why do projectiles keep moving?'' It was Aristotle, more than any other individual, who first phrased the basic questions of physics.

5. Criticisms and Commentaries on Aristotle's Physics

With the decline of Greek political power and the rise of Rome, scientific progress in the ancient world slowed to a crawl. Although the Roman Empire gave much to Western civilization in certain areas, it reverted largely to the earlier practice of explaining physical phenomena in religious terms. What scientific progress there was (and there was little compared with the large contributions of the Greek philosophers) came from the eastern Mediterranean. Alexandria, in Egypt, was the site of a famous library and a center for scholars for several hundred years. It was in Alexandria that Claudius Ptolemy (about 500 years after Aristotle) compiled his observations of the wandering stars and devised his complex geocentric system, as discussed in Chapter 2.

Another 350 years after Ptolemy, in the year 500, also in Alexandria, John Philoponus (''John the Grammarian'') expressed some of the first recorded serious criticism of Aristotle's physics. John Philoponus criticized Aristotle's explanation of projectile motion, namely, the notion of antiperistasis. Philoponus argued that a projectile acquires some kind of motive force as it is set into motion. This is much closer to our modern understanding, as we shall see, that the projectile is given momentum, which maintains it in motion. Philoponus also questioned Aristotle's conclusion that objects of different weights would fall with the same speed *only* in a vacuum (which Aristotle rejected as being impossible). Philoponus demonstrated that objects of very different weights fell with essentially the same speeds in air. It is remarkable, however, that these first serious criticisms came more than 800 years after Aristotle's death.

With the fall of the Roman Empire and the beginning of the ''dark ages'' in Europe, and in particular with the destruction of the library and scholarly center in Alexandria, essentially all scientific progress ended. Some work, especially in astronomy and in optics, was carried on by Arab scholars. Finally, another 700 years later, the Renaissance began and European scholars rediscovered the early Greek contributions to philosophy and art. Much of the early work had to be obtained from Arabic texts. In the thirteenth century, St. Thomas Aquinas (1225–1274) adroitly reconciled the philosophy of Aristotle to the Catholic Church. By the beginning of the fourteenth century, Aristotle's philosophical system was church dogma and thereby became a legitimate subject for study. Fundamental questions regarding the nature of motion were finally raised again, particularly by various scholars at the University of Paris and at Merton College in Oxford, England. William of Ockham (1285–1349), and Jean Buridan (1295–1358) were noted scholars of motion and became serious critics of some of Aristotle's ideas. Buridan resumed the attack started 800 years earlier by Philoponus on Aristotle's explanations of projectile motion. Buridan provided examples of objects for which, he emphasized, the explanation of antiperistasis clearly would not work; for example, a spear sharpened at both ends (how could the air push on the back end?), and a millwheel (with no back end at all). Buridan also rejected the alternate explanation that a column of moving air created in front of and alongside a projectile kept the object moving. He noted that if this were true, one could start an object in motion by creating the moving air column first—and all attempts to do this fail. Buridan, like Philoponus, concluded that a moving object must be given something which keeps it moving. He referred to this something as **impulse,** and felt that as an object kept moving it continually used up its impulse. When the impulse originally delivered to the object was exhausted, it would then stop moving. He likened this process to the heat gained by an iron poker placed in a fire. When the poker is removed from the fire, it clearly retains something that keeps it hot. Slowly, whatever is acquired from the fire is exhausted, and the poker cools. As is discussed in Section B3 below, Buridan's impulse is close to the correct description of the underlying "reality," eventually provided by Galileo and Newton.

The level of criticism of Aristotle's physics increased. In 1277, a religious council meeting in Paris condemned a number of Aristotelian theses, including the view that a vacuum is impossible. The council concluded that God could create a vacuum if He so desired. Besides the specific criticisms by Buridan, scholars of this era produced, for the first time, precise descriptions and definitions of different kinds of motion and introduced graphical representations to aid their studies. These contributions, chiefly by the Scholastics in Paris and the Mertonians in England, provided the better abstractions and idealizations that made possible more precise formulations of the problems of motion; for example, Nicole Oresme (1323–1382) invented the velocity–time graph (see below) and showed a complete understanding of the concept of relative velocity. This ability to idealize and better define the problem was essential for further progress, as discussed below.

B. Galilean Mechanics

1. Graphical Representations of Motion

A familiar example of motion is shown in Fig. 3.1, as a graph of speed versus time. The example is a hypothetical trip in an automobile. The vertical axis (the ordinate) indicates the auto's speed at any time in miles per hour (mi/h). The

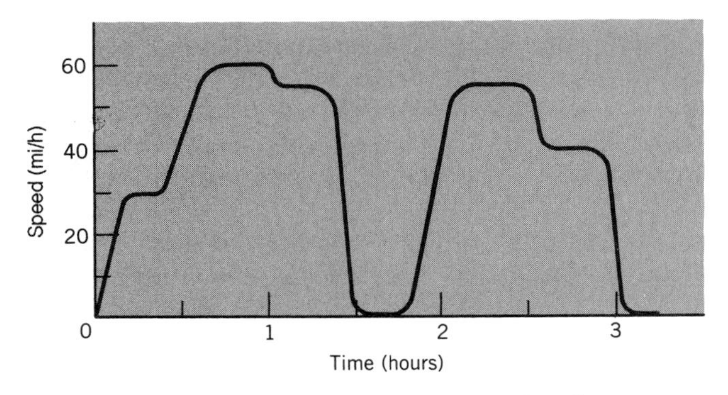

Figure 3.1. Hypothetical automobile trip.

horizontal axis (the abscissa) indicates the elapsed time in hours. The car starts the trip by accelerating (speeding up) to about 30 mi/h and then proceeding at this speed until reaching the edge of town, about one-half hour after starting. At the edge of town, the driver accelerates up to about 60 mi/h, on the open road. After about 1 hour from the beginning of the trip, the driver notices a State Police car and slows down (decelerates) to the speed limit of 55 mi/h. About one-half hour later, the driver pulls into a rest stop, decelerating to zero speed. After a short rest of about 15 minutes, the driver starts again, immediately accelerating to 55 mi/h, and continues until reaching the town of his destination. At the town limits the driver slows down to 40 mi/h until reaching the exact spot of his final destination, where he comes to a complete stop.

The example is simple and easy to follow. The important point, however, is that the entire trip can be recorded on the speed versus time graph, and the graph can be used to determine the motion of the car at any time during the trip. The basic kinds of motion are demonstrated in the hypothetical trip. Motion at a constant (nonchanging) speed is called **uniform motion.** The times spent at constant speeds of 30, 60, and 55 mi/h are all examples of uniform motion. The rest stop is also an example of uniform motion (with zero speed). Simple uniform motion is shown in the graph of Fig. 3.2. The graph is a horizontal straight line. The example shown is for a uniform speed of 30 mi/h. If the automobile were traveling at 60 mi/h, the graph would also be a horizontal straight line, but twice as high. If the automobile were not moving, then the graph would also be a horizontal line, but at zero mph; that is, rest is also uniform motion. (As will be discussed in Section D2 below, uniform motion must also be constant in direction.)

The times spent speeding up and slowing down in Fig. 3.1 are all examples of **accelerated motion.** Although slowing down is often called deceleration, it is more useful to consider it as negative acceleration. From the graph, one can see that some of the accelerations involve a rapid change of speed, whereas others correspond to a slower change of speed. Thus there are different kinds of accel-

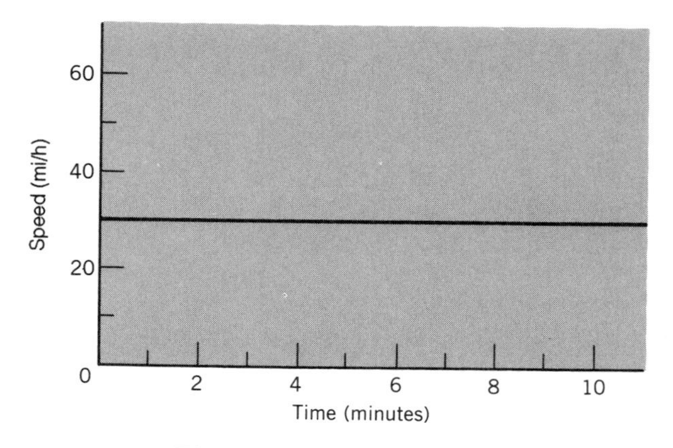

Figure 3.2. Uniform motion.

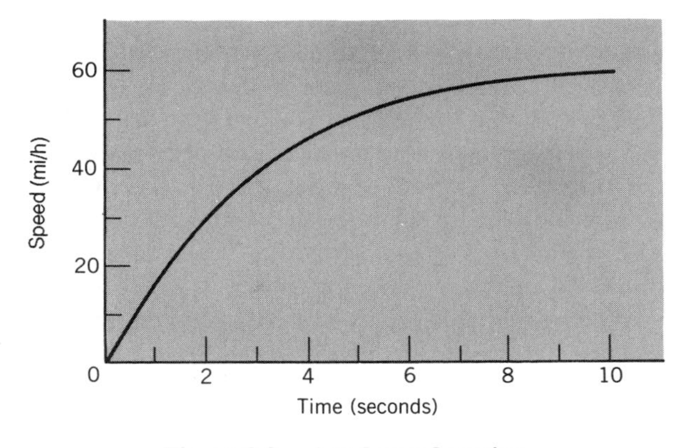

Figure 3.3. Accelerated motion.

erations. A specific example of accelerated motion is shown in Fig. 3.3. It corresponds to a car (for example) which is accelerating from rest to more than 50 mi/h in several seconds. The speed is not constant, but continually increases during the time shown on the graph. The "steepness" of slope of the graph indicates the magnitude of the acceleration. Thus at 1 second the slope of the graph is greater than at 5 seconds, and therefore the acceleration is greater at 1 second than at 5 seconds, in this particular case.

If the speed changes at a steady or constant rate (increase or decrease), then the motion is **uniformly accelerated motion.** Examples of uniformly accelerated motion are shown in Fig. 3.4. The essential characteristic of uniformly accelerated motion is that the graph is a sloping straight line. The actual speed does not matter. What is important is that the speed is changing by a constant amount during each time interval. Thus, for example, the line in Fig. 3.4 that represents a car starting from rest (0 mi/h) indicates that its speed increases by 10 mi/h in each second: from 0 to 1 seconds the speed increases from 0 to 10 mi/h, from 1

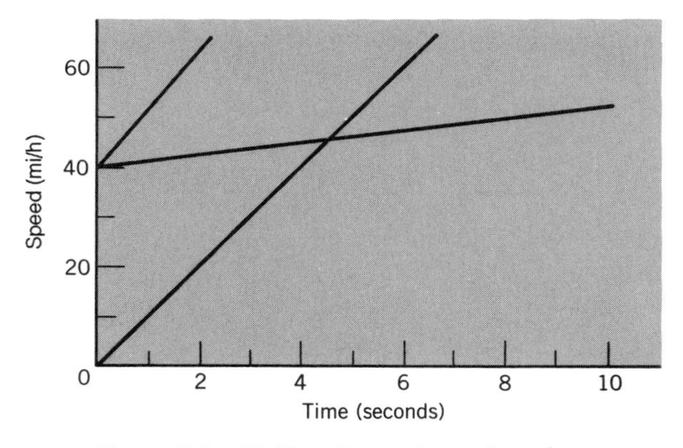

Figure 3.4. Uniformly accelerated motion.

to 2 seconds the speed increases from 10 to 20 mi/h, from 2 to 3 seconds the speed increases from 20 to 30 mi/h, and so on. By contrast in Fig. 3.3, the increase of speed is less in each successive one-second interval; the motion is accelerated but not uniformly accelerated.

The graph of speed versus time for the motion of an object tells us a great deal about the motion at any one instant, but it does not directly indicate the result of the motion—namely, the distance traveled. Frequently, one is as interested in knowing the distance traveled as in knowing the speed at any one instant. Furthermore, in studies of motion, it is often much easier to measure distances traveled than to measure speeds at all times. Thus we are interested also in obtaining a distance versus time graph for the motion of an object and relating the information on the distance versus time graph to the information on the speed versus time graph.

It is actually quite easy to determine the distance traveled from a speed versus time graph. The distance traveled by a car with a constant speed of 30 mph in one hour of driving is 30 miles. In two hours, the same car would travel a total of 60 miles. We obtain these simple results, by using (in our head) the standard formula for distance traveled by an object with speed v (for velocity) over a time t; namely, distance = speed × time, or symbolically

$$d = v \times t$$

Now if we plot on a speed versus time graph the motion of an object with uniform motion (speed) v, we obtain a simple horizontal straight line, with zero slope, as previously discussed for Fig. 3.2. In Fig. 3.5 we show that the formula $d = v \times t$ corresponds simply to the area under the horizontal line, from time = 0, up to time = t. The light area in this case represents the distance traveled in two hours. This correspondence is true in general, that is, whether the speed is uniform or not. Consider the graph of Fig. 3.6.

The actual speed of the object is shown by the curved line on the graph. The distance traveled in six hours is represented by the light area. Shown also on this

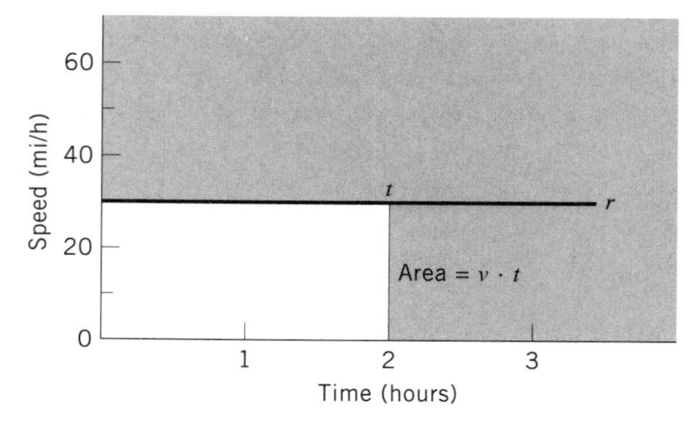

Figure 3.5. Distance traveled as area under a speed versus time plot for uniform motion.

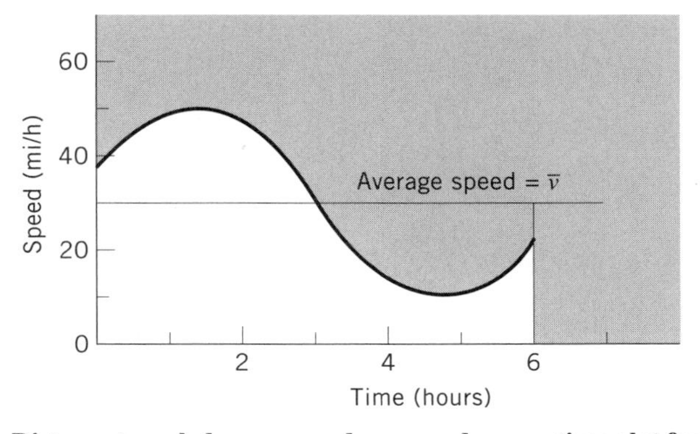

Figure 3.6. Distance traveled as area under a speed versus time plot for nonuniform motion.

graph is the straight horizontal line corresponding to the *average* speed of the motion represented by the curved line. The average speed is that constant speed for which the object will travel the same distance in the same time interval; that is, this is the average of all of the individual speeds, taken at regular intervals (e.g., every one second or every one minute) during the actual motion. Now the total distance traveled by an object with this average speed (say $\bar{v}$) over the time t is (just as in Fig. 3.2) equal to $d = \bar{v} \times t$ and is the area under the straight line $\bar{v}$ in Fig. 3.6. Notice that in some regions the light area rises above the straight line corresponding to the average speed, and in other regions it falls below. Because $\bar{v}$ is the average height (speed) of the actual motion curve, the two areas are exactly equal; that is, any excess of one over the other in one region is canceled by a deficit in another region.

Thus it is always possible to determine the actual distance traveled from a speed versus time graph, no matter what kinds of motion are involved, simply by finding the area under the curve. This area is very easy to calculate if the speed is uniform, but the area can always be determined, even for very complicated curves, by plotting on fine graph paper and simply counting the squares under the curve. Specific mathematical techniques exist for finding the area under a curve for which a definite mathematical formula exists (i.e., the curve is described exactly by a formula). These techniques are well known and form part of the branch of mathematics known as calculus. There are two important special cases that can be discussed without the use of calculus.

Fig. 3.7a shows the graph of speed versus time and Fig. 3.7b shows the graph of distance versus time for the case of uniform motion (constant speed). The specific case considered is that of a car traveling at 30 mi/h. As the graph indicates, at the end of one hour, the car has traveled 30 miles; at the end of two hours, it has traveled 60 miles, and so forth. The graph is seen to be a straight line with slope. Note that a faster speed (say 60 mi/h) would produce a straight line with a greater slope (it would rise faster), so that greater distances would be covered in the same time.

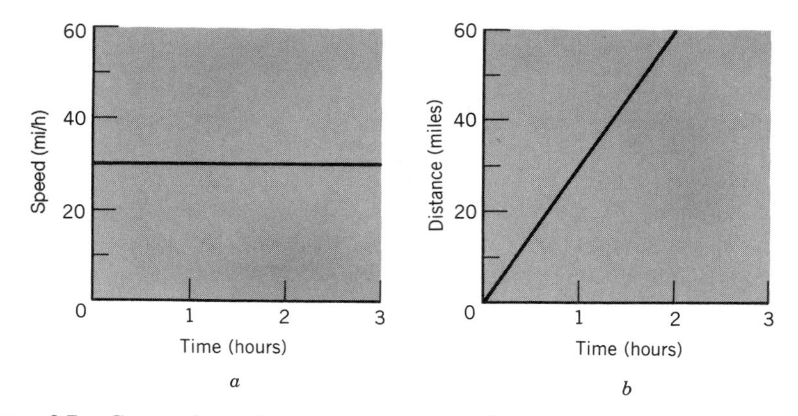

Figure 3.7. Comparison of speed versus time and distance versus time graphs for uniform motion. Note that the distance traveled at any time is equal to the area under the left-hand graph up to that time.

Now consider the case of uniformly accelerated motion shown in Fig. 3.8. The speed versus time curve (Fig. 3.8a) is a straight line with slope. As discussed earlier, the speed is increasing at a uniform (constant) rate. The distance traveled in any time interval is, as discussed above, equal to the area under the line over that time interval. Note that if we consider two equal time intervals, say Δt_1 and Δt_2, at different times (early and later), the corresponding areas under the line are quite different. The area corresponding to the later time interval is larger, simply reflecting the fact that the speed is greater.

The corresponding distance versus time graph is shown in Fig. 3.8b. The distances traveled during the time intervals Δt_1 and Δt_2 are represented by the sep-

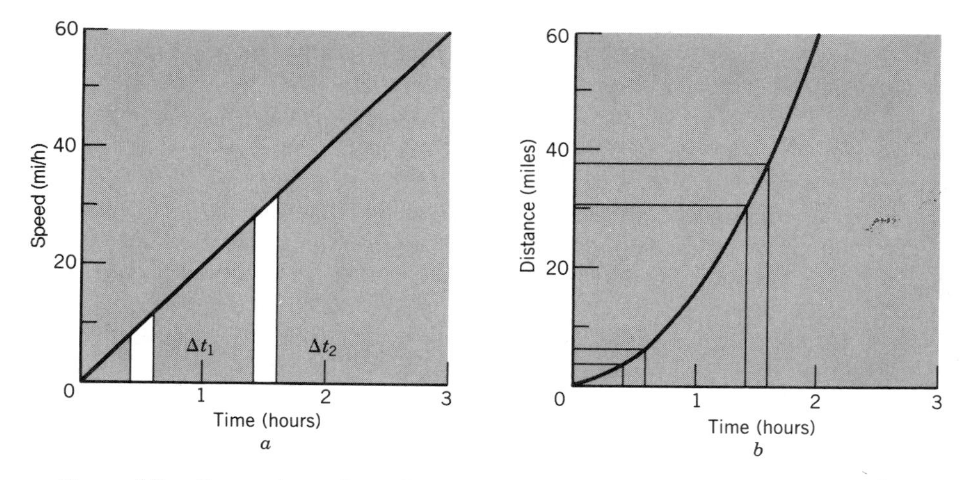

Figure 3.8. Comparison of speed versus time and distance versus time graphs for uniformly accelerated motion.

arations between the pairs of horizontal lines. The distance traveled per unit time increases with time. Note that the slope (i.e., the "steepness") of the distance versus time curve is just the speed. The distance versus time graph for uniformly accelerated motion is seen to be an upward curve rather than a straight line. The precise shape of the curve can be found by carefully finding the area under the speed versus time curve as time increases. For example, suppose the time intervals Δt_1 and Δt_2 are both taken to be 12 minutes, or $\frac{1}{5}$ hour. In the time interval Δt_1, the average speed is seen to be 10 mph, so that the car would travel about 10 mi/h $\times \frac{1}{5}$ hour $= 2$ miles. In the time interval Δt_2, the average speed is seen to be 30 mi/h, so that the car would travel about 30 mi/h $\times \frac{1}{5} = 6$ miles. These respective distances are shown in Fig. 3.8b, corresponding to these two time intervals.

The curve shown in Fig. 3.8b is called a quadratic; that is, it turns out to be precisely described by the mathematical formula

$$d \propto t^2$$

where $\propto$ is the mathematical symbol meaning *is proportional to.* Thus the formula (and the curve of Fig. 3.8b) says that for uniformly accelerated motion, if the time traveled is doubled, the total distance covered is quadrupled (increases four times), or if the time is tripled, the distance increases nine times, and so on. The mathematics of calculus confirms this result.

These various developments were extremely important because they made it possible, for the first time, to analyze motion precisely and in detail, whether it be falling bodies or projectiles, in terms of specific *kinds* of motion. The definitions of uniform and uniformly accelerated motions were critical to properly describe motion, so that the underlying nature of the motions could be discovered. The graphical representations are needed in order to properly identify these different types, as we have seen above. Thus the stage was finally set for someone to determine precisely the kinds of motion involved with falling bodies and projectiles.

2. Falling Bodies

Galileo Galilei (1564–1642) first determined exactly *how* objects fall, that is, with what kind of motion. Galileo was born and raised in northern Italy just after the height of the Italian Renaissance. He demonstrated unusual abilities in several areas as a young man (he probably could have been a notable painter), and studied medicine at the University of Pisa, one of the best schools in Europe. But Galileo early became more interested in mathematics and "natural philosophy" (i.e., science), and displayed so much ability that he was appointed professor of mathematics at Pisa while still in his mid-20s. Following some difficulties with the other faculty at Pisa (he was aggressive, self-assured, and arrogant), Galileo, at age 28, moved to the chair of mathematics at Padua, where he stayed for nearly 20 years. He quickly established himself as one of the most prominent scholars of his time and made important contributions to many areas of science and technology.

Galileo Galilei. (Courtesy of Editorial Photocolor Archives.)

Galileo became convinced fairly early in his career that many of the basic elements of Aristotle's physics were wrong. He apparently became a Copernican rather quickly, although he did not publish his beliefs regarding the structure of the heavens until he used the telescope to study objects in the night sky about 1610, when he was almost 46. More important for this chapter, Galileo also became convinced that Aristotle's descriptions of how objects fall and why projectiles keep moving were quite incorrect.

Apparently, Galileo became interested in how objects fall as a young man, while still a student at Pisa (although it is not true that he performed experiments with falling objects from the Leaning Tower of Pisa). He realized that it was necessary to "slow down" the falling process so that one could accurately measure the motion of a falling object. Galileo also realized that any technique devised to solve this problem must not change the basic nature of the "falling" process.

Aristotle's solution to the problem of how to slow down the falling process (so as to be able to study it) was to study falling objects in highly resistive media, such as water and other liquids. From these studies, Aristotle observed that heavy objects always fall faster than lighter ones and that objects fall with uniform speeds. He did observe that heavy and light objects fall at more nearly the same speeds in less resistive media, but felt they would have the same speeds only in a vacuum, which he rejected as being impossible to obtain. Therefore he felt that the medium was an essential part of the falling process. He concluded that a basic characteristic of the falling process was that heavy and light objects fall at different speeds.

Galileo, on the other hand, felt very strongly that *any* resistive effects would mask the basic nature of falling motion. He believed that one should first determine how objects fall with no resistive effects, then consider separately the resistive medium effects, and add the two together to get the net result. The important point here is that Galileo felt the basic nature of the falling process did not include these resistive medium contributions. Thus he wanted to find some other way to slow down falling without changing the basic nature of the process.

Galileo devised two methods that he felt satisfied his criteria. The first, which he began to study as a student, was a pendulum. The second method was that of a ball rolling down an inclined plane. In both of these phenomena, he believed that the underlying cause of the motion was the same as that responsible for falling (i.e., gravity). Galileo argued that the motions involved should have the same basic characteristics, but they should be somewhat diminished for pendulums and rolling balls on inclined planes. He obtained the same results from both methods.

Let us consider his second method, the inclined plane, in some detail. By choosing an inclined plane without too much slope, one can slow down the process in order to observe it. Galileo did more than just observe the motion of the rolling ball, however; he carefully studied how far the ball would travel in different intervals of time (i.e., he determined the distance versus time graph for the rolling ball). Galileo did this because he knew how to identify the kind of motion involved from such a graph. (Galileo claimed to be the first to define uniform and uniformly accelerated motions, although he was not.)

Such measurements were not easy for Galileo. For example, there were no accurate stopwatches at that time, so he had to devise ways to measure equal intervals of time accurately. He reportedly started measuring time intervals simply by counting musical beats maintained by a good musician. His final technique was to use a ''water'' clock, which measured time by how much water (determined by weighing) would accumulate when allowed to drain at a fixed rate.

The results of Galileo's inclined-plane experiments are represented (in a simplified form) in Table 3.1, and are plotted in Fig. 3.9. The distance traveled at the end of the first unit of time is taken to define one unit of distance. The succeeding distances are seen to follow a quadratic dependence on the time, that is,

$$d \propto t^2$$

Table 3.1. Total Distance Traveled After Different Times for the Inclined Plane Experiment

Time Elapsed	Distance Traveled
0	0
1	1
2	4
3	9
4	16
⋮	.

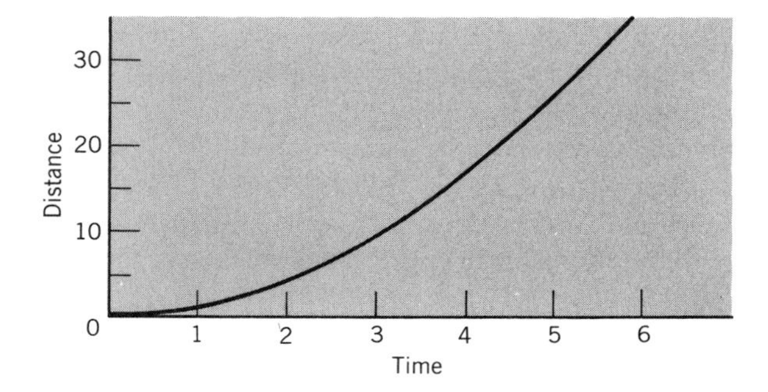

Figure 3.9. Graphical representation of distance versus time for Galileo's inclined-plane experiments. The data from Table 3.1 fit this graph.

which we saw in the previous section indicates uniformly accelerated motion. The graph in Fig. 3.9 has the same characteristic upward curving as that shown in Fig. 3.8b. Thus Galileo was able to determine experimentally *how* objects fall; namely, with uniformly accelerated motion. This is the kind of motion a falling object has when the effects of the resistive medium can be eliminated and is the kind of motion that characterizes the basic falling process.

Galileo certainly was aware that falling objects, in air or in liquids, do not keep falling faster and faster; that is, he knew that the resistance of the medium will eventually cause the acceleration to cease, and that the falling velocity will reach a constant value. This final, maximum velocity is now generally referred to as the *terminal velocity* for a falling object. It was precisely because of the resistive effects of the air (or a liquid) that Galileo decided he needed to study falling with pendulums and inclined planes, thereby keeping the speed low enough to minimize the effects of the medium. In order to understand the net result of the falling process, let us consider Fig. 3.10.

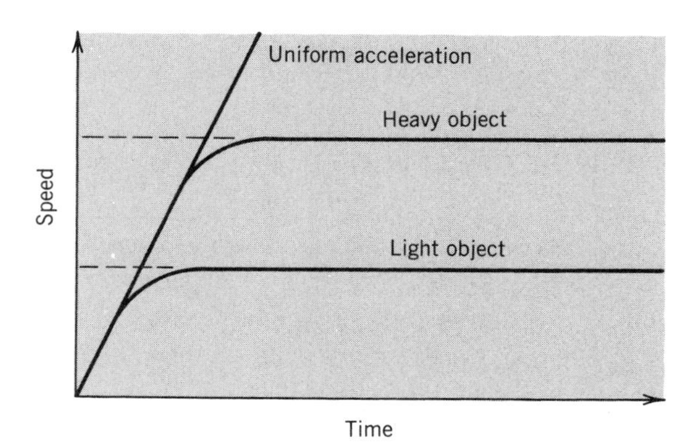

Figure 3.10. Graphical representation of motion of light and heavy falling objects.

The figure represents the velocity (or speed) of both a heavy and a light object, as a function of time, and thus should be compared with Figs. 3.7a and 3.8a. The diagonal line represents the motion of an object falling with a constant (uniform) acceleration, which is the motion an object would have when falling in a vacuum. In anything but a vacuum, resistance of the medium increases as the speed increases, until the resistive force equals the downward force of gravity, and the acceleration stops. The resulting *terminal velocity* for an object will depend not only on its weight but on its size and shape as well. The terminal velocity for a falling person (a skydiver, for example) may be as much as 130 mph, depending on body position, clothing, and so on. The horizontal lines in Fig. 3.10 indicate the kind of motion predicted by Aristotle for a light and heavy object. These lines represent constant (although different) speeds, with no indication of a time during which the object speeds up from zero speed to its final terminal velocity.

Although it is important to understand the actual motion of a falling object, as shown in Fig. 3.10, from a scientific viewpoint, it is more important to understand the true nature (the ''true reality'' in Plato's ''Allegory of the Cave'') of falling, disregarding the resistance of the medium. Galileo asserted that this true nature is that objects fall with uniformly accelerated motion. This was a very important step toward understanding the falling process. Once the *how* of falling motion is known, it is then necessary to know *why* they fall. The understanding of why was provided by Isaac Newton, who was born in the same year that Galileo died. Before discussing Newton's contributions, it is necessary to consider some other important contributions to the science of mechanics by Galileo and also the general relationship of mathematics to science.

Optional Section 3.1 *Motion of Falling Bodies*

As discussed above, Galileo determined that objects fall with uniformly accelerated motion. The mathematical formulae that express this relationship are

$$v \propto t$$

and

$$y \propto t^2$$

where v is the velocity (speed), y is the vertical drop, and t is the time. In order to make these proportionality relationships into equations, we need to include the acceleration of gravity constant, g. Then we have

$$v = gt$$

and

$$y = \frac{1}{2}gt^2$$

Table 3.O.1. Falling Object

Time (sec)	Vertical Speed (m/sec)	Distance Fallen (m)
1	9.8	4.9
2	19.6	19.6
3	29.4	44.1
4	39.2	78.4
⋮	⋮	⋮

The acceleration of gravity constant g has been determined from experiment to be $g = 9.8$ m/sec^2, or 32 ft/sec^2, where m/sec^2 means meters per second per second. Thus, if an object is dropped from a large vertical height, we can calculate the speed and total vertical distance fallen using these equations. These equations tell us that the velocity (speed) of a falling object will increase 9.8 m/sec every second, so that the object keeps falling faster and faster. Because of this increased speed, the distance fallen increases faster and faster with time, at a rate proportional to the square of the time elapsed. The result is given in Table 3.O.1.

These results are correct if one neglects the resistance of the air for the falling object and are strictly valid only in a vacuum. A small dense object (e.g., a rock) will follow the speeds and distances of Table 3.O.1 for a few seconds before the resistance will slow down the rate of increase and the object reaches its terminal (final) velocity.

3. Projectile Motion

As part of Galileo's studies of objects rolling down inclined planes, he was able to arrive at two very important generalizations regarding motion. Both of these new insights applied directly to Aristotle's old problem of how to explain projectile motion.

Galileo noted that a ball rolling down an inclined plane would continually experience uniform acceleration (see Table 3.1), even for only a very small angle of inclination. He then wondered what would happen to a ball rolling *up* an inclined plane. He discovered that a ball started along a flat surface with uniform motion (constant speed) would experience uniform deceleration (i.e., *negative* acceleration) if it rolled up an inclined plane. Because these two conclusions appeared always to be true, even for very small angles of inclination, he concluded that as the angle of the plane was changed from being slightly downward to slightly upward, the acceleration must change from being slightly positive to slightly negative, and an angle must exist for which there is no acceleration. This angle, he concluded must correspond to a horizontal plane, with no slope whatsoever.

Galileo's analysis was extremely important, because, as he realized, if a ball rolling on a flat plane experienced no acceleration (or deceleration), then it would *naturally* roll along forever. That is, Galileo realized that a ball rolling along a perfectly level surface would naturally keep rolling with uniform speed, unless something (e.g., friction or upward inclined slopes) acted on the ball to cause it to slow down. Thus Galileo recognized that Aristotle's question: "Why do projectiles keep moving?" although addressing a very important issue, was actually the wrong question. The question should be: "Why do projectiles *stop* moving?" Galileo realized that the natural thing for a horizontally moving object to do is to keep moving with uniform motion.

He then seized on this result to explain how the Earth could behave like other planets without requiring an engine to drive it (Chapter 2, Section C4). Galileo reasoned that horizontal was parallel to the surface of the Earth, and therefore circular. Thus it was natural for objects to travel on a circular path, without requiring an engine, unless there were some resistance due to a medium. But if the Earth were traveling in "empty space" (i.e., in a vacuum), then there would be no resistance, and the Earth could travel in a circular orbit forever, just as Copernicus had postulated. The Earth's atmosphere would also travel along with it, and thus there would also be no problem of high winds as a consequence of the Earth's motion. Unfortunately, Galileo's reasoning was somewhat fallacious—he had drawn the wrong conclusion from this aspect of his rolling ball experiments. The ball rolling on the flat plane was rolling in a *straight line,* not on a level surface parallel to the Earth's surface.[1]

The property of a body by which it keeps moving we now call **inertia.** Mass is a measure of this inertia. Perhaps the most dramatic example of inertia is the motion of a spacecraft in "outer" space. Recall how the astronauts traveled to the Moon. First, their spacecraft "blasted-off" from Cape Canaveral and put them into orbit around the Earth. Next, after determining that everything was "A-OK," they fired their rocket engines for several minutes in order to leave their Earth orbit and start toward the Moon. Once they reached the proper speed and direction, they shut off the engines. From there, they were able to simply "coast," for about 3 days and 240,000 miles, to a rendezvous with the Moon. Sometimes (but not always), they needed to fire their engines for a few minutes at about the halfway point, not to change their speed but only to apply a "mid-course correction" to their direction. The point is that the trip used the inertia of the spacecraft to keep it moving at a large constant speed (about 3000 miles per hour) all the way from the Earth to the Moon. It worked so well because there was essentially no resistance in the form of friction, which acted upon the craft in space.[2]

Thus Galileo answered Aristotle's question by pointing out that it is natural for a moving object to keep moving. Of course, Galileo did not address *why* this

[1]It might be speculated that this misunderstanding was the reason that Galileo failed to recognize the significance of Kepler's elliptical orbits. Galileo believed in circular orbits and that his experiments had proven their validity. It would be difficult to explain a noncircular orbit. Not until the time of Isaac Newton was the proper explanation given of the physics of planetary orbits.

[2]Actually the spacecraft slowed down somewhat until the Moon's gravitational attraction exceeded that of the Earth, and then the spacecraft speeded up somewhat.

is the natural state. But it was extremely important to first determine *what* nature does before it was possible to address questions of why things happen the way they do. Thus Galileo (just as he did for falling objects) had for the first time properly recognized what happens in horizontal motion. In fact, Galileo went on to specifically consider projectile motion, the example that so perplexed Aristotle.

Galileo realized that the tricky thing about projectile motion (i.e., thrown or struck objects) was that both horizontal and vertical motions were involved. He had already determined that "pure" horizontal motion was uniform motion in a straight line. But how should these be combined in order to describe a projectile? Galileo was also faced with this question in his use of inclined planes and pendulums to study falling. Both of these systems involve combined horizontal and vertical motions. Because he found that balls rolling down inclined planes always exhibited uniformly accelerated motion, no matter what the slope of the plane (although the magnitude of the acceleration increases as the slope increases), Galileo guessed that the effects of falling were independent of the horizontal motion. Because he found that all of his experiments with both inclined planes and pendulums gave results consistent with this hypothesis, he eventually concluded that it was always true. Thus Galileo was led to what is known as the **superposition principle,** which states that for objects with combined vertical and horizontal motions, the two motions can be analyzed separately and then combined to yield the net result.

According to the superposition principle, projectile motion involves both horizontal motion and vertical motion, which can be analyzed separately. Thus the vertical motion is just falling motion—the same falling motion that one would expect from an object with no horizontal motion. Similarly, the horizontal motion is uniform motion with whatever horizontal velocity the object has initially. The resulting combined motion for an object originally thrown horizontally from some height is represented in Fig. 3.11.

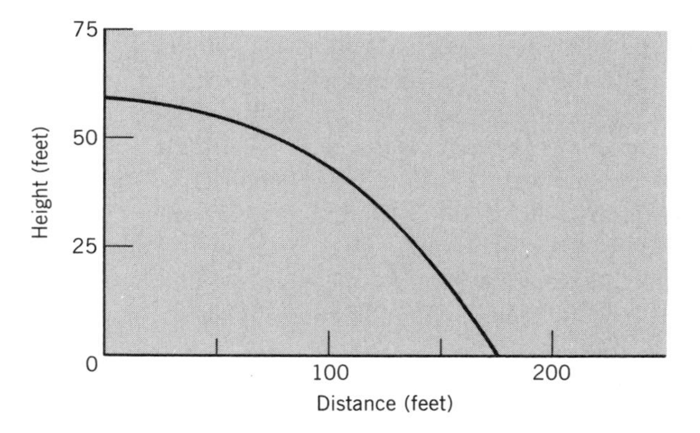

Figure 3.11. Graphical representation of height versus distance for object thrown horizontally from a height.

The object will strike the ground in the same amount of time as an object simply dropped from the same height. The thrown object will move horizontally at its initial horizontal speed until it strikes the ground. Thus for example, if the horizontal speed of the object is 60 mi/h (or 88 feet per second), and it takes 2 seconds to fall to the ground (which would be the case if it starts from a height of 64 feet), then the object will travel 176 feet horizontally (88 ft/sec × 2 sec). Note that the superposition principle says that a bullet shot horizontally from a high-powered rifle, and a bullet dropped from the same initial height will hit the ground at the same time! The bullet fired from the rifle simply has a very large horizontal velocity and is able to move a rather large horizontal distance in the short amount of time available. (This result clearly does not apply to an object that "flies" in the air or in any way derives some "lift" from its horizontal motion. It is strictly valid only in a vacuum.)

Galileo also applied the concept of projectile motion to an apple (for example) falling from a tree fixed to the moving surface of the Earth. The apple maintains its "forward motion" while falling and thus does not get left behind, as had been claimed by some of the opponents of the heliocentric theory.

The net result of all of Galileo's contributions to the development of the science of mechanics is great. He determined exactly how objects fall and introduced the idea of inertia for moving objects. He was able to correct Aristotle's conceptual analysis of falling bodies and was able to "answer" his question of why projectiles keep moving. However, Galileo did not give the why's of the motions. As discussed below, it was Isaac Newton who provided the "why" for falling motion.

C. Logic, Mathematics, and Science

Before discussing Newton's contributions to the science of mechanics, it is necessary to discuss the difference between inductive and deductive logic, to consider the value and use of mathematics in science, and to indicate what is basically involved in the branches of mathematics known as analytical geometry and calculus. This will provide a better understanding of Newton's contributions and how he made them. These discussions will be useful also for the material in subsequent chapters.

1. Inductive and Deductive Logic

There are two principal kinds of logic used in science. When one performs, or collects the results from, a series of experiments and extracts a general rule, the logic is referred to as **inductive.** For example, if it is concluded from several carefully performed surveys that people prefer to watch mystery shows above all others, inductive logic is being used. The general rule was obtained directly from experiment and not through any line of reasoning starting from certain basic principles.

The famous English philosopher Francis Bacon (1561–1626) was a rather extreme proponent of the inductive method. He believed that if one were to carefully list and categorize all available facts on a certain subject, the natural laws governing that subject would appear obvious to any serious student. While it is certainly true that Bacon made several important contributions to science in this manner in some of his particular studies and experiments, it is also true that he did not accomplish one of his stated goals in life—namely, to rearrange the entire system of human knowledge by application of the power of inductive reasoning. He tried to rely too heavily on this one kind of logic.

The other principal type of logic is **deductive** reasoning. In this style of logic one proceeds from a few basic principles or theories and logically argues that certain other results must follow. Thus if one theorizes that all people love to read a good mystery and argues that therefore mysteries will be popular television shows, then one is using deductive logic. Perhaps the strongest proponent of the deductive method was the French philosopher René Descartes (1596–1650). He believed that one could not entirely trust observations (including experiments) to reveal the truths of the universe. Descartes believed that one should start from only a very few, irrefutable principles and argue logically from them to determine the nature of the universe. Although Descartes made many important contributions to philosophy and mathematics (he developed much of the subject of analytical geometry, which is discussed below), because of his insistence on starting with a few basic principles not obtained from observation, he was not able, as he wished, to obtain a consistent description of the physical universe.

The proper use of the inductive and deductive methods of logic is in combination. Newton was quite skilled at this combined technique. He started with certain specific observations and generalized from these observations to arrive at a theory indicating the general physical law that could explain the phenomenon. Once he arrived at a theory, he would deduce consequences from it to predict new phenomena that could be tested against observation. If the predictions were not upheld, he would try to modify the theory until it was able to provide successful descriptions of all appropriate phenomena. His theories always started from observations (the inductive method) and were tested by predicting new phenomena (the deductive method). Newton properly saw both methods as necessary in determining the physical laws of nature.

2. Value and Use of Mathematics in Science

Before discussing some of the basic mathematical concepts that were developed about the time of Galileo and Newton (some by Newton himself) and that were necessary for Newton's synthesis of the science of mechanics, it is appropriate to discuss in a general way why mathematics is often used and even needed to describe physical laws.

As described succinctly by Schneer (see reference at the end of this chapter), mathematics may be said to be the science of order and relationship. Thus to the extent that the physical universe is orderly and has related parts, mathematics can be applied to its study. It is, perhaps, somewhat surprising that the universe is ordered and has related parts. The universe might not have been ordered and may have had no related parts. Such a universe would be hard to imagine. Nevertheless, it is still amazing that the more we learn about the universe, the more we know that it is ordered and appears to follow certain rules or laws. (As Einstein said, "The most incomprehensible thing about the universe is that it is comprehensible.") The more ordered and related the universe is found to be, the more mathematics will apply to its description.

Developments in physics often have followed developments in mathematics. Occasionally some areas of mathematics owe their origin to the need to solve a particular problem, or class of problems, in physics. We have already discussed how Kepler was able to deduce the correct shape of the orbits of the planets because of his knowledge of conic sections, including the ellipse. He also made extensive use of logarithms, which had just been invented, in his many calculations. Similarly, in order for Galileo to determine the true nature of falling, he needed proper definitions of motion and certain simple graphical techniques. We will see increased dependence on mathematics as we continue, and must now introduce two new mathematical developments in order to discuss Newton's works.

3. Analytical Geometry

In order to use mathematics to describe physical motions and shapes, it was necessary to find ways to combine algebra and geometry. Describing geometrical figures with mathematical formulas (and vice versa) is known as analytical geometry. As indicated above, this field was largely developed just before the time of Newton. The French philosopher-mathematician René Descartes was one of the principal persons involved in this work.

Most simple and many rather complex figures can be described by mathematical formulas. Figure 3.12 provides several examples. The circle is described by a rather simple formula, as is the ellipse (see Optional Section 3.2, below). Some of the more complicated patterns involve the use of trigonometric functions (sine, cosine, etc.) in their algebraic descriptions. It is not necessary here for us to understand how these algebraic descriptions are obtained—or even how they work. What is important, though, is to be aware that such algebraic descriptions of figures often exist and are known.

A particularly important class of geometrical figures for the discussions of this chapter are those called **conic sections.** A conic section figure is obtained by slicing through a cone, as shown in Fig. 3.13. If one cuts through the cone parallel to the base, one obtains a circle. If one cuts through the cone parallel to the side of the cone, a parabola is obtained. A cut across the cone but not parallel to the base or the side yields an ellipse, and finally a cut vertically but not parallel to the side yields an hyperbola. All of these figures are found in nature, as water ringlets in a pool, planetary orbits, comet trajectories, and so on.

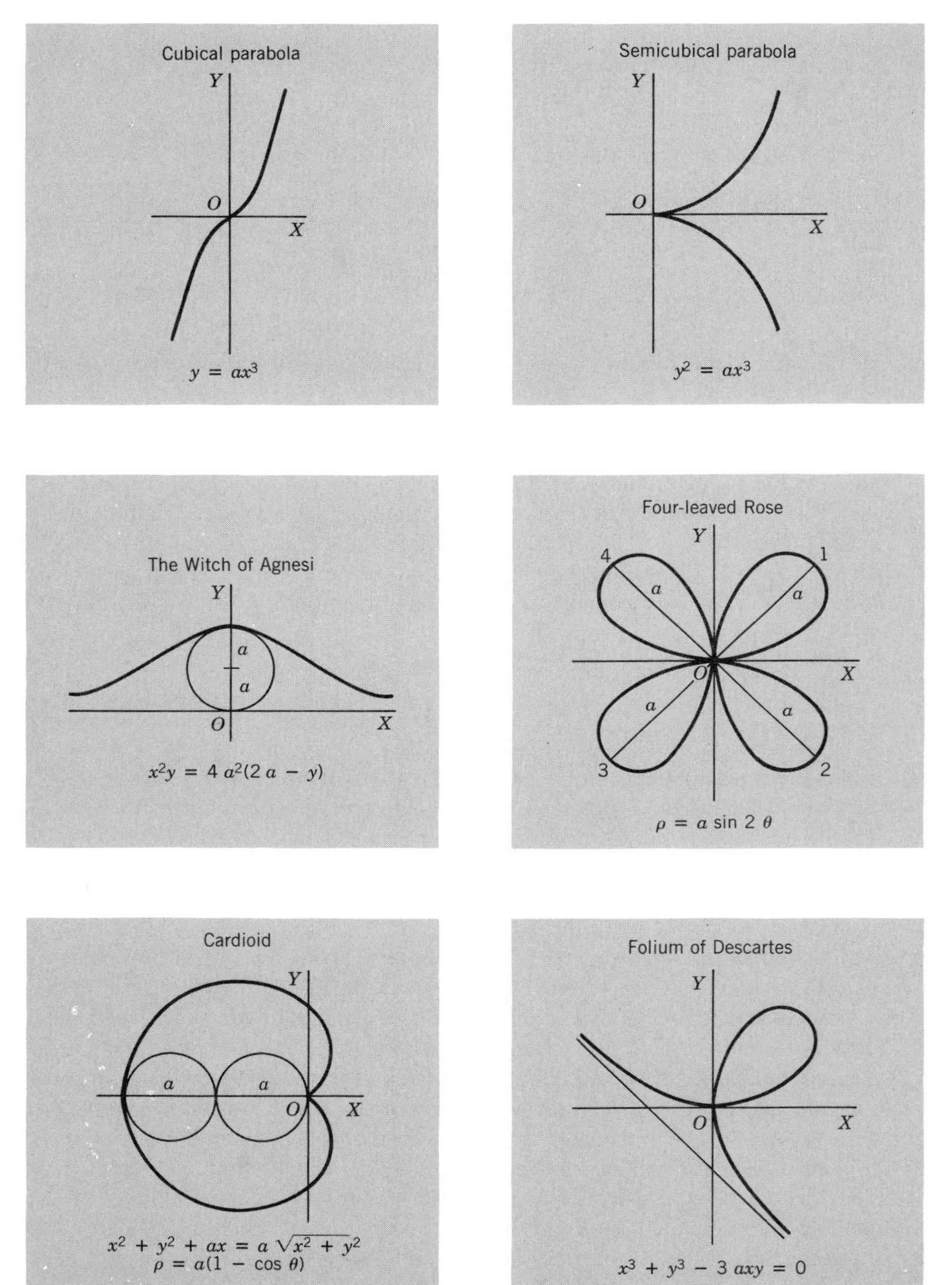

Figure 3.12. Various mathematical formulae and their graphical representations. (By permission from W.A. Granville, P.F. Smith, and W.R. Longley, *Elements of the Differential and Integral Calculus,* John Wiley & Sons, New York, 1962.)

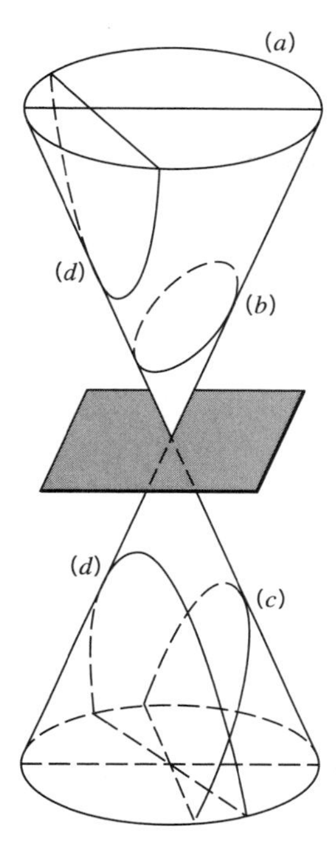

Figure 3.13. Conic sections: (a) circle, (b) ellipse, (c) parabola, and (d) hyperbola.

OPTIONAL SECTION 3.2 *The Formulae for a Line, a Circle, and an Ellipse*

Perhaps the simplest example of analytic geometry is the formula for a line. This formula is just

$$y = mx + b$$

This equation describes a straight line with slope m and that intercepts the y axis at value b.

Another simple example is the formula for a circle. This formula is

$$x^2 + y^2 = R^2$$

This formula describes a circle of radius R centered at the origin, that is, at $x = y = 0$. Let us consider a simple case. Let $R = 4$. We can assume various values for x, and then using this equation, determine the values for y. These values are shown in Table 3.O.2. If we plot these values on a simple x–y grid, we will get a circle. (Note that neither x nor y can be greater than 4, i.e., x or y cannot be greater than R.) Note also because of the squares, $\pm x$ yields the same value of y (which can also be $\pm y$). This circle is plotted in Fig. 3.O.1.

Table 3.O.2. A Circle

$\pm$x	$\pm y = \sqrt{4^2 - x^2}$
0.0	4.00
0.5	3.97
1.0	3.87
1.5	3.70
2.0	3.46
2.5	3.12
3.0	2.64
3.5	1.94
4.0	0.00

Finally, let us consider the formula for an ellipse. This example is important also since it is the figure that describes planetary orbits. This formula is

$$\frac{x^2}{a^2} + \frac{y^2}{b^2} = 1$$

This ellipse has its center at the origin and semi-axes a and b. One could map out this figure in the same way as for the circle, that is, by solving the equation for y and then substituting in values for x. If one has $a = 4$ and $b = 3$, one obtains the ellipse shown in Fig. 3.O.2. Note that if $a = b$, the ellipse becomes a circle with R = a.

An ellipse can be drawn in the following simple fashion. Drive two nails into a board a few inches apart. Cut a string somewhat longer than

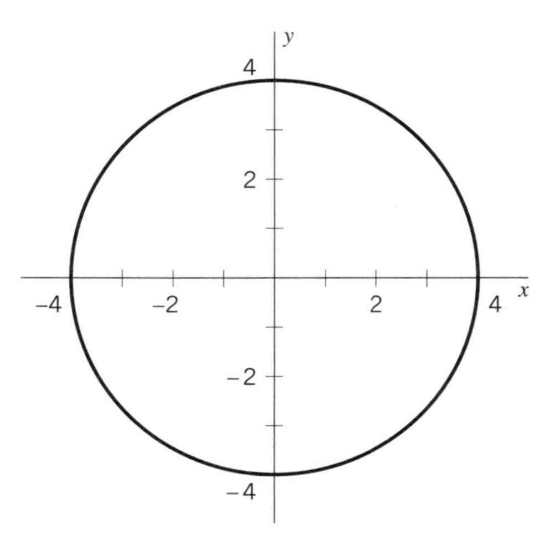

Figure 3.O.1. A circle with radius = 4.

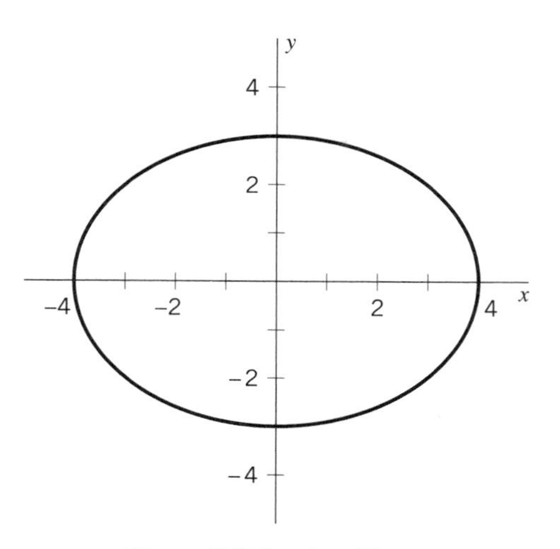

Figure 3.O.2. An ellipse.

the distance between the nails and tie the ends to the nails. Place a pencil against the string until it is taut and then make a figure by moving the pencil along the string always keeping it taut. The mark will form an ellipse.

4. Calculus

In our discussion of types of motion and Galileo's studies of falling objects, we sometimes found it necessary to describe the motion on a speed versus time graph, and sometimes on a distance versus time graph. We indicated how some of these graphs were related, for example, how uniform motion would appear on both kinds of graphs. In general, the mathematical techniques for relating curves on these two kinds of graphs are part of the subject known as calculus. For example, we discussed how the area under a speed versus time curve corresponds to the total distance traveled.

The subject of **integral** calculus deals with mathematical techniques for finding the area under any curve for which a mathematical formula exists (analytical geometry). Similarly, the slope of a distance versus time graph indicates the speed of an object, and the slope of a speed versus time graph gives the acceleration. The mathematical techniques for determining slopes of curves described by mathematical formulas are known as **differential** calculus.

Newton and the German mathematician-philosopher Gottfried Leibnitz (1646–1716) are credited as independent inventors of the calculus. They performed their work separately, but arrived at the same basic results on differential calculus. That Newton developed the ideas and methods of calculus certainly enabled him to arrive at some of his conclusions regarding mechanics and the universal law of gravity. He used calculus not only to translate from speed versus time graphs to distance versus time graphs, and vice versa, but also to find areas and volumes of figures and solids.

5. Vectors

Finally, it is necessary to discuss the fact that some quantities have *direction,* as well as magnitude (i.e., size). For example, if we indicate that an automobile is traveling at 30 mph, we have not fully described its motion. We need to indicate also in what direction the car is traveling. Quantities that logically require both a direction and a magnitude are called **vector** quantities. The vector quantity associated with motion is called *velocity.* If we indicate the magnitude of the motion only, we refer to its *speed.* Thus if we say that a car is traveling 30 mph in a northerly direction, we have specified its velocity.

Velocity is not the only quantity that is a vector; force is another example. Whenever a force is impressed on an object, it not only has a certain strength, but also a specific direction, which must be indicated in order to describe it completely. Similarly, acceleration (positive or negative) is always at a certain rate and in a specific direction, so that it also is a vector quantity. Thus circular motion, even with constant speed, is accelerated motion.

Some quantities do not require a direction, but only a magnitude. Such quantities are called **scalar** quantities. The mass of an object (in kilograms, for example) needs no direction and is only a scalar quantity. The length of an object is also a scalar quantity. Other examples include time, electric charge, temperature, and voltage. Speed is a scalar quantity, but velocity is a vector. Because a proper discussion of motion, including the actions of impressed forces, must include directions, it is appropriate to use vector quantities. As we will see shortly, Newton was well aware of the need to use vector quantities in his descriptions of motion, and made a special point to introduce them carefully.

Because vector quantities have both magnitude and direction, they can be represented graphically by arrows, as shown in Fig. 3.14. The length of the arrow is proportional to the magnitude of the vector, and the orientation of the arrow and position of the arrowhead represent the direction. The simple mathematical operations carried out with ordinary numbers are only somewhat more complicated when carried out with vector quantities—including addition, subtraction, and multiplication. For the purposes of this and subsequent chapters, it is sufficient to describe graphically how vectors are added. To add two or more vectors,

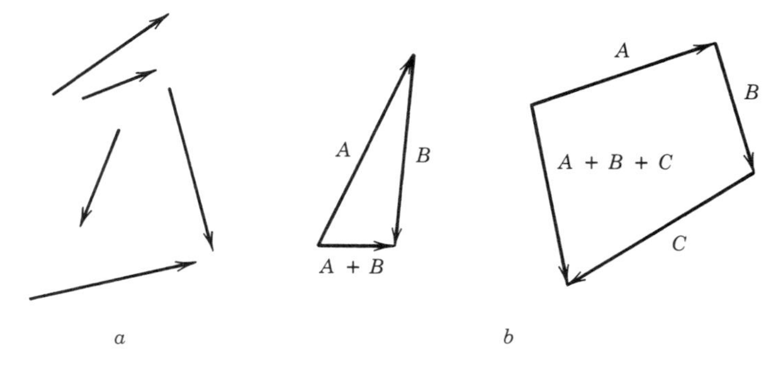

Figure 3.14. Vectors. (a) Examples of vectors represented by arrows. (b) Examples of vector addition in graphical representation.

they are simply placed head to tail, maintaining their orientation and direction; and a final arrow or resultant vector is drawn from the tail of the first arrow to the head of the last arrow. This final arrow represents the vector sum of the individual vectors, as shown in Fig. 3.14b. (One interesting result of vector addition is that a vector sum can have a smaller magnitude than each of the magnitudes of the vectors contributing to the sum.)

OPTIONAL SECTION 3.3 *Addition of Vectors*

As an example of vector addition, let us consider an airplane flying in a diagonal cross wind. Let us suppose that the airplane is flying at 100 mi/h in a due east direction and that there is a cross wind of 50 mi/h moving due north. The resultant motion will be somewhat north of east and can be determined quantitatively from Fig. 3.O.3.

As shown, triangle *ABC* is a right triangle, hence by the Pythagorean theorem,

$$R^2 = x^2 + y^2$$

or

$$\begin{aligned} R &= \sqrt{x^2 + y^2} \\ &= \sqrt{100^2 + 50^2} \\ &= 112 \text{ mi/h} \end{aligned}$$

The direction is given by θ. We see that $\tan \theta = y/x = 50/100 = 0.5$, or $\theta = 27°$. (If one is not familiar with the trigonometric functions, the angle can be found graphically, by plotting the triangle to scale and measuring the angle with a protractor.) The resultant motion of the airplane is that it travels at an angle of 27° north of due east at 112 mi/h.

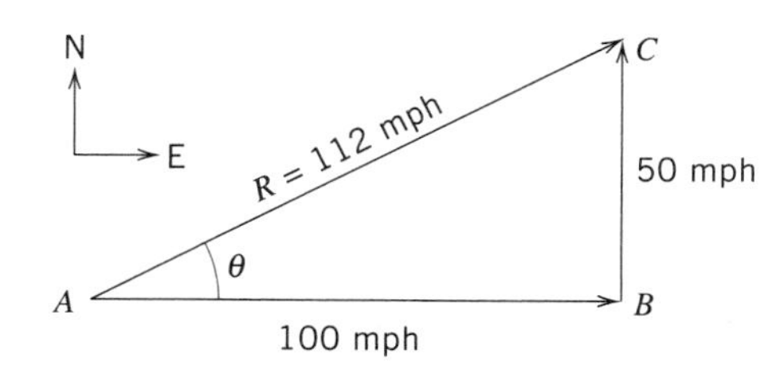

Figure 3.O.3. The vector triangle for airplane in a cross wind.

D. NEWTONIAN MECHANICS

We are now ready to discuss the contributions of Isaac Newton (1642–1727) to the science of mechanics. His work presents fairly complete answers to the questions first raised by Aristotle, which have been the subject of this chapter. In fact, Newton's work represents one of the greatest contributions made by an individual to human understanding of the physical universe. It is difficult to overestimate the impact of his work on Western thought, as we will discuss further below. He developed a picture of the universe as a subtle, elaborate clockwork, working according to well-defined rules.

1. Historical Background

Isaac Newton was born in England in 1642, the same year that Galileo died. Newton continued the work of Galileo in several important ways, especially with regard to mechanics. Like Galileo, Newton was the dominant scientist of his generation. Newton made important contributions in mathematics, optics, wave phenomena, mechanics, and astronomy. Any of his several most important contributions would have established him as one of the most important scientists in history.

Newton became interested in experimental devices as a child and demonstrated an ability to make original designs of windmills, water clocks, and sundials. Because of his great potential, he was able to attend Cambridge University, where he showed exceptional mathematical ability. Following his graduation in 1665, Newton returned to his boyhood home at Woolsthorpe in Lincolnshire, where he lived with his widowed mother. This was the time of the Great Plague throughout Europe, including England. Some 31,000 people in London alone perished from the Black Death over a two-year period. (This was also the time of a great fire that devastated much of London.) Newton spent the two years from 1665 to 1667 essentially in seclusion at Woolsthorpe. It is clear now that during those two years he formulated the well-known binomial theorem of mathematics, developed calculus, studied the decomposition of white light into its spectrum (using a prism), and began his study of mechanics, including the universal law of gravitation. None of this work was published immediately, and some of it not for 30 years. But these were surely two of the most productive years ever spent by a single scientist.

In 1667 Newton returned to Cambridge as a lecturer. Newton's master at Cambridge, a noted mathematician named Isaac Barrow, became so impressed with Newton's work and abilities that he resigned his own chair of mathematics (in 1669) so that it might go to Newton. Newton was thus put into a position to freely follow his various studies. Unfortunately, Newton's first publications in the area of optics were not well-received by other English scientists. Newton quickly became disillusioned and was reluctant to publish at all. He retreated into studies of alchemy, theology, and Biblical prophecy (Newton was an ardent Christian).

Later, following some suggestions from Robert Hooke and at the urging of an astronomer friend named Edmund Halley (the discoverer of the famous comet), Newton finished his work on mechanics begun nearly 20 years earlier. In 1687 he published his most important work, *Principia Mathematica Philosophia Naturalis* or in English, *The Mathematical Principles of Natural Philosophy,* or simply the *Principia.* (It was written in Latin, as were most scientific works at that time.) The *Principia* represents one of the greatest accomplishments of the human mind, and perhaps the single most important work in the history of physics.

Certainly Newton's work was the continuation of the earlier work of Aristotle, Kepler, and especially Galileo. Newton himself said, "If I have been able to see so far, it is because I have stood on the shoulders of giants." Nevertheless, Newton's work was more than the great synthesis of all the thinking that had been accomplished earlier, and as a result of the publication of the *Principia,* he was established as the outstanding natural philosopher (i.e., physicist) of his time. He later served in Parliament (as the representative of Cambridge University) and became Warden of the Mint in London. Eventually Newton completely overshadowed all his early denigrators and even became president of the prestigious Royal Society, with unparalleled power in the realm of science.

As a person, Isaac Newton was jealous, egotistical, complex, and troubled. An absent-minded, confirmed bachelor, he was cared for by his niece. He was recognized as the leading scientist of his time and was the first scientist to be knighted. He had heated rivalries with other scientists, especially John Flamsteed, an English astronomer, and Gottfried Leibnitz, the German philosopher who co-invented calculus. Newton was admired by Voltaire and Alexander Pope, despised by Jonathan Swift and William Blake. He made some of the most important contributions to science ever and his work ushered in a new wave of optimism regarding the ability of man to understand his universe.

2. The *Principia Mathematica*

Newton's *Principia* was written in a rigorously logical, axiomatic style, following the example set by the ancient Greek mathematician Euclid in his book on geometry. Newton presented a number of basic definitions and assumptions, defined what he called "fundamental" and "derived" quantities, and then presented his three laws of motion, and a few other laws, including his universal law of gravitation, all based on induction from experiment. Then by logical deduction he showed that all the observed motions of objects in the physical universe, including celestial motion, natural local motion, and projectile motion, were simply consequences of his definitions and laws of motion. The *Principia* stands as a simple but truly monumental exposition of the nature of motion. Only in modern times, with the development of relativity and quantum mechanics, have any limitations on Newton's system been substantiated. These limitations are important only for systems involving extreme conditions: high speeds, small size, or very low temperatures.

Newton's definitions begin with the **quantity of matter,** which he described as the product of the density and volume of an object. This quantity describes "how much stuff" there is in an object, and is now called **mass.** The **quantity of motion** is defined as the product of the quantity of matter of an object and its velocity—that is, its mass times its velocity (**mv**)—and is a vector quantity called

momentum. Note that both mass and velocity contribute to momentum. A 10-ton truck traveling at 20 mph has a greater magnitude of momentum than a subcompact automobile traveling at 50 mph, because the truck is about 10 times more massive than the automobile, whereas the automobile has a speed only 2.5 times greater than the truck.

Inertia is introduced as an inherent property of mass that describes its resistance to a change of its state of uniform motion. This will be discussed more fully below. Finally, Newton defined **impressed force** as being an action that may change the state of motion (i.e., the momentum) of a body. Normally, one thinks of an impressed force as likely to speed up or slow down an object. Newton recognized that it is also possible for a force to change the direction of motion of an object, without changing its speed. Such a force is called a *centripetal force.* An example is an object spun in a circle fastened to a string. The object may swing around with a constant speed, but its direction is continually changing. This change of direction is a form of acceleration and is caused by the string.

Newton next introduced his ''fundamental quantities,'' which were required to be measurable and objective, that is, independent of the state of mind of the individual observer. Moreover, he desired that there be only a small number of these quantities, so that the complete description of science would be based on as few basic ideas as possible. He needed only three such fundamental quantities: time, length, and mass. These were to be measured in terms of fundamental units. The modern scientific units are the *second,* the *meter* (39.37 in.), and the *kilogram* (equivalent to 2.2 lb), respectively. The studies of electricity and heat have necessitated the introduction of two additional fundamental quantities. Other so-called derived quantities are formed as combinations of the fundamental quantities, for example, kinetic energy is half the product of momentum and velocity (hence its units are mass times the square of the ratio of length to time).

Having defined various concepts and quantities carefully, Newton was able to introduce his three laws of motion very simply. These laws were intended to specify the relationship between impressed forces and the *changes* in the motions of an object.

FIRST LAW OF MOTION—LAW OF INERTIA

In the absence of a net external force, an object will continue in a state of uniform motion (including rest) in a straight line.

As already discussed, this law had been recognized by Galileo before Newton was born, and represented a recasting of Aristotle's question, ''Why do objects keep moving?'' into the form, ''Why do objects stop moving?'' This law simply states that all objects with mass have a common property, called inertia, which ''keeps them doing what they have been doing.'' (In fact, mass is a measure of that inertia.) Newton, however, stated the law correctly in recognizing that inertial motion is straight-line motion, not circular motion.

SECOND LAW OF MOTION—LAW OF ACCELERATION

The time rate of change of motion (momentum) of an object is directly proportional to the magnitude of the impressed force and is in the direction of the impressed force.

This law relates acceleration of an object to the impressed force. Note that momentum is mass times velocity, so that if the mass of an object is not changing, then change of momentum implies change in the velocity, which is called acceleration. Hence, this law says that the acceleration of an object (with fixed mass) is proportional to the impressed force. Doubling the force doubles the acceleration, and tripling the force triples the acceleration. Mathematically, the second law becomes

$$\text{mass} \times \text{acceleration} = \text{impressed force}$$

or symbolically,

$$m \cdot \vec{a} = \vec{F}$$

where the arrows over the a and F are just to remind us that both acceleration and force are vector quantities, with specific directions, which, according to the law, must be in the same direction. If we divide both sides of this equation by the mass m, we obtain the expression for the acceleration,

$$\vec{a} = \frac{\vec{F}}{m}$$

In this form, Newton's second law serves to tell us how the acceleration of an object depends on both the impressed force and the mass of the object. In words, it says that the acceleration of an object is directly proportional to the net impressed force (and in the same direction) and *inversely* proportional to the mass of the object.

It is important to understand what "inversely proportional to" means. Symbolically, if A is inversely proportional to B we write

$$A \propto \frac{1}{B}$$

and it means that if B gets larger, A gets smaller, and vice versa. Note that if A were directly proportional to B, then if B gets larger, A also gets larger. An example of "directly proportional to" is the change in length of a stretched spring that increases as the stretching force increases. An example of "inversely pro-

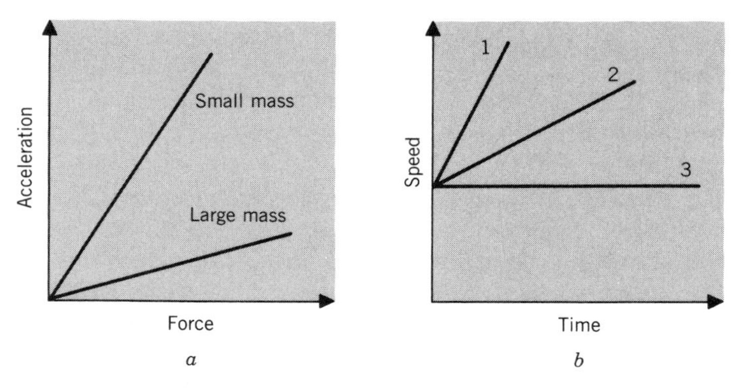

Figure 3.15. Newton's second law of motion. (a) Effect of a force on acceleration for a small- and large-mass object. (b) Effect of application of constant force on a speed-time graph: 1—small mass or large force; 2—large mass or small force; 3—zero net force.

portional to'' is the volume of a balloon full of air, which decreases as the pressure of the surrounding atmosphere increases. We will need to use each of these two different kinds of ''proportional to'' several times as we continue in this text.

Thus Newton's second law tells us that the acceleration of an object is directly proportional to the impressed force and inversely proportional to the mass of the object. This is reasonable, of course, for we know that if we push or pull harder on an object, it will speed up faster; and that if the object is heavier (more massive), the same amount of push or pull will not be as effective. Larger engines give greater accelerations, but more massive cars reduce the acceleration. But Newton's second law goes beyond just what we know must be true in a *qualitative* way and tells us how acceleration is related to impressed force and mass in a *quantitative* way. The law indicates that the acceleration is not directly proportional to the square root of the force, for example, or the cube, or whatever; but simply is proportional to the magnitude of the force (i.e., to the *first* power). Similarly, the second law indicates that the acceleration is inversely proportional to the mass, in a specific way.

This relationship of acceleration to force for large and small masses is shown graphically in Fig. 3.15. Newton's second law gives one the ability to calculate how much the acceleration of an object will be for a given force, if the mass is known. Scientists and engineers use this formula possibly more than any other in their various calculations. It applies to cars, spaceships, rockets, rubber balls, and subatomic particles. It is probably the most frequently used single equation in physics.

OPTIONAL SECTION 3.4 *Examples of Newton's Second Law of Motion*

Suppose we have a car of mass 1400 kg (about 3000 lbs) and we wish to be able to accelerate it from 0 to 100 km/h (62 mi/h) in 10 seconds. We can use Newton's second law to determine how powerful the engine must be.

We start with

$$F = m\,a$$

The desired acceleration is

$$\frac{100 \text{ Km/h}}{10 \text{ sec}} = \frac{2.8 \text{ m}}{\text{sec}^2},$$

and the necessary force is

$$F = (1400 \text{ kg}) \left(\frac{2.8 \text{ m}}{\text{sec}^2}\right) = 3890 \text{ N}.$$

Here N stands for newton, which is the unit of force in the metric system. One newton is defined as the amount of force required to accelerate one kilogram of mass by an acceleration of one meter per second per second (m/sec^2). One newton is about 0.22 pounds. For our example here, an engine of about 72 hp would be required, which is a reasonable-sized modern car engine.

THIRD LAW OF MOTION—LAW OF ACTION AND REACTION

The mutual actions of two bodies on each other are always equal and directed toward contrary parts.

Implicit in this law is the idea that forces represent the interaction between bodies. Whereas the first and second laws focus attention on the motion of an individual body, the third law says that forces exist because there are other bodies present. Whenever a force is exerted on one object or body by a second object, the first object exerts an equal and opposite force on the second object. The force relationship is symmetric.

For example, when a horse pulls a wagon forward along a road, the wagon simultaneously pulls back on the horse, and the horse feels that reaction force. The horse's hooves push tangentially against the surface of the road, if it is not too slippery, and simultaneously the surface of the road pushes tangentially against the bottom of the horse's hooves. It is important to remember which forces are acting on the horse. They are the reaction force of the road surface on the horse's hooves, which "pushes" forward, and the reaction force of the wagon as transmitted through the harness to the horse's shoulders, which "pulls" backward. In order for the horse to accelerate, the force of the road surface on the horse's hooves must be greater than the force of the harness on the horse's shoulders. But the force of the road surface is a reaction force and must be the same magnitude as the force exerted by the horse's hooves on the road. If the road is slippery there will not be enough friction between the road surface and the bottom of the horse's hooves, and the horse will slip. The horse will be unable to use its

Figure 3.16. Force pairs for a horse pulling a wagon. (a) Horse pulls on wagon (through the harness). (b) Wagon pulls back on horse. (c) Horse pushes against road surface. (d) Road surface pushes back against horse's hooves.

powerful muscles to push sufficiently hard against the road surface to generate a reaction force that will push the horse forward. These various forces, which are all involved in the horse pulling a wagon, are shown (by arrows) in Fig. 3.16. It is essential to recognize, strange as it may seem, that it is the *reaction force* of the road surface against the horse's hooves that propels the horse forward.

A further example of the role of reaction forces and the third law of motion is supplied by a rocket engine. In a rocket, high-speed gases are expelled from the exhaust of a combustion chamber. The rocket in effect exerts a force on the gases to drive them backward. Simultaneously the gases exert a reaction force on the rocket to drive it forward. It is this reaction force that accelerates the mass of the rocket. Indeed, the corporate name of one of the major American manufacturers of rocket engines in the 1950s was Reaction Motors, Inc.

2.1 Law of Conservation of Mass. In addition to his three laws of motion, Newton considered it important to indicate two **conservation laws** that he concluded must be true. In physics, a quantity is said to be conserved if the total amount is fixed (always the same). Newton's simplest conservation law is that for mass (quantity of matter). He believed that the total amount of mass in a carefully controlled, "closed" system is a fixed amount. A closed system simply means that nothing is allowed to enter or leave the system being studied. Certainly mass is not conserved in a system if more material is being added. The real import of this law is that Newton believed that mass could be neither created nor destroyed. He recognized that matter can change in form, such as from a solid to a gas in burning. But Newton believed that if one carefully collected all the smoke and ashes from the burning of an object, one would find the same total amount of matter as there was in the beginning. We now recognize that this conservation law is true only if we also capture all of the heat and light which escape.

2.2 Law of Conservation of Momentum. Using his second and third laws of motion, Newton was able to show that whenever two bodies interact with each other, their total momentum (i.e., the sum of the momentum of the first body and the momentum of the second body) is always the same, even though their individual momenta will change as a result of the forces they exert on each other. Thus if two objects (e.g., an automobile and a truck) collide with each other, their paths, speeds, and directions immediately after the collision will be different than they were before the collision. Nevertheless, if the momentum of the first object *before* the collision is added to the momentum of the second object *before* the collision, and compared to the sum of the momentum of the first object and the momentum of the second object *after* the collision, the results will be the same. This is called the law of conservation of momentum, and it can be generalized to apply to a large number of objects that interact only with each other.

Although Newton derived the law of conservation of momentum as a consequence of his laws of motion, it is possible to use this law as a postulate, and derive Newton's laws from it. In fact, the modern theories of relativity and of quantum mechanics show Newton's laws to be only approximately correct, particularly at high speeds and for subatomic particles, but the law of conservation of momentum is believed to be always exact.

OPTIONAL SECTION 3.5 *Conservation of Momentum*

As an example of how total momentum is conserved, let us consider the collision between a large and small automobile. Suppose the large car has a mass of 2000 kg (weight ~ 4400 lb) and the small car a mass of 1000 kg (weight ~ 2200 lb). Suppose that they are each traveling at 65 km/h (~ 40 mi/h) and collide head-on in an intersection and stick together. From the fact that the total momentum is fixed, we can calculate the velocity (i.e., speed and direction) of the two cars immediately after the impact.

We recall that momentum is quantity of motion, or mass times velocity. Mathematically, we write this as

$$\vec{p} = m \times \vec{v}$$

If we call the original direction of the large car the positive x-direction, then the smaller car was originally moving in the negative x-direction. Hence the total momentum just before the collision was

$$\begin{aligned} P_i &= m_1 v_i - m_2 v_i \\ P_i &= (2000 \text{ kg})\,(65 \text{ km/h}) - (1000 \text{ kg})\,(65 \text{ km/h}) \\ P_i &= 65{,}000 \text{ kg} \cdot (\text{km/h}) \end{aligned}$$

Immediately after the collision, the momentum must be

$$\begin{aligned} p_f &= m_1 v_1 + m_2 v_2 \\ &= m_1 v_f + m_2 v_f \\ &= (m_1 + m_2) v_f \\ P_f &= (3000 \text{ kg}) v_f \end{aligned}$$

The law of conservation of momentum requires

$$P_f = P_i$$

or

$$(3000 \text{ kg})v_f = 65{,}000 \text{ kg (km/h)}$$

or

$$v_f = \frac{65{,}000}{3{,}000} \text{ km/h} = 22 \text{ km/h } (\sim 13 \text{ mi/h})$$

in the positive x-direction.

3. The Universal Law of Gravitation

As we have discussed, Galileo showed that if the effects of buoyancy, friction, and air resistance are eliminated, all objects fall to the surface of the earth with exactly the same acceleration, regardless of their mass, size, or shape. He showed also that even projectiles (i.e., objects that are thrown forward) also fall *while* they are moving forward. Their forward, or horizontal, motion is governed by the law of inertia; but their simultaneous falling motion occurs with exactly the same acceleration as any other falling object. It was left to Newton to analyze the nature of the force causing the acceleration of falling objects.

Newton considered that falling objects accelerate toward the Earth because the Earth exerts an attractive force on them. It is said that the basic idea for the law of gravity occurred to Newton when an apple fell on his head. This story is almost certainly untrue; but Newton did remark that it was while thinking about how gravity reaches up to pull down the highest apple in a tree, that he first began to wonder just how far up gravity extended. He knew the Earth's gravity still existed high on the mountains and concluded it likely had an effect in space. Finally, he wondered if gravity reached all the way up to the Moon, and if one could, in fact, demonstrate that gravity was responsible for keeping the Moon in its orbit. He recognized that the Moon is in effect also "falling" toward the Earth, because its velocity is always changing, with the acceleration vector perpendicular to its circular path and therefore pointing to the Earth at the center of the Moon's orbit. This force, because it changes the direction of the Moon, is a *centripetal force,* as discussed earlier. Similarly, the planets are "falling" toward the Sun because in their curvilinear elliptical paths their acceleration vectors point toward the Sun. Because accelerations can exist only when there are impressed forces, the Sun must exert a force on the planets. Therefore Newton concluded that the same kind of force, the force of gravity, might be acting throughout the universe.

By the use of Newton's laws of motion, as well as experimental measurements of the orbits of the Moon and the planets, it is possible to infer (reason by induction) the exact form of the gravitational force law. According to Newton's second law, if the force were constant, more massive objects should fall to Earth with a smaller acceleration than less massive objects. Therefore the Earth must exert a greater force on more massive objects—they weigh more (weight is a force). In fact, the gravitational force on an object must be exactly proportional to the mass of the object in order that all objects have the same acceleration. Therefore the Earth exerts a force on an object proportional to the mass of the object. But according to Newton's third law, the object must exert an equal and opposite force on the Earth and this force must be proportional to the mass of the Earth, because from the object's "point of view" the Earth is just another object. Both proportionalities must apply, and therefore the force of gravitation must depend on *both* the mass of the Earth and the mass of the object.

Newton knew that the acceleration of the Moon, while "falling" toward the Earth (he considered the Moon to be like a projectile) was much less than the acceleration of objects falling near the surface of the Earth. (Knowing the dimensions of the Moon's orbit, and the length of time it takes the Moon to go around its orbit once—$27\frac{1}{3}$ days—he was able to calculate the acceleration of the Moon.) He eventually concluded that the Earth's force of gravitation must depend on the distance between the center of the Earth and the center of the object (the Moon in this case). The distance from the center of the Earth to the center of the Moon is 60 times greater than the distance from the center of the Earth to an object on the surface of the Earth. The calculated "falling" acceleration of the Moon is 3600 times smaller than the falling acceleration of an object near the surface of the Earth. But 3600 equals 60 times 60. Therefore it is a good guess that the force of gravity decreases with the *square* of the distance from the center of the attracting object. Putting all these considerations together, Newton proposed his law of gravitation.

THE UNIVERSAL LAW OF GRAVITATION

Every material particle in the universe attracts every other material particle with a force that is proportional to the product of the masses of the two particles, and inversely proportional to the square of the distance between their centers. The force is directed along a line joining their centers.

In mathematical form, the magnitude of the force is represented by

$$F_{\text{gravity}} = \frac{Gm_1m_2}{r^2}$$

where F_{gravity} is the magnitude of the force, m_1 is the mass of one of the objects, m_2 is the mass of the other object, r is the distance between the centers of the two objects, and G is a proportionality constant (which is needed just to use units already defined for mass, distance, and force).

The universal law of gravitation is referred to as an *inverse square law,* because the magnitude of the force varies inversely with the square of the distance between the objects (see the discussion of Newton's second law regarding inversely proportional to). Now, in truth, Newton was not the first person to consider that the law of gravitation might be an inverse square law. A few other European scientists had or were also considering such a dependence, especially the English physicist Robert Hooke, and the Dutch physicist Christian Huygens. Because they lacked Newton's mathematical skills (especially the ideas of calculus), however, they were unable to demonstrate that an inverse square law works to describe falling objects and the motion of the Moon.

With his three laws of motion and the universal law of gravitation, Newton could account for all types of motion: falling bodies on the surface of the Earth, projectiles, and the heavenly bodies in the sky. In particular, he was able to account for Kepler's laws of planetary motion (see Chapter 2). Newton was able to show that the three laws of planetary motion are logical mathematical consequences of his laws of motion and of his universal law of gravitation. For example, by using the mathematical form of the universal law of gravitation for the impressed force in his second law of motion, he was able to obtain the mathematical formula for an ellipse (using analytic geometry) as per Kepler's first law of planetary motion.

Newton's law of gravitation applies not only to the Earth and the Moon, but also to the Earth and Sun, the other planets and the Sun, Jupiter and its moons, and so on. An analysis similar to that for the Earth and Moon works also for these other examples. The law of gravitation, as well as the laws of motion, apply throughout the solar system. Moreover, astronomical evidence shows that they apply also to the universe on an even larger scale. These laws apparently are truly "universal."

Newton's laws of motion, together with the universal law of gravitation, provided a great synthesis. With just a few laws, Newton was able to explain within the framework of a single theory, phenomena that previously had been considered to be entirely unrelated, viz., falling bodies, projectiles, and planetary orbits. This synthesis represents the first real "unification" of forces in physics. Another unification of this type was achieved later when Maxwell combined the laws of electricity and magnetism into an electromagnetic force. More recently, electromagnetic forces and the "weak" nuclear force have been explained in terms of the "electroweak" interaction. Ever since this first tremendous success by Newton, physicists have searched hopefully for further developments that can unify different phenomena. There is talk and hope for a "grand unified theory" that would account for all physical phenomena within the framework of a single theory. This subject is discussed more fully in Chapter 8.

OPTIONAL SECTION 3.6 *The Universal Law of Gravitation*

We can use Newton's universal law of gravity to calculate the gravitational force between various objects.

Suppose the weight of a person is 150 lb. What is the force of gravity on this person? Weight *is* merely the force of gravity, so the force is 150 lb. But this weight is just the force of gravity on the person due to the Earth if the person is on the surface of the Earth. Suppose the person is an astronaut and travels in a spaceship to the Moon. What is the ''weight'' of the person then?

If we mean by weight the force of gravity due to the Earth, it is very small. Newton's universal law of gravity has the mathematical form

$$F_{\text{gravity}} = \frac{GmM}{r^2}$$

where m is the mass of the smaller object (the person), M is the mass of the larger object (the Earth), r is the distance between the centers of the two objects, and G is the gravitational constant. On the surface of the Earth we have

$$F_{\text{gravity}} = \frac{GM_E m_{\text{person}}}{r^2_{\text{Earth}}}$$

At the distance of the Moon, we have

$$F_{\text{gravity}} = \frac{GM_E m_{\text{person}}}{d^2_{\text{Moon}}}$$

So the only difference is that r changes from r_{earth} ($\sim$ 6000 mi) to d_{moon} (240,000 mi). Hence the force of gravity due to the Earth will **decrease** by the ratio of

$$Ratio = \frac{r^2_{\text{Earth}}}{d^2_{\text{Moon}}} = \frac{(6000\ \text{mi})^2}{(240{,}000\ \text{mi})^2}$$
$$Ratio = 0.000625.$$

The person's ''weight'' will be less than 1/1000 of what it was on the surface of the Earth.

If the astronaut actually lands on the Moon, we might want to know what is his or her ''weight'' on the moon. What we now are interested in is the force of gravity due to the moon, that is,

$$\text{F}_{\text{gravity}} = \frac{GM_{\text{Moon}} m_{\text{person}}}{r^2_{\text{Moon}}}$$

Now

$$\frac{M_{\text{Moon}}}{M_{\text{Earth}}} \simeq 0.012$$

and

$$\frac{r_{\text{Moon}}}{r_{\text{Earth}}} \approx 0.27$$

Hence,

$$F_{\text{gravity}} \text{ (on Moon)} = \frac{G(0.012M_{\text{Earth}})m_{\text{person}}}{(0.27r_{\text{Earth}})^2}$$
$$\simeq 0.17\frac{GM_{\text{Earth}}m_{\text{person}}}{r_{\text{earth}}^2}$$

so that the force of gravity on the surface of the Moon, due to the Moon, is about $\frac{1}{6}$ of the force of the gravity on the surface of the Earth, due to the Earth. Thus the astronaut would ''weigh'' about 25 lb on the Moon.

OPTIONAL SECTION 3.7 *The Acceleration of Gravity*

Let us consider quantitatively one simple example of how Newton's second law of motion and his universal law of gravitation can be combined to yield a familiar result. Recall that Newton's second law appears mathematically in the form:

$$\vec{a} = \frac{\vec{F}}{m}$$

Now if we consider the application of this law to a falling object on the surface of the Earth, then the impressed force is just the force of gravity, given by

$$F = F_{\text{gravity}} = \frac{GMm}{R^2}$$

where M is the mass of the Earth, m is the mass of the falling object, R is the radius of the Earth (recall that the law of gravity says that the force is inversely proportional to the square of the distance *between the centers* of the two objects), and G is the universal gravitational constant. If we use the expression of F_{gravity} in the first formula, we have

$$a = g = \frac{F_{\text{gravity}}}{m} = \frac{GM}{R^2}$$

and we see that the acceleration of a falling object is predicted to be independent of the mass (m) of the object; that is, all objects will fall at the same rate. This is, of course, just what Galileo had concluded. Furthermore, if we use modern values for the mass and the radius of the Earth and the known value of G, this equation predicts $g = 9.8$ meters per second per second (m/sec^2), or equivalently, $g = 32$ ft/sec^2, the known acceleration of gravity.

4. Tests and Criticisms

Newton's laws allow one to go beyond Kepler's laws of planetary motion. Because Newton's law of gravitation indicates that *all* material objects exert a gravitational force on all other bodies, we know that the planets should experience gravitational attraction not only from the Sun, but also from each other. Thus the planets should disturb each other's orbits so that the resulting orbits are not perfectly elliptical. Newton was concerned about the effects of these perturbations. He thought that the planetary orbits might be unstable, so that the gravitational pulls of the other planets might cause a planet to leave its orbit. Newton speculated that divine intervention might be required to maintain the planets in their orbits. Nearly 100 years later, a famous French mathematician and astronomer, Pierre Laplace (1749–1827), showed that the planetary orbits actually are stable, so that although the orbits are not elliptical, the planets will not leave their orbits if perturbed.

Eventually, the perturbations caused by the planets' gravitational attractions for each other provided some of the most convincing evidence for the correctness of Newton's laws. In 1781, a new planet was discovered and named Uranus. Using improved mathematical techniques to calculate the effects of all the other planets on the orbit of Uranus, by about 1820 it became clear that there were discrepancies between the observed orbit and the calculated orbit. Suspecting that there was another planet in the solar system, the French astronomer U. J. J. Leverrier and the English astronomer J. C. Adams eventually were able to calculate just where, and how massive, the new planet must be in order to explain the discrepancies. In 1846, Neptune was discovered almost exactly where they had predicted. Careful study of the orbit of the new planet Neptune indicated that yet another planet must exist. In 1930, Pluto was discovered and most of the remaining unexplained perturbations were removed. A debate continues about whether there are now any residual discrepancies in the orbits of all the planets, after taking the gravitational attractions of all the presently known ones into account. The remaining discrepancies are small, and they may be only slight errors in the difficult observations (or there may be yet another planet).

The important point of the planetary perturbations is that, in the end, they yielded a remarkable test of Newton's laws. The effects of all the planets on each other are predicted and observed, and the agreement is remarkable. Only in the case of the planet Mercury have the calculations been significantly off the mark. For Mercury, Einstein's general theory of relativity is required to resolve the discrepancy, and this will be discussed in Chapter 6.

In addition to planetary orbits, Newton's laws have been shown to be highly accurate in many other physical situations. Calculations similar to those for the planets are made for artificial satellite orbits around the Earth. The appropriate combinations of Newton's three laws of motion and the universal law of gravitation explain all projectile motions. Newton's laws are used daily by engineers and scientists to describe the motions of everything from trains to subatomic particles. Lawyers often study Newton's laws in order to be able to analyze automobile accidents. Although we will see that relativity and quantum mechanics will require some modification and extension of Newton's laws, they are continually being demonstrated to be accurate descriptions of how our physical universe behaves in motion.

It is important to recognize how Newton combined the use of inductive and deductive logic. He always started with a critical observation. In the case of gravity, for example, he began by analyzing how objects fall and then extended his analysis to the motion of the Moon. Once he had a general theory, arrived at from his observations (inductive logic), he then proceeded to use the general theory to explain or predict other phenomena (deductive logic). In the case of gravity, he showed that his inverse square law could explain Kepler's laws of planetary motion. Newton set a clear example, followed since by nearly all scientists, of combining inductive and deductive logic to obtain and test a theory.

Before we consider the philosophical consequences and implications of Newton's work, it is necessary to point out that his laws do not answer all questions with regard to motion and gravity. Although Newton explained *why* objects fall with uniformly accelerated motion using an inverse square law for gravity, he did not indicate *why* gravity has such a dependence on distance, or *how* gravity acts over a distance. This latter question is known as the *action-at-a-distance* problem and has puzzled scientists for a very long time. Note that gravity can act across (apparently) empty space. The Earth attracts the Moon (and vice versa) and the Sun attracts the Earth (and vice versa). The Earth and the Moon are separated by about 240,000 miles of almost perfectly empty space. It is quite remarkable that, somehow, these bodies interact with each other. On a small scale, it can be quite intriguing to play with two small bar magnets. It is easy to feel them start to attract (or repel) each other rather strongly before they actually touch. How do they do it? It is another version of the action-at-a-distance problem.

We are not ready yet to fully discuss the answer to this problem, which we will consider again in Chapter 6 and then discuss in more detail in Chapter 8 on the fundamental building blocks of nature. Briefly, it appears that bodies interact, whether by gravity or electromagnetic forces or other forces, by exchanging tiny subatomic particles that are invisible to the human eye. Newton discussed the action-at-a-distance problem, but unable to solve it, concluded by saying, "I make no hypothesis."

Summarizing, we may say that Galileo showed exactly *how* objects fall, namely, with uniformly accelerated motion. Newton then explained that objects fall this way because of certain laws of motion and the inverse square law of gravity. We now know how forces are transmitted, but we still do not know *why* objects have inertia, or why gravity exists at all. We will make further progress on these questions as we discuss relativity, quantum mechanics, and elementary particles, but we will see that every answer to a question always seems to raise a new question.

E. Consequences and Implications

Newton's great synthesis of the laws of motion and the universal law of gravitation to explain all the kinds of motion had a great philosophical and emotional impact on the scientists of the time. His work created an image of a great clockwork-like universe. Because it is only in relatively recent times that this image has been questioned, it is important to understand what this image was and some of the philosophical consequences it implied.

1. The Mechanical Universe

Newton's laws indicated that the motions and interactions of all the material bodies of the universe obey a few certain, relatively simple rules (i.e., laws). According to these rules, if one knows the locations, masses, and velocities of a set of objects at some one instant, then one can determine how the objects interact with each other and what will be the result of collisions. For example, one uses the law of gravitation to determine what net force is impressed on each object because of the presence of all the other objects. Next, one can use the second law of motion to determine the magnitude and direction of the consequent acceleration of that object, which allows one to know the velocity of the object. The law of conservation of momentum will enable one to determine the result of any possible collisions. Now, of course, this can be quite involved, but it is often performed for certain sets of bodies by engineers and scientists. And it works!

For example, the motions of a set of billiard balls on a table, or of subatomic particles in nuclear reactions, can be accurately predicted if one knows at some one time where each object is and what its velocity is. Forensic physicists can reliably calculate backward to determine the initial directions and speeds of automobiles involved in an accident if the final directions and speeds can be determined (from skid marks, etc.). Newton's laws indicated that the future motions of any set of objects can be determined if we know their motions initially.

If one takes this realization to its apparent logical conclusion, it implies that the future motions of all the objects of the universe could be predicted if we could only determine what they are right now. Of course, we cannot determine the locations and velocities of all the bodies of the universe at any one moment (not even with the world's largest computer). But philosophers after Newton were quick to realize that the important point was that it could be done *in principle.* Whatever the positions and velocities are now completely determines what they will be in the future. It does not matter that we cannot carry out the calculations of the future of the entire universe; it is still completely determined. Newton's laws indicate a universe evolving in time, according to his laws, in a completely predetermined fashion. The universe is like a giant, elaborate clockwork, operating in a prescribed way.

Many philosophers cited Newton's work as proof of predestination. If it is all evolving in a (theoretically) predictable way, it is predetermined. Newton's universe was evoked to challenge the view of self-determination of individuals—and even of divine election of kings. Such conclusions may appear valid. As we will see later with regard to the apparent consequences of other physical theories, however, such grand conclusions applied to the universe as a whole are very questionable. Specifically, we will see that the results of quantum mechanics directly challenge this predetermined universe. More generally, it is hard to be sure of the extension of the specific laws of nature determined in our small region of the universe to the net course of the universe as a whole.

2. The Rational Universe

Newton's laws were a great success for science. So much of physics had appeared to be separate, unrelated phenomena with no simple explanations. Newton explained all kinds of motion in terms of only a few simple laws. Scientists were extremely encouraged. They became confident that our universe was indeed a rational system, governed by simple laws. It appeared that perhaps there were good reasons for everything.

Scientists in all areas set out anew to discover the "basic" laws of their fields. In fact, areas not regarded previously as sciences were declared to be so. The "social sciences" were created and scholars started to search for the laws that govern human behavior. There was a movement to make a science out of every field of human endeavor. Adam Smith in his *Wealth of Nations* tried to define the laws of economics so that scientifically sound economic policies could be adopted. Auguste Comte tried to do the same in sociology. Faith in the ultimate power of the "scientific mind" was tremendous. The kings of Europe established academies dedicated to solving all human problems. Frederick the Great of Prussia and Catherine of Russia founded scientific academies in imitation of those of Charles II in England and Louis XIV in France. They spoke of the time as an Age of Reason and of Newton as a bearer of light. The point cannot be made too strongly: Newton's work unified a very diverse range of phenomena by explaining all motion with only a few laws. If the physical universe was so simple, the thinking proceeded, why not all branches of knowledge?

The idea that there are simple, logical rules for everything extended beyond mere scholarly pursuits. The American and French Revolutions were both dedicated to the idea that people had certain *natural* rights that must be honored. The designers of the new governments proceeded to try to establish laws and governing principles that would be consistent with these natural rights and tried to argue logically from the assumed rights to deduce the kind of government that should exist. The Declaration of Independence has the same kind of axiomatic structure, with deductive logic supporting the final conclusions, as does Newton's *Principia.* Benjamin Franklin, who was one of the best scientists of his time, wrote an essay entitled, "On Liberty and Necessity; Man in the Newtonian Universe." Jefferson called himself a scientist and began the American tradition of governmental encouragement of the sciences. The Age of Reason was in full bloom.

Study Questions

1. What is the subject of the science of mechanics?
2. What is the main point of Plato's "Allegory of the Cave"?
3. Name the four prime substances according to Aristotle.
4. Name the four basic kinds of motion according to Aristotle.
5. Explain why only vertical motion was "natural motion" according to Aristotle.
6. According to Aristotle, why is the Earth "imperfect"?
7. What was the question Aristotle attempted to answer with his process known as antiperistasis?
8. How does Buridan's example of a millwheel provide difficulty for the process of antiperistasis?
9. Why did Aristotle believe a perfect vacuum couldn't exist?
10. What is the difference between uniform motion and uniformly accelerated motion?
11. Give an example of an object with negative acceleration.
12. How many "accelerators" are there in a car? (Remember that velocity is a vector quantity.)
13. If the distance versus time graph is a straight line, what kind of motion does it indicate?
14. If the distance versus time graph is a curved line, what kind of motion does it indicate?
15. Why did Galileo study "falling" by using pendulums and inclined planes?
16. What kind of motion did Galileo determine that falling objects would have, ignoring resistive effects?
17. Describe the falling process of an object in air, indicating the kinds of motion the object will have at different times.
18. How did Galileo rephrase Aristotle's question about projectile motion?
19. Define inertia. With what characteristic of a body is inertia associated?
20. Why did Galileo need the superposition principle in order to use pendulums and inclined planes to study falling objects?

21. Is predicting the outcome of presidential elections from polls an example of inductive or deductive logic?
22. Why is mathematics useful in science?
23. Finding the area under a curve will likely involve what two branches of mathematics?
24. The slope of a speed versus time curve corresponds to what quantity? Finding the slope of a curve is the subject of what branch of mathematics?
25. Indicate whether the following quantities are scalar or vector quantities: (a) speed, (b) velocity, (c) force, (d) length, (e) mass, and (f) momentum.
26. Give a few examples of the addition of two or more forces whose sum is zero.
27. Why is Newton's *Principia* said to be in axiomatic style?
28. Explain why momentum is reasonably referred to as "quantity of motion."
29. Distinguish between a centrifugal and a centripetal force. Give an example of each.
30. Name Newton's fundamental quantities.
31. What is a "derived" quantity? Give an example.
32. Give your own examples of directly and inversely proportional quantities.
33. According to Newton's second law of motion, how does acceleration depend on the impressed force? How does it depend on the mass of the object?
34. Which object can have the larger acceleration: (a) a 2-ton car with a 100-horsepower engine, or (b) a 5-ton truck with a 300-horsepower engine (assume the impressed force available is directly proportional to the horsepower of the engine)?
35. What is meant by a conservation law?
36. What is the reaction force to the weight of a book sitting on a table? To a baseball bat "hitting" a ball? Indicate what would happen if the reaction forces "failed" in these two cases.
37. How does a jet engine work? Why can't a jet engine be used in outer space? Why can a rocket engine be used?
38. In what way is Newton's universal law of gravitation "universal"?
39. What part of the universal law of gravitation incorporates Newton's third law of motion?
40. How did the discoveries of the planets Neptune and Pluto confirm Newton's law of gravitation?

41. What was Newton's concern about the planetary orbits and how was it eventually alleviated?

42. What did Newton say regarding the origin of gravity?

43. How can Newton's laws of motion be used to determine the initial velocities of the cars in an accident from the skid marks? Which laws must be used?

44. How do Newton's laws imply that the universe is predetermined?

45. What was the goal of the Age of Reason?

PROBLEMS

1. Suppose a car starts from rest and accelerates to 60 mi/h with uniformly accelerated motion in 10 sec. What is the acceleration? What is the average speed? What is the distance traveled?

2. A rock is dropped from the top of the Sears Tower in Chicago, which has a height of 1200 feet. Ignoring air resistance, how long does it take for the rock to hit the ground? How fast is the rock falling at impact? Would either of these answers change if the rock is thrown 90 mph horizontally from the top? How far would the latter rock go horizontally before impact?

3. A rock is thrown upward at 90 mi/h. How high does the rock go and how long does it take to hit the ground? (*Hint:* This problem is symmetrical about the top point of the path of the rock.)

4. A car is to make a trip over a hill 1 mile up and 1 mile down. Suppose that the car is only able to go up the hill at 30 mi/h. How fast must the car travel down the hill to average 60 mi/h over the hill? (Be careful!)

5. Show that if in the formula for an ellipse, $a = b$, then the equation for a circle is obtained.

6. As in the examples of Optional Section 3.2, make graphs of the two equations

$$y = 4x^2$$

$$\frac{x^2}{4} - \frac{y^2}{9} = 1$$

   The former is called a parabola and the latter a hyperbola. What can you say about the difference in the shapes of these two curves?

7. A motorboat can travel at 25 mi/h in still water. The boat is to traverse a river flowing at 10 mi/h. At what angle must the boat point upstream in order to travel directly across the river? (Graphical solution is satisfactory.)

8. Two people push on a large rock on a frozen lake. One person pushes with a force of 50 lb toward the north and the other person with a force of 30 lb toward the east. In what direction does the rock move? (Graphical solution is satisfactory.)

9. An object of mass 10 kg traveling at 40 km/h collides with a stationary object of 10 kg and the two objects stick together. How fast does the combined object move after the collision?

10. An object of mass 10 kg traveling north at 40 km/h collides with another object of mass 10 kg traveling east at 30 km/h. The two objects collide and stick together. How fast and in what direction does the combined object move after the collision? (Graphical solution is satisfactory.)

11. Calculate the acceleration of an object with weight 500 lb subject to a constant force of 50 lb.

12. Calculate the acceleration of an object of mass 250 kg subject to a constant force of 200 newtons.

13. Calculate the acceleration of a 4000-lb car with a 250-hp engine. How long would the car take to accelerate from zero to 60 mph, assuming constant acceleration?

14. Calculate the force of gravity between the Earth and the Moon. $M_{\text{Earth}} = 6 \times 10^{24}$ kg. $M_{\text{Moon}} = 7.35 \times 10^{22}$ kg. $d_{\text{Moon}} = 3.84 \times 10^{5}$ km. $G = 6.7 \times 10^{-11}$ newton · m²/kg². Calculate the force of gravity between the Sun and the Moon. $M_{\text{Sun}} = 2 \times 10^{30}$ kg. $d_{\text{Earth-Sun}} = 1.5 \times 10^{8}$ km. The Moon is very slowly moving away from the Earth. How far will it have to be from the Earth for the Sun to "capture" our Moon (i.e., for the gravitational force from the Sun to be greater than the gravitational force from the Earth)?

REFERENCES

Herbert Butterfield, *The Origins of Modern Science,* Free Press, New York, 1957.
An introduction to the historical background of classical physics. Does not discuss "modern" physics.

Douglas C. Giancoli, *The Ideas of Physics,* Harcourt Brace Jovanovich, New York, 1978.
An introductory text, similar to this one, with somewhat different emphases and choice of topics.

Eugene Hecht, *Physics in Perspective,* Addison-Wesley, Reading, Mass., 1980.
An introductory text, similar to this one, more on the anecdotal historical details and somewhat different topics.

Gerald Holton and Stephen G. Brush, *Introduction to Concepts and Theories in Physical Science,* 2nd ed., Addison-Wesley, Reading, Mass., 1973.
An introductory text with major emphasis on the historical background of physics.

Stuart G. Inglis, *Physics: An Ebb and Flow of Ideas,* John Wiley & Sons, New York, 1970.
An introductory text, similar to this one, with good, clear development of the major ideas of physics.

Cecil J. Schneer, *The Evolution of Physical Science,* Grove Press, New York, 1964 (a reprint of *The Search for Order,* Harper & Row, New York, 1960).
See comment in references for Chapter 1.

William Thompson (Lord Kelvin). (Bettmann Archive.)

Chapter 4

THE ENERGY CONCEPT

Energy is what makes it go

Although Newton's laws of mechanics were extremely successful in explaining many natural phenomena, there still remained many processes in nature that could not be understood solely by the application of those laws. One of the best known of these other phenomena was heat. Although various attempts were made to understand heat from at least as long ago as the time of Aristotle, most of our present knowledge of heat was developed in the nineteenth century. The studies performed during the nineteenth century in order to better understand heat and its production eventually led to the modern concept of energy. Both because this concept is most significant for understanding the nature of our physical universe, and because energy has become such an important topic in our lives, it is worthwhile to retrace the important steps in the development of the energy concept. It is important to know what energy is, how it is characterized and measured, what forms it can have, and how it can be converted from one form to another.

A. Interactions and Conservation Laws

1. Collisions

The first real step in the development of the energy concept was the realization that there are conservation laws in nature. If there is a conservation law for some quantity, then the total amount of that quantity is maintained at a constant value for all time in an isolated system. Newton developed the law of conservation of momentum in his *Principia.* Newton showed that the total momentum (he called it quantity of motion) of a system, which was characterized by the sum of the products of the mass of each object and its velocity, remained a constant, even if the various objects were allowed to collide with each other. By "isolated system" is meant a collection of objects that can interact with each other, but not with anything else. As the objects interact, say by collisions, their individual velocities and therefore their momenta may change; however, the vector *sum* of all their momenta will always have the same value. (The idea for this conservation law was not original with Newton, but had been developed by the English scientists Hooke, Wallis, and Wren and proposed as a law by the Dutch physicist Christian Huygens in 1668.)

For example, this conservation law allows one to predict what will happen when one object with a known amount of momentum (mass times velocity) collides with another object of known momentum. The momentum of each of the objects after the collision can be calculated if one other fact about the collision is known; for example, whether the two objects stick together or whether the collision is elastic, as discussed in the next paragraph. If we can discover more "laws" like this that nature always obeys, we can increase our ability to predict exactly what will happen in increasingly complex situations. This is because every conservation law can be expressed as an equation. Because equations can be solved mathematically, exact results can be predicted.

Christian Huygens also recognized that in certain collisions another quantity besides momentum is conserved (i.e., its total value remains constant). Originally,

Leibnitz (1646–1716) called this quantity **vis viva,** or *living force.* Mathematically it was calculated as the product of the mass of an object times the *square* of its velocity (mv^2). Some time later a factor of one-half was included and the quantity was renamed kinetic energy. (The factor of one-half follows when this quantity is derived from Newton's laws of motion.) The kind of collision in which kinetic energy is conserved is known as an *elastic collision.* A billiard-ball collision is a familiar example of an essentially elastic collision.

The significance of the concept of kinetic energy of an object was enhanced when it was recognized to be the result of applying a force to the object. If one measures the amount of a force and the distance over which the force is applied to the object, the gain in kinetic energy for the object will be exactly equal to the product of the force times the distance (in the absence of other forces), that is,

$$\text{change in kinetic energy (K.E.)} = \text{force } (F) \times \text{distance } (d)$$

We can represent this graphically as shown in Fig. 4.1.

The force need not be a constant force as shown in the figure, but might vary in any manner over the measured distance. The kinetic energy acquired by the object would still be equal to the area under the force versus distance curve. An increase of kinetic energy is thus the integrated effect of an accelerating force acting over a distance.

In fact, the converse of this relationship shows why kinetic energy was first known as *vis viva.* If an object has velocity, it has the ability to exert a force over a distance; for example, a large moving ball colliding with a spring can compress the spring more than the "dead weight" of the ball alone. (Note that the kinetic energy of the object will decrease as it compresses the spring.) This ability to exert additional force is associated with the motion of the object and is measured by its kinetic energy. This force due to motion alone was referred to as *vis viva,* or living force.

Similarly, the change in momentum of an object (mass times velocity) can be demonstrated to be the effect of a force on an object multiplied by the length of

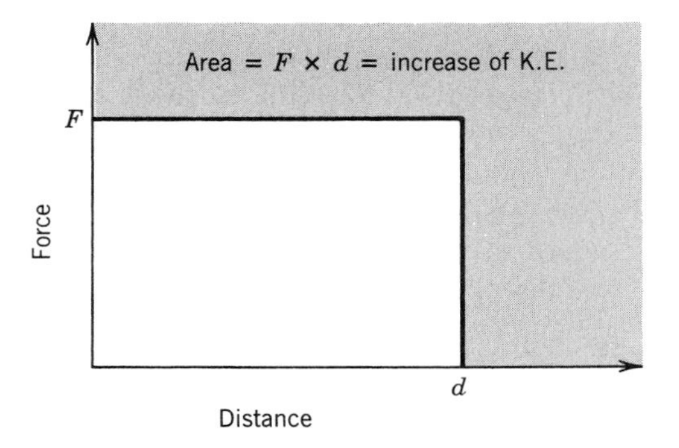

Figure 4.1. Force versus distance.

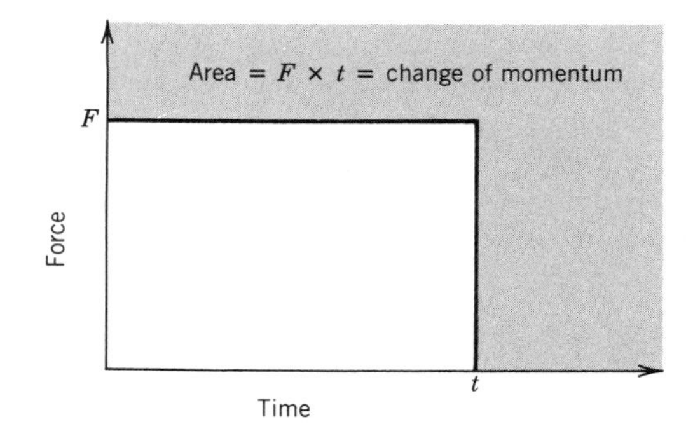

Figure 4.2. Force versus time.

time the force is allowed to act. So we have (with only one force acting on the object)

$$\text{change in momentum} = \text{force } (F) \times \text{time } (t)$$

This is shown graphically in Fig. 4.2.

Here again, the force need not be a constant force. The momentum added to the object will be equal to the area under the force versus time curve, whatever shape the curve might display.

Both kinetic energy and momentum are related to the velocity of an object. Kinetic energy is a scalar quantity, whereas momentum is a vector quantity (see Chapter 3, Section C5). Both quantities are determined by the integrated (or cumulative) effect of the action of net force. However, kinetic energy is the integrated effect of net force acting over a distance as contrasted with momentum, which is the integrated effect of a force acting over time.

2. Significance of Kinetic Energy—Work

The product of force times distance is a much more general concept than indicated so far. It is an important quantity and is called work. From the discussion above, work $(W) = F \times d =$ K.E., for a force that accelerates an object.[1] One must be very careful when using the word *work,* because it now means something very specific. A physicist would say that if you pushed very hard against a wagon full of bricks, and the wagon did not move, then you did no work on the wagon! Only if the wagon moved as a result of the applied force would you have done some work on it. Note that one can double the amount of work done either by

[1]Actually, it is only the component of the force parallel to the direction of motion that does work. Moreover, it is the net force that results in the acceleration and produces a change in the kinetic energy.

doubling the applied force or by doubling the distance over which the force is applied. In either case, the net effect will be the same—twice the amount of kinetic energy will have been imparted to the object.

Kinetic energy is our first example of one of the forms energy may have. Whenever we do work (in the physicist's sense) on an object, we change the energy of the object.

One can work on an object and not change its *kinetic* energy, however. For example, the force acting over the distance may just be moving an object slowly up an inclined plane. When the force is removed, the object is left higher up the ramp, but not moving. If work has been done on the object, what kind of energy has been imparted to it? The answer is *potential energy* because the object, in its higher position, now has the *potential* to acquire kinetic energy. If the object were nudged off the side of the ramp to make it fall, it would accelerate and gain kinetic energy as it fell. **Potential energy** is energy associated with an object because of its position or configuration, as will be discussed in the next section.

3. Conservative Mechanical Systems

There are certain common systems not involving collisions, but nevertheless involving movement, for which some aspect is unchanging. Something is being conserved, physicists prefer to say. One such system is a pendulum. A pendulum, if constructed properly, will continue to swing for a very long time. If we watch such a pendulum for only a little while, we can analyze it as if it were not going to stop at all. More precisely, we can measure just how high the pendulum bob rises on every swing and see that it comes back to almost the same height every time. It would not be hard to convince ourselves that if we could just get rid of the effect of air resistance and put some extremely good lubricant on the pivot, the pendulum bob would always return to the exact same height on every swing. This is an interesting observation. What is it that makes nature act this way for a pendulum?

Is there something being conserved in the pendulum's motion that requires it to behave this way? If so, *what* is being conserved? It is not kinetic energy. The pendulum bob speeds up and slows down, even coming to a stop, momentarily, at the top of each swing. When the bob passes through the bottom position, it is moving quite fast. Hence, because kinetic energy (K.E.) is one-half of mass $\times$ (velocity)^2, its K.E. is zero at the high points of a swing and a maximum at the low point.

Mathematical analysis shows that the quantity being conserved is the sum of the kinetic energy and the *potential energy* of the bob. (Potential energy was originally called latent *vis viva* because it was considered to have the potential of being converted into *vis viva.*) The potential energy is said to be zero when the bob is at its lowest position and increases directly with the height above that position. It reaches its maximum when the bob is at the top of the swing where the kinetic energy has gone to zero. It works out just right. The **total mechanical**

energy (i.e., the sum of these two kinds of energies) is always the same. This is an example of what we call a **conservative system.**

Another example of a conservative system is a ball rolling up and down the sides of a valley. If the terrain is very smooth, and one starts a ball at a certain height on a hill on one side of the valley, when released it will roll down into the valley and up the hill on the other side to the same height as it had on the first side. As it moves, it will possess the same combination of potential and kinetic energies as a pendulum bob. The sum, the total mechanical energy, will again always be a constant.

It can be seen why this new form of energy is called *potential* energy. When the ball is high on one hill, it has a great amount of "potential" to acquire kinetic energy. The higher up the hill the ball is started, the faster it will be moving, and therefore the more kinetic energy it will have when it reaches the valley. Note that this potential energy is actually **gravitational potential energy.** It is against the force of gravity that work is done to raise the ball higher, and the force of gravity will accelerate the ball back down into the valley.

Another system that also involves only motion and the gravitational force is the system of a planet moving around the Sun. This is also a conservative system. Kepler's first law of planetary motion (see Chapter 2) states that the planet's orbit is an ellipse, with the Sun at one focus. Thus the planet is sometimes closer, then sometimes farther from the Sun. From Newton's universal law of gravitation, it can be shown that when the planet is farther from the Sun it has more gravitational potential energy (it has further to "fall" into the Sun, so it is "higher"). Kepler's second law of planetary motion states that the radius vector from the Sun to a planet sweeps out equal areas in equal times. This law accounts for the known fact that the planet moves faster when it is closer to the Sun. Again, one can see that things occur in just the right way: The planet moves faster when it is closer to the Sun, so the planet has more kinetic energy when its potential energy is less. If the mathematics is done carefully, it can be seen that the sum of the potential and kinetic energies will remain exactly constant. (Kepler's second law also illustrates another conservation principle—conservation of angular momentum, discussed in Chapter 8).

Conservative systems thus appear to be operating in our universe. These systems always seem to maintain the same value for the total energy when taken as the sum of the kinetic energy plus the potential energy. But these systems are actually only approximately conservative. The pendulum will slowly come to a stop, as will the ball in the valley. The total energy appears to go slowly to zero. The planets are slowly moving out into larger-sized orbits. Is there or isn't there an exact conservation law for these systems?

The answer to the preceding question is that the total energy actually is conserved, but there exists yet a third kind of energy, and the kinetic and potential energies are slowly being converted into this third form. The new form of energy to be considered is **heat.** The systems have frictional forces (between the pendulum bob and the air and at the pivot for the first example); and friction (as people who have rubbed their hands together to warm them know) generates heat.

The heat so generated is gradually lost to the system. It escapes into the surrounding air and slightly warms it. That all of the lost energy appears as heat, and that heat is another form of energy, was established by nineteenth-century physicists. It is worthwhile to discuss now, in some detail, the evolution of our understanding of heat as a form of energy. In fact, historically, physicists seriously considered that there was a conservation law for total energy only after heat was shown to be a form of energy.

OPTIONAL SECTION 4.1 *Work and Mechanical Energy*

As discussed above, work can produce mechanical energy (i.e., either kinetic energy or potential energy). Work is calculated as force times distance, that is,

$$\text{work} = F \times d$$

Kinetic energy is calculated as one-half of the mass times the square of the velocity

$$\text{K.E.} = \frac{1}{2} mv^2$$

Gravitational potential energy is calculated as the weight of an object ($W = mg$) times the height (i.e., distance) of the object above some reference level,

$$\text{P.E.} = Wh = mgh$$

We can use these mathematical expressions to calculate the amount of work done and kinetic and potential energies present in a simple pendulum. Suppose we have a pendulum bob of mass 1 kg suspended by a thin wire. If we pull the pendulum up (keeping the wire tight) to a height 10 cm above its bottom (lowest) position, then the work done is

$$\text{work} = F \times d$$

where F is equal to the force of gravity that we must overcome in lifting the bob and is just the weight, $W = mg$. (Actually, the work done is the product of the distance moved times the component of the force along that distance. In this case, the force is vertical and so only the height matters for the distance moved.) So,

$$\begin{aligned} \text{work} = W \cdot h &= mgh \\ &= (1\ \text{kg}) \left(9.8\ \frac{\text{m}}{\text{sec}^2}\right) (0.1\ \text{m}) \\ &= 0.98\ \frac{\text{kg}\cdot\text{m}^2}{\text{sec}^2} \end{aligned}$$

Now 1 kg·m²/sec² = 1N·m (see Chap. 3) which is called 1 joule, and is actually a unit of energy. So the work done is 0.98 joules.

At this point, before we release the pendulum, all the work done went into increasing the potential energy.

$$\text{P.E.} = \text{work} = mgh = 0.98 \text{ joules}$$

Now we release the pendulum and it begins to swing. Using our expression for kinetic energy and assuming that the system is conservative, so that the total energy is the sum of potential energy and kinetic energy, we can calculate the velocity of the pendulum bob as it passes through its lowest point. At this lowest point, the potential energy (mgh) goes to zero since the height (h) goes to zero. Hence all the energy must now be in the form of kinetic energy and this must be equal to the original potential energy, which is equal to the work done.

$$\text{K.E.} = \frac{1}{2} mv^2 = mgh = 0.98 \text{ joules}$$

or

$$\frac{1}{2} mv^2 = 0.98 \frac{\text{kg·m}^2}{\text{sec}^2}$$

or

$$v^2 = (2)\left(0.98 \frac{\text{kg·m}^2}{\text{sec}^2}\right)\left(\frac{1}{\text{m}}\right)$$

Now $m = 1$ kg, so

$$v^2 = (2)\left(0.98 \frac{\text{m}^2}{\text{sec}^2}\right)$$
$$= 1.96 \frac{\text{m}^2}{\text{sec}^2}$$

or

$$v = 1.4 \frac{\text{m}}{\text{sec}}$$

It is important to realize that this result is obtained by merely recognizing the relationship between work and energy and by assuming the pendulum system is a conservative system.

B. Heat and Motion

1. Heat as a Fluid

Heat has been studied from very early times. Aristotle considered fire to be one of the five basic elements of the universe. Heat was later recognized to be something that flowed from hot objects to colder ones and came to be regarded as a type of fluid. About the time of Galileo this heat fluid was known as **phlogiston** and was considered to be the soul of matter. Phlogiston was believed to have mass and could be driven out of, or absorbed by, an object when burning.

In the latter part of the eighteenth century the idea that heat was a fluid was further refined by the French chemist Antoine Lavoisier and became known as the **caloric theory of heat.** The caloric theory was accepted by most scientists as the correct theory of heat by the beginning of the nineteenth century. The fluid of heat was called the caloric fluid and was supposed to be massless, colorless, and conserved in its total amount in the universe. (Lavoisier developed this theory while proving that mass is conserved when the chemical reaction known as oxidation, or burning, takes place.) The caloric fluid, with its special characteristics, could not actually be separated from objects in order to be observed or studied by itself. The caloric fluid theory thus presented a growing abstractness in the explanations of heat phenomena. Several phenomena were accurately understood in terms of the caloric theory, even though it was an incorrect theory.

As an example of a process that was thought to be understood by analysis with the caloric theory of heat, consider the operation of a steam engine. This is a good example both because improved understanding of the nature of heat came from attempts to improve steam engines and because it introduces several important concepts needed for the study of heat. The first working steam engine was constructed about 1712 by Thomas Newcomen, an English blacksmith. Very rapidly, the Newcomen engine was installed as a power source for water pumps in coal mines throughout England. It replaced cumbersome and costly horse-team powered pumps.

A simplified Newcomen steam engine is shown in Fig. 4.3. The fire under the boiler is continually producing steam from the water in the boiler. The operation of the engine is accomplished with essentially four steps. These steps are performed by using the two valves labeled *steam valve* and *water valve.* The steps are (1) open the steam valve to let steam into the piston chamber, with the pressure of the steam causing the piston to rise and the pump rod to descend; (2) close the steam valve; (3) open the water valve, which allows cool water to spray into the chamber, condensing the steam, creating a partial vacuum, and causing the piston to come down and the pump rod to be lifted; and (4) close the water valve. The whole cycle is then repeated.

The early ideas regarding the essentials of the steam engine were very crude by today's standards. Although it is called a *steam* engine, it is the fuel being burned under the boiler that actually provides the power for the engine. Early experimenters were not entirely convinced of this, however. The power source

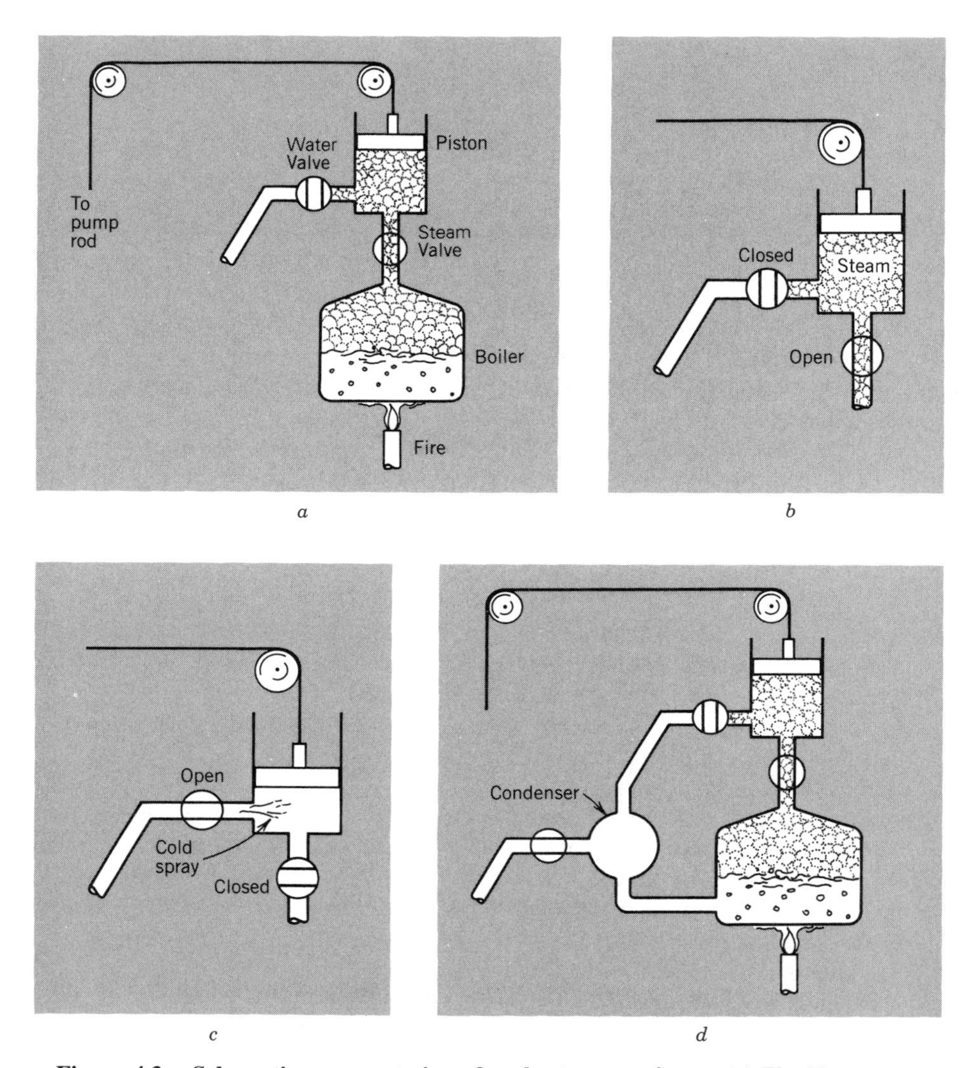

Figure 4.3. Schematic representation of early steam engines. (a) The Newcomen engine. (b) Steam pressure pushes piston up. (c) Condensing steam pulls down. (d) Watt's improved steam engine with separate condensing chamber.

for the steam engine was considered to be the steam, and the efficiency of the engine was measured in terms of the amount of steam it consumed. Many of these early ideas did improve the steam engine somewhat, especially those of the English inventor James Watt, who patented the first really efficient steam engine in 1769.

In the Newcomen engine, the walls of the cylinder are cooled during the condensation or "down-stroke" step of the cycle. When steam is admitted into the cylinder for the expansion or "up-stroke" step of the cycle, about two-thirds of the steam is consumed just to reheat the walls of the cylinder so that the rest of the steam can stay hot enough to exert sufficient pressure to move the piston up. Watt realized that the condensation step was necessary for the cyclic operation of the engine, but he also realized that the actual cooling of the steam could take place in a *different location* than the hot cylinder. He therefore introduced a separate chamber called the condenser into his engine, to carry out the cooling for the condensation step, as shown in Fig. 4.3d. The condenser was placed between the water valve and the piston chamber, and between steps (2) and (3) above, the steam was sucked into the condenser by a pump (not shown in the diagram) and step (3) was modified appropriately. Also the water from the condenser was returned to the boiler.

Watt also recognized that the necessary condensation step resulted in the loss of some steam, which was capable of pushing the piston further. In order to decrease the relative amount of steam lost in the condensation step, he decided to use higher pressure (and, therefore, higher temperature) steam in the initial stages of the expansion step. Thus even after the steam inlet valve was closed, the steam in the cylinder would still be at a high enough pressure to continue expanding and pushing the piston upward. While it continued expanding, its pressure would drop and it would "cool" off, but it would still be doing work. Only after it had done this additional work would the condensation step begin.

With these and other improvements, Watt's steam engine was so efficient that he was able to give it away rather than sell it directly. All the user of the engine had to pay Watt was the money saved on fuel costs for the first three years of operation of the engine. Watt and his partner Matthew Boulton became wealthy, and the Industrial Revolution in England received a tremendous impetus from a new source of cheap power.

In 1824 a young French military engineer named Sadi Carnot published a short book entitled *On the Motive Power of Fire,* in which he presented a very penetrating theoretical analysis of the generation of motion by the use of heat. Carnot was considerably ahead of his time, and so his contribution went essentially unnoticed for about 25 years.

In considering Watt's work, Carnot realized that the real source of the power of the steam engine was the *heat* derived from the fuel, and that the steam engine was simply a very effective means of using that heat to generate motion. He therefore decided to analyze the fundamental way in which heat could generate motion, believing that once this was understood it might be possible to design even more efficient engines than Watt had developed. He made use of the then accepted caloric theory of heat, and arrived at some remarkable conclusions despite using an erroneous theory. In time, after the publication of his book, he realized that the caloric theory might be in error, but unfortunately he died of cholera before he could pursue the matter much further. Nevertheless, Carnot explained the performance of a steam engine in terms of an analogy that is extremely useful for understanding the motive power for heat.

According to the caloric theory, heat is a fluid that flows from hot objects to colder ones. The steam engine is understood to be the heat equivalent of the mechanical waterwheel. The waterwheel derives its power from the flow of water from a higher elevation to a lower one. The temperatures of the objects can be regarded as analogous to the elevations of the pond and the exit steam for the waterwheel. It is commonly known that for a constant amount of water flowing over the waterwheel, the larger the wheel, the more work that can be done (i.e., the farther the water must fall). Hence, Carnot suggested that a steam engine (he said it should more properly be called a *heat engine*) could do more work with the same amount of heat (caloric fluid) if the heat was made to flow over a larger temperature difference.

One must be somewhat clever to utilize this suggestion. The temperature difference that exists in a steam engine is between the heat of the steam and the temperature of the surroundings. The steam is at the boiling point of water. At first it appears that we cannot control either the high temperature or the low temperature. But Carnot pointed out that if the steam is produced at a higher pressure, the temperature of the steam will be higher than the normal boiling point of water. This is just the same as the principle of a pressure cooker. Thus there is a way to raise the higher temperature and obtain a larger temperature difference. Carnot thus explained the improvement Watt found by using higher pressures. All modern steam engines operate at high pressures and realize an increase in efficiency accordingly.

The caloric theory thus seemed to explain the steam engine very nicely. Some other aspects of heat flow also appeared to be accounted for by the caloric theory. Some concepts and definitions that will be needed again later can be introduced in terms of the caloric theory.

One of the first things noticed about heat flow is that it is always from hot to cold. When two objects at different temperatures are placed in contact with each other, the colder one becomes warmer and the warmer one becomes colder. One never observes that the warmer one gets warmer still and the colder one, colder. The caloric theory accounts for the observed fact that heat always flows from hot to cold by simply stating that temperature differences cause heat flow. Heat flows "down temperature." In terms of the waterwheel analogy, it is just the same as saying that water always flows downhill. This will be discussed further in Chapter 5.

Another characteristic of heat is that it takes a certain amount of heat to raise the temperature of a given object a certain number of degrees. For example, a large object is much harder to "warm up" than a small one; that is, it takes more heat to do it. In terms of the caloric theory, the temperature corresponds to the "height" of the caloric fluid within the object. The amount of heat needed to raise the temperature of an object one degree Celsius is called the **heat capacity** of the object. Similarly, objects made of different materials are harder or easier to warm up than some others might be. This characteristic of a material is given the name **specific heat** and its numerical value is determined by the amount of heat necessary to raise one gram of the material by one degree Celsius.

It is important to understand the difference between **heat** and **temperature.** Heat refers to the **amount,** while temperature refers to the degree, or concentration. A bathtub full of slightly warm water can easily contain more heat than a thimbleful of nearly boiling water. The total amount is greater in the former, but the concentration is greater in the latter.

Water has been chosen to be the standard to which other materials are compared and its specific heat is defined to be unity. The amount of heat required to raise the temperature of one gram of water one degree Celsius is called one **calorie** (one "unit" of caloric fluid). Eighteenth-century physicists believed that materials with different specific heats possessed different abilities to hold caloric fluid. The heat capacity of an object is calculated by multiplying its mass by its specific heat.

Another characteristic of materials that was explained within the framework of the caloric theory of heat is called **latent heat.** Water provides a good example of this concept. As heat is added to water, its temperature rises relatively quickly until it reaches its boiling point. If a thermometer is placed in the water, it will be seen that the temperature quickly rises to the boiling temperature (212° F, 100° C) and then remains there as the evaporation (boiling) begins. Heat is continually added at the boiling temperature while the water vaporizes or turns into steam. Because there is no temperature rise, it may be asked: Where is the heat going that is being added to the water? The calorists (supporters of the caloric theory) answered that it was going into a hidden, or latent, form in the water. They did not suggest that it was disappearing.

There is no question that some heat is being added during this time. It is simply going into an invisible form and can be completely recovered by cooling the water. This hidden heat is called the **latent heat of vaporization.** A parallel phenomenon is observed when heating ice to melt it. Additional heat is required after the ice first reaches its melting temperature to actually melt the ice. This hidden heat is called the **latent heat of melting.** These latent heats have been observed for all materials as they change from solid to liquid or liquid to gas (i.e., change from one state to another) and are different for different materials. These latent heats are determined as the amount of heat required to melt or vaporize one gram of the material.

It was regarded as significant that exactly all of the heat put into latent heat is released when the material is later cooled down. For example, water in the form of steam is more scalding than the same amount of hot liquid water at the same temperature. The latent heat of the steam will be released as it condenses on skin. The point is, however, that something appears to be conserved—that is, not destroyed. This conserved quantity was believed to be the caloric fluid.

There were other phenomena, however, that were not explained sufficiently by the caloric theory; some of these "problem phenomena" led a few physicists to perform investigations that ultimately demonstrated that there is no caloric fluid at all. Perhaps the best known of these "problem phenomena" was heat generated by friction. We can warm our hands just by rubbing them together. There does not appear to be any heat source for the warmth one feels. There is no fire or even a warm object to provide a flow of heat into our cold hands. Where does the heat come from? If somehow heat is being generated where there was none before, then we must be (according to the caloric theory) producing caloric fluid. But caloric fluid was believed to be something that cannot be created or destroyed; it can only flow from one object to another. Where then, does the caloric fluid come from when friction occurs?

The calorists' answer was that a *latent* heat is released when heat is produced by friction. The claim was that a change of state is involved, just as in changing a liquid to gas, when small particles are ground off an object by friction—during machining, for example. Some calorists said that the material was "damaged" and would "bleed" heat. These answers were never entirely satisfactory. Any small pieces ground off by friction appear to be just tiny amounts of the original material. Eventually it was shown that such grindings have the same characteristics, including the same specific heat, as the original material, and they certainly had not undergone a change of state.

One of the first people to be impressed by this difficulty with the caloric theory was Count Rumford.[2] In a famous experiment performed in 1798, he measured the amount of heat developed from friction during the boring of a cannon at the Bavarian royal arsenal in Munich. He was greatly impressed with the large amount of heat produced and the small amount of metal shavings that resulted. Count Rumford was convinced that the tremendous amount of heat being generated could not be from some latent heat. He measured the specific heat of the shavings and found it to be the same as the original metal. Count Rumford proceeded to demonstrate that, in fact, one could produce any amount of heat one desired, without ever reducing the total amount of metal. The metal was simply being slowly reduced to shavings with no change of state involved. He even proceeded to show that heat could be generated without producing any shavings at all. Count Rumford seriously questioned whether the calorists' interpretation could be correct and suggested that the heat being produced was *new* heat.

Actually, the calorists readily accepted the results of Rumford's experiments and felt they provided important clues regarding the nature of the caloric fluid. It was said that these experiments showed that caloric fluid must be composed of small, nearly massless particles whose total number in an object is vastly greater than the number ever released by friction. This interpretation is very close to our modern understanding of electricity in terms of atomic electrons. In many ways, Rumford's experiments were simply taken to further the understanding of the caloric fluid.

Besides friction, a second class of phenomena that caused difficulties for the caloric theory was the expansion and contraction of gases. In a little-known (then or now) experiment in 1807, the French physicist Gay-Lussac measured the temperature of a gas allowed to expand into an evacuated chamber (i.e., one with all the air removed). He measured the temperature of the gas both in its original chamber and in the chamber that had originally been evacuated. The temperatures were found to be the same and equal to the original temperature of the gas. The caloric theory predicted that half of the original caloric fluid should be in each chamber (if the two chambers were of equal size). Because temperature was believed to be determined by the concentration of caloric fluid, this should have resulted in the temperature being considerably *lower* than it was originally. This

[2]Rumford, originally named Benjamin Thompson, was a former American colonist and Tory who fled to England during the American Revolution. Interestingly enough, Rumford, who helped overthrow the caloric theory, married the widow of Lavoisier, who first proposed the caloric theory.

experiment and others similar to it simply were not understood by the calorists and, in fact, were not understood by anyone for over 30 years.

The final blow to the caloric theory was provided by the careful experiments of British physicist James Joule, in the 1840s. It was well-known that heat could do work and this was understood by the calorists with their waterwheel analogy. Heat did work as it flowed down-temperature just as water does when it flows downhill. But can work create heat? This was one of the most important questions to arise in the entire development of our present-day understanding of heat and energy. The caloric theory answer was a very definite no. This would be creating new caloric fluid; it would be like creating water by mechanical means, and could not be done—except that is just what Joule's experiments demonstrated.

Joule, encouraged by the work of Count Rumford and others, began to study whether mechanical work could produce heat. His experimental apparatus consisted of a paddle wheel inside a cylinder of water (see Fig. 4.4). The vanes placed inside the cylinder just barely allowed the paddles to pass and kept the water from swirling with the paddles. Joule wanted to see if one could raise the temperature of the water just by rotating the paddle. The vanes ensured that the rotating paddle could only ''agitate'' the water, and microscopic constituents (i.e., molecules) would be set into faster motions. Additionally, if the temperature of the water rose, it must have been because of the mechanical work expended in turning the paddle wheel. Thus if Joule measured a temperature rise in the water, he would have direct evidence that work can produce heat. Finally, one certainly cannot talk about damaging or bruising water to make it ''bleed'' heat.

Although the temperature rise was rather small, and Joule had to construct his own very sensitive thermometer, he observed a definite increase in the tempera-

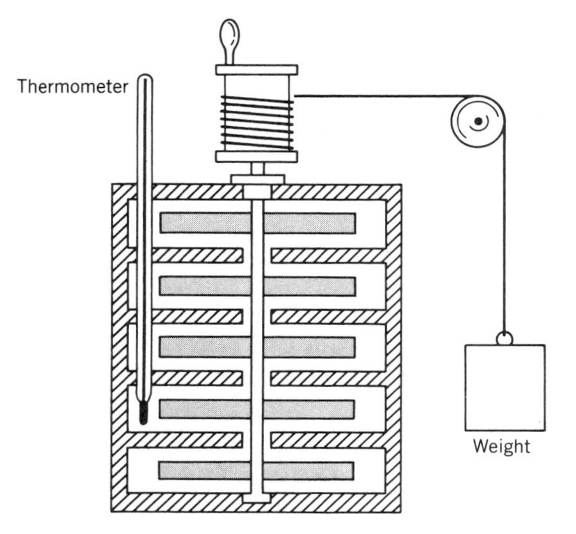

Figure 4.4. Joule's churn. Apparatus used to determine the mechanical equivalent of heat.

ture of the water just by rotating the paddle. Joule even set up the experiment to measure just how *much* heat is produced by a certain amount of mechanical work. Instead of just turning the paddle by hand, he set it up so that a metal cylinder suspended by a rope over a pulley could be allowed to turn the paddle as it fell a measured distance. In Section A2 of this chapter, work is defined as the product of force multiplied by distance. For Joule's experiment, the force is just the weight of the metal cylinder and the distance is how far it is allowed to fall, so he could measure the mechanical work exactly. He could determine the amount of water in his experiment and, from the temperature rise and the specific heat of water, deduce the amount of heat that must have been generated. Thus Joule determined the mechanical work equivalent of heat. He did this for a number of different materials such as various types of oils, and he always found the same value for the mechanical equivalent of heat. Joule eventually determined a value that varies less than 1 percent from the modern accepted value. These experiments represented a landmark in the history of science.

There are two extremely important consequences of Joule's experiments. The first is that mechanical work can produce heat. Heat is not conserved. The caloric theory cannot be correct and in fact there is no caloric fluid that cannot be created or destroyed. We need another theory of heat. Heat must be a form of energy. Recall that mechanical work can produce kinetic energy or potential energy. Joule demonstrated that work can also produce heat. It is clear that heat must also be a form of energy.

The second important consequence of Joule's experiments is the clue about how to construct a new theory of heat. Because of the vanes in the cylinder of water, the rotation of the paddles can do nothing to the water except agitate it. This restriction was an intended aim by Joule. Thus it must be concluded that agitated water is warmer water. But what is being agitated in the water? The answer is the molecules of which the water is comprised. These molecules have the familiar chemical symbol H_2O, meaning they consist of two atoms of hydrogen and one of oxygen. It must be molecules that are being agitated, and apparently the more agitated they are, the warmer the substance. Temperature must be related to the energy of the microscopic motions of these molecules. Thus temperature can be understood only in terms of molecular motions, as described by the molecular model of matter.

Optional Section 4.2 *Heat Capacity and Latent Heat*

Let us consider quantitatively the example of heating water in order to make it boil. Suppose we have a pot of water with 1 liter of water in it. As discussed above, the unit of heat, one calorie, is defined as the amount of heat necessary to raise the temperature of one gram of water by one degree Celsius. Suppose the water starts at about room temperature, 25° C (77° F). If the pot is placed on a burner that can produce 30,000 cal/min of heat and if about one-half of the heat produced is transferred to the water, then about 15,000 cal/min of heat goes into the water. The total amount of heat (in calories) required to heat the water up to its boiling point (100° C) is

$$\text{heat} = \text{specific heat} \times \text{mass} \times \text{change in temperature}$$

or symbolically,

$$Q_1 = c \cdot m \cdot \Delta T$$

and for this example,

$$Q_1 = \frac{1 \text{cal}}{(\text{gm})\ (^\circ\text{C})}\ (1000\ \text{gm})\ (100^\circ\ \text{C} - 25^\circ\ \text{C})$$
$$Q_1 = 75{,}000\ \text{cal}$$

The amount of time required to bring the water up to the boiling point is then

$$t_1 = \frac{Q_1}{R} = \frac{75{,}000\ \text{cal}}{15{,}000\ \text{cal/min}} = 5\ \text{min}$$

Now the latent heat of water, L, has been determined to be 539 calories per gram. This is the amount of heat (per gram) required to convert the water into steam (often called the **heat of vaporization**). If we wish to completely boil the water away, using the same burner, the total heat required would be

$$\begin{aligned} Q_2 &= L \cdot m \\ &= \frac{539\ \text{cal}}{\text{gm}}\ (1000\ \text{gm}) \\ &= 539{,}000\ \text{cal} \end{aligned}$$

and the time required to do this would be

$$t_2 = \frac{Q_2}{R} = \frac{539{,}000\ \text{cal}}{15{,}000\ \text{cal/min}} = 36\ \text{min}$$

So we see that actually it would take about seven times as long to boil the water away as to raise the temperature to the boiling point. The amount of energy (i.e., heat) required to overcome the mutual attractive forces between the molecules in the water is actually quite significant.

2. The Kinetic-Molecular Model of Matter and the Microscopic Explanation of Heat

As early as 400 B.C., Democritus, a Greek philosopher, suggested that the world was made up of a few basic building blocks, too small for the eye to distinguish, called atoms. It was originally believed that there existed only a few different kinds of atoms. Later, it was erroneously thought that there was a different kind of atom for every different material such as wood, rock, air, and so on. We now know that there are hundreds of thousands of different chemical substances, which are now called compounds. The smallest amount of a compound is called a molecule, not an atom. Molecules, however, are made up from atoms, of which there are only about 100 chemically different kinds. Each naturally occurring atom is associated with one of the elements from hydrogen to uranium. Molecules and their constituent atoms obey the laws of mechanics and also the laws of conservation of energy and momentum. The dominant force is electrical and not gravitational.

It should be noted that we now know that even the objects identified as atoms are not the fundamental building blocks of nature as originally supposed. Atoms are themselves built up from protons, neutrons, and electrons; furthermore, the protons and neutrons are built up of even smaller units called quarks. In order to understand heat, however, one need not be concerned with the structure of a molecule.

Consider what happens to the molecules of a substance, such as water, as the substance goes from a gas (steam) to a liquid (water) and finally to a solid (ice). When the substance is a gas, the molecules are essentially not bound to each other at all. They are each free to move in any direction whatsoever, and to spin and vibrate. For a gas at normal pressures (such as the pressure of the atmosphere around us), molecules will not proceed very far in any one direction before colliding with other molecules. They will bounce off each other, obeying Newton's laws of motion and the laws of conservation of momentum and kinetic energy. If we add one gas to another (such as by spraying perfume in one corner of a room), the molecules of the new gas will, after many collisions with the molecules of the original gas, diffuse out in all directions. Soon the perfume can be detected across the room, as the perfume molecules become completely mixed with the original air molecules.

As the gas is cooled, the molecules move more slowly, and usually do not collide with each other with as much speed as when the gas was warmer. When the gas is cooled enough, we know it will condense into the liquid state. In the liquid state, the molecules are still fairly free to move about, colliding with each other as they move. However, there exists a small force of attraction between all the molecules (which is electrical in origin), and this small force from all nearby molecules adds up, preventing an individual molecule from escaping all the others and leaving the liquid. Thus liquids will remain in a container, even without a lid, for a long time. We do know that most common liquids will slowly evaporate. This is explained by noting that occasionally a molecule will be struck by two or more other molecules in rapid succession and driven toward the surface of the liquid. This gives the molecule an unusually large velocity, which allows it to overcome the net force acting on it from all the other molecules, and it will (if there are no intervening collisions) escape from the liquid.

If the liquid is cooled enough, it will finally "freeze" into a solid form. In a solid, the molecules are moving so slowly that the small forces on each one from the nearby molecules hold it in one place. The molecule is no longer free to move around among all the other molecules. The only motion it can have is a vibrational motion about its fixed place in the solid.

The molecular description of the states of matter has been verified repeatedly by many careful experiments. No "problem phenomena" are known to exist that question the accuracy of the account of how a material passes from the gaseous to the solid state as the temperature is lowered. (Many of the details of these transitions are not yet fully understood, however.) Thus gases, liquids, and solids are all composed of particles (the molecule) that are moving and are subject to the laws of mechanics.

Many of the earlier discussions can now be understood in detail. Heat input causes the molecules to move faster; heat removal causes them to move more slowly. If the molecules are moving very rapidly, the material is "hot." In fact, the temperature of a gas (whose molecules interact only by elastic collisions) is simply a measure of the average kinetic energy of random translational motion for all the molecules (recall that kinetic energy is $\frac{1}{2}mv^2$). The average speed of an air molecule at normal room temperature is found to be greater than 1000 mph! Because there are about a million billion molecules in one cubic foot of air, gas molecules collide with each other frequently (each molecule is struck by other molecules about 100 billion times per second).

It is interesting to note that one of the first direct verifications of the molecular nature of matter was provided by Albert Einstein, who explained the observed motion of very small solid particles suspended in a liquid solution. This motion, called Brownian movement, is very irregular, with the particle moving a short distance in one direction and then quickly in another direction and quickly again in another, and so forth. Einstein explained (and showed quantitatively) that this motion was caused by bombardment of the small particle from the even smaller individual molecules of the liquid.[3] The same kind of motion can be seen for the dust particles that make a "sunbeam" visible.

One can now also understand why a gas can exert a pressure such as it does in an automobile tire. The individual molecules are very light, but they are moving very fast and there are a large number of them hitting the wall of the tire every second. The pressure represents the average force exerted by molecules on the wall as they hit and rebound. Of course, the air molecules on the outside of the tire are also exerting a pressure on the other side of the walls of the tire at the same time that the molecules inside exert their pressure. There is a greater density of molecules on the inside, however, and thus the inside pressure is greater than the outside pressure by the amount necessary to hold up the vehicle. This extra pressure inside is accomplished by forcing (from an air pump) more molecules per unit volume inside the tire than there are in the air outside of the tire. Note that if the temperature of the gas (air) inside the tire were to be raised, the pressure would also increase, because the molecules would have more kinetic energy. Such an increase in pressure is known to occur when a tire becomes warmer, on a long

[3]The particles observed in Brownian motion can be seen with a low-powered microscope. Molecules, however, are so small that they cannot be seen in any significant detail with the most powerful microscopes presently available. Until Einstein's work, many scientists felt that there were no convincing proofs of the existence of molecules. Einstein published this work in the same year (1905) in which he published his first paper on relativity theory (see Chapter 6, Section D3).

drive on a hot day, for example. The flexing of the tire results in more agitation of the molecules, just as in Joule's experiment.

Specific heat can now be understood in detail. Recall that the specific heat of a material is the amount of heat required to raise the temperature of one gram one degree Celsius. This means increasing the average kinetic energy of the molecules. Depending on the actual geometric arrangements of the atoms in a molecule or solid, different modes of motion are possible (such modes may include internal vibrations). Thus the specific heats of various compounds will be different because the averaging will be different. Similarly, one can understand latent heats. In order to make a liquid boil, enough heat must be supplied to overcome the net force on an individual molecule produced by all the other molecules still in the liquid state, so we do not see a rise in temperature. Once enough heat has been absorbed to overcome the bonds between the molecules and all the liquid boils into a gas, then any additional heat *will* serve to increase the kinetic energy of the molecules and a rise in temperature will occur.

The kinetic-molecular model of matter reduces the problem of understanding heat and the states of matter to understanding systems of particles that have certain forces between them and that obey the laws of mechanics. The success of this model made it appear that Newton's laws of mechanics permit a unified description of the behavior of all objects, large and small, and in complete detail. Only the development of quantum and relativity theories in the twentieth century (which are discussed in Chapters 6 and 7) placed limitations on the universality of Newton's laws.

C. Conservation of Energy

1. Precursors, Early Forms

Enough has now been learned to recognize that heat surely is another form of energy, together with kinetic and potential energies. In the discussion of conservative systems (such as a pendulum), it was concluded that the total energy of the system was *almost* maintained at a constant value. The fact that the pendulum does eventually come to a stop indicates that the total energy calculated as the sum of the kinetic plus potential energies is not exactly conserved. It was speculated that the friction at the pivot and between the pendulum bob and the air made it slowly stop.

We have now seen that friction generates heat (Count Rumford's experiment) and that heat is a form of energy. It seems only reasonable to see whether or not the total amount of heat generated by the pendulum is exactly equal to the energy it loses. But this is a very difficult experiment to perform accurately. One would need to have a way to measure the amount of heat produced and know that it represented the amount of energy that the pendulum had lost. Joule measured how much heat was produced by a certain definite amount of mechanical work and, of course, work produces energy. But how can it be established that some of the energy is not lost? If the heat produced in Joule's experiment could be all converted back into mechanical energy (i.e., kinetic or potential), then energy conservation would be proved. But it turns out that heat energy can never be totally converted into another form of energy. (This will be discussed in detail in the next chapter.) So then, how did a few scientists originally come to believe that by including heat as another form of energy the total energy of an isolated system is conserved?

Perhaps the first person to become thoroughly convinced that energy is conserved was a German physician, Julius Robert Mayer (1812–1878). Mayer served as a ship's surgeon on a voyage to Java in 1840. He noticed, while treating the members of the crew, that (venal) blood is redder in the tropics than in Germany. Mayer had heard of a theory by Lavoisier that body heat came from oxidation in body tissue using the oxygen in blood. Mayer reasoned that because the body needs to produce less heat in the tropics than in Germany, the blood was redder because less of its oxygen was being removed.

Mayer was intrigued by this phenomenon and continued to think about what it implied. He noted that body heat must warm the surroundings. This heat is, of course, from the oxidation of the blood going on inside the body. But there is another way a body can warm the surroundings—by doing work involving friction. Mayer wondered what the source of this kind of heat was. He concluded that it must also be from the oxidation of blood, originally. Thus the oxidation process produces heat into the surroundings both directly by radiation of body heat and indirectly through the link of mechanical work (i.e., by the action of muscles). Because the heat in both cases must be proportional to the oxygen consumed, so must be the amount of mechanical work that temporarily takes the place of one part of it. Therefore, Mayer believed that heat and work must be equivalent; that is, so many units of one must exactly equal so many different units of the other. Mayer reasoned further that energy originally stored in the food was converted in the oxidation process into energy in the forms of heat and mechanical work. This was the beginning of Mayer's idea of conservation of energy.

It should be noted how relatively easy all this was for Mayer. It was probably easy because he ignored the caloric theory. He did not consider that heat could not be created where there was none before. He tacitly assumed that all this heat from the body was new heat. This ignorance of an accepted theory was certainly an asset at first, but it later caused him great difficulty in making his ideas well-known. After considerable investigation and thinking, Mayer published a private pamphlet entitled "Organic Motion in Its Connection with Nutrition," in which he recognized various forms of energy. The forms included mechanical, chemical, thermal, electromagnetic, heat, and food. He even proposed, 80 years ahead of his time, that the Sun was the ultimate source of all these energies—something every grade-school student now learns.

Mayer's proposal that energy is conserved was ridiculed or simply ignored because of his archaic language and use of metaphysics, or sometimes even just incorrect physics. Although he spent years trying to convince scientists of his conclusions, he saw others get credit for his ideas and temporarily lost his sanity from his extreme frustration.

Although Julius Mayer may have been the first scientist (for that is what he became) to recognize the law of conservation of energy, it was the work of Joule that led to the scientific community's acceptance of this idea. Lord Kelvin (William Thompson), a well-known British physicist of the mid-nineteenth century, recognized the importance of Joule's work and gained its acceptance by the Royal Society of London. Kelvin was aware of Carnot's extensive work on heat engines and realized that Joule's work provided a vital clue to the correct analysis of such engines. The amount of heat transferred to the low-temperature reservoir (i.e., the surrounding atmosphere) is *less* than the amount of heat removed from the high-temperature reservoir (i.e., the steam in the condenser chamber) and the difference is found in the mechanical work done. Heat and mechanical work must be two different forms of the same thing—namely, *energy.* Kelvin's recognition of this fact for heat engines was a very important step in the development of the energy concept.

Joule continued to perform careful experiments and was able to show that electrical energy could be converted to heat with a definite relationship, and then eventually that electrical energy and heat could be converted from one to the other. By this time, it was already known that certain chemical reactions could generate electrical current, and the stage was finally set for recognizing that energy has several forms and can be converted from one of these forms to another.

By the time Joule had finished many of these now-famous experiments, he apparently became convinced that energy was something that could not be destroyed, but could only be converted from one form to another. For the first few years (starting about 1845) that Joule expounded a theory of the conservation of energy, almost no one would believe he was right. It was not until Kelvin became convinced that Joule's experiments were correct, saw the connection with the new interpretation of heat flow in a steam engine, and then supported Joule's idea, that the world began to accept the idea of conservation of energy.

There was also a third "co-discoverer" of the idea of conservation of energy, and that was the German physician Hermann von Helmholtz. In 1847, when only 26 years old, he presented a paper entitled "On the Conservation of Force." Although unaware of Mayer's work and hearing of Joule's experiments only near the end of his own work, Helmholtz concluded that energy must have many different forms and be conserved in its total amount. He arrived at this conclusion because he believed that heat, light, electricity, and so on, are all forms of *motion* and therefore forms of mechanical energy (i.e., kinetic energy). It must have seemed obvious to Helmholtz that kinetic energy was always conserved, not just in elastic collisions but in all collisions, because it went into other forms that were kinetic energy also. Much of what Helmholtz so boldly suggested was eventually found to be correct, although not *all* forms of energy can be reduced to kinetic energy. The immediate response to his paper was discouraging, and most historians agree that it was the experiments of Joule, backed by the reputation and influence of Kelvin, that eventually convinced the physicists of the mid-nineteenth century that there was indeed a law of conservation of energy.

2. The Forms of Energy

Let us review what has been discussed thus far. It has been said that heat is a form of energy and that when an account is kept of energy, in all its forms, the total amount in an isolated system is conserved. This is the law of conservation of energy. What, then, are all the possible forms of energy?

Basically, for the present discussion, energy can be considered to be in one or more of the following five major forms:

1. Kinetic
2. Potential
3. Heat
4. Radiant
5. Mass

Kinetic energy is one-half of Huygens' *vis viva,* or $\frac{1}{2}mv^2$. It is the energy associated with a body due to its (organized) motion. **Heat** has been discussed in some detail, and is energy added to (or taken from) the microscopic modes of (disorganized) motion of the individual molecules of matter. The amount of heat energy is determined by the change in *temperature* of the object (for example, in the gas state the average kinetic energy of random microscopic motion of the molecules), and the *amount of matter* in the object (i.e., the number of molecules). **Potential energy** is energy associated with the shape, configuration, or position of an object in a force field. It was first introduced as gravitational potential energy, but because of the work of Joule and others is now known to have several forms. Some of these forms are:

1. Gravitational
2. Electrical
3. Chemical (electrical on a microscopic scale)
4. Nuclear

Whenever energy appears in one of these forms of potential energy, it is possible (although not always practical) to convert the energy completely into kinetic energy or heat energy (which is microscopic random kinetic energy). The fourth major form of energy, **radiant energy,** is electromagnetic radiation such as light. This is a form that has not been discussed here. For a long time, radiant energy was believed to be a form of heat, but it is also a form of energy that is completely convertible to one of the other forms.

That **mass** is a form of energy was first indicated by Einstein in 1905 (see Chapter 6, Section D5) and verified some 30 years later when certain nuclear reactions were carefully observed experimentally. Until Einstein's work, it was believed that there were separate conservation laws for energy and mass. Newton had included the law of conservation of mass in his *Principia.* In fact, mass and all the other forms of energy *are* separately conserved to a very good approximation, unless a process including a nuclear reaction is involved. Ordinary chemical reactions, such as oxidation, involve only an extremely small transformation of mass into energy or vice versa. Because nuclear reactions do not occur naturally on the surface of the Earth (except in radioactive materials, which were not noticed until about 1896), the early workers who first established the law of conservation of energy had no reason to recognize mass as another form of energy.

3. Energy Concept for Systems

One of the quantities that needs to be determined to characterize a system completely is now seen to be the total energy of the system. One should specify not only the size, mass, temperature, and so on of a system, but its total amount of energy as well. This clearly is more important than just indicating the heat content, which was often done before the energy concept was fully developed.

The workings of a system can be analyzed to keep an account of the total amount of energy and how it flows from one form to another. For an isolated system, we now know that if some energy disappears in one form, it must reappear in another form. This fact can provide an important clue in analyzing a system. Consider a heat engine as an example. If work is done *on* the engine, then energy must have been supplied from somewhere to accomplish this. Conversely, if the engine *does* work, then it supplies a very definite amount of energy to something else. This energy supplied could be kinetic energy (increasing the speed of a car) or potential energy (lifting a weight). In the next chapter we will discuss how Rudolf Clausius (1822–1888) was able to reanalyze the workings of a heat engine with an awareness of this law of conservation of energy. He demonstrated that although heat definitely is a form of energy, it is a very special form of energy. As applied to systems, the law of conservation of energy is called the *first law of thermodynamics.*

4. What Is Energy?

Perhaps we should consider for just a moment what this energy, whose total amount in all forms is conserved in an isolated system, actually is. The first thing to be noted is that it is strictly an abstract quantity. Energy is not a kind of material. Each kind of energy is defined by some mathematical formula. Moreover, energy has only a relative nature. Gravitational potential energies are normally measured with respect to the surface of the Earth. Sometimes, however, one might measure how much potential energy an object has with respect to the floor of a room, or to the top of a table, and so on. One is only interested in how much potential energy the object has relative to other objects around it. Kinetic energy also has a relative nature, because kinetic energy is measured as the quantity $\frac{1}{2}mv^2$ where v, the velocity, is measured with respect to an arbitrary reference. The chosen reference might be the Earth, the floor of a room, a moving car, the Moon, or whatever seems appropriate for the situation. All potential and kinetic energies are relative.

The next thing to be understood about energy is that only energy changes are significant. There is no need to keep track of the potential or kinetic energies of an object unless those energies are going to change. It is the *change* in potential energy that shows up as a *change* in kinetic energy (or vice versa), for example. These changes in energy are especially important when one considers interacting systems. In fact, it is by energy changes that interactions proceed. It can be important to know that if one system loses a certain amount of energy, then the other (for only two systems involved) must have gained exactly that amount. Scientists and engineers use the law of conservation of energy continually in analyzing or designing systems involving energy transfers.

The relationship between energy and the physicists' concept of *work* cannot be overemphasized. Work is defined to be a force acting over a distance. Work is *in fact* how things are accomplished in the world. Except that thinking is also considered (by most people) to be "work," the physicists' work is involved in all the forms of work of which we are commonly aware. Work produces energy (either kinetic or potential, for example) and energy can do work. From a utilitarian viewpoint perhaps this is our best understanding of what energy is. Energy has the ability to do work; it can cause a force to be exerted over a distance. Energy is like stored work.

Heat was originally thought to be a fluid, but now it is recognized to be one of the several different forms of energy. Energy is not (as heat was originally thought to be) any actual substance. Energy is strictly an abstract idea. It is *associated* with motion, or position, or vibrations, and is determined from mathematical expressions for each of its different forms. Energy can be converted into forces acting on objects (Helmholtz referred to the law of conservation of energy as the law of conservation of *force,* before a careful distinction was made between force and energy.) It is because energy is conserved that it is such an important concept. That nature "chooses" to maintain the amount of this strictly abstract quantity at a constant value in the many possible transformations of energy from one form to another is quite amazing.

To conclude this chapter, consider an analysis of a common energy system—the heating of a home by an electrical resistive system—as a demonstration of the concepts of energy. In most parts of the United States, electrical power is produced by the burning of fossil fuels. The energy, therefore, was originally stored as chemical potential energy in the fossil fuel. This energy is converted into heat energy by the burning of the fuel. The heat energy is converted to the kinetic energy of a directed stream of hot gas, which exerts pressure on the vanes of a turbine engine, causing it to do mechanical work. The turbine engine, in turn drives the shaft of an electrical generator, which converts the work into electrical energy. This electrical energy is then transmitted to the house to be heated. In the house, the electrical energy is converted into both heat energy and radiant energy in a resistive heater unit. This emitted energy finally then is absorbed by the air molecules in a room, giving them more random motion (i.e., heat) and thereby raising the temperature of the air in the room.[4]

The law of conservation of energy does *not* require that the amount of energy finally delivered to the home as heat be exactly equal in amount to the energy originally stored in the fossil fuel. This whole process involves ''loss'' of energy at every step. The amount finally delivered may be considerably less than originally available. The law of conservation of energy *does* require that all of the energy will go *somewhere.* Most often some of it escapes from the network as heat and raises the temperature of the surroundings at that point. The law of conservation of energy also requires that more energy never be released from the system at any point than is inserted at that point.[5] Many schemes have been suggested that would try to do just this; that is, release more energy than is put in. Any device that would violate this first law of thermodynamics is known as a perpetual motion machine (of the first kind), because some of the energy released could be used to drive the machine itself. No such scheme has ever been successfully executed, and the first law of thermodynamics stands as one of the most important principles known describing the way nature behaves.

OPTIONAL SECTION 4.3 *The First Law of Thermodynamics*

The first law of thermodynamics is the law of conservation of energy, as discussed above. The real significance of this law is that it allows us to do ''bookkeeping'' on a system. Let us consider an example. Suppose we wish to accelerate a 2000-kg ($\sim$ 2-ton) car up to 100 km/h ($\sim$ 62 mi/h), and then let it coast up a hill. We would like to determine how much gasoline will be required, and how far up the hill the car will go. We will neglect any friction involved.

In order to solve this problem, we need to know how much heat is released when burning one gallon of gasoline and the efficiency of the automobile engine. We also need to know the mechanical equivalent of heat; this was first determined by James Joule, as discussed above. The heat of combustion of gasoline is 3.4×10^7 cal/gallon. Let us assume that the engine is 50 percent efficient (a ''good'' engine). The mechanical equivalent of heat is 1 cal = 4.19 joules.

[4]Strictly speaking, heat is best defined as energy that is transferred between two objects as a result of a temperature difference between the two objects. However, the temperature of an object can be related to the random motion of its constituents, at least for the purposes of this book. This is an example of the need, for the purposes of explanation of a complex subject, to oversimplify matters somewhat.

[5]Actually, it is possible to release more energy than inserted for a while by using the internal energy of the system itself; however, this can only occur for a limited time and cannot be performed continuously.

First we need to know how much kinetic energy is associated with the car moving at 100 km/hr (= 28 m/sec). We calculate this from the expression for kinetic energy

$$\begin{aligned} \text{K.E.} &= \frac{1}{2} mv^2 \\ &= \left(\frac{1}{2}\right) (2000 \text{ kg}) \left(28 \frac{\text{m}}{\text{s}}\right)^2 \\ &= 7.8 \times 10^5 \text{ joules} \end{aligned}$$

So we see that we must burn enough gasoline to produce this amount of kinetic energy. Since 1 cal = 4.19 joules, the amount of heat required is

$$\begin{aligned} Q &= \frac{7.8 \times 10^5 \text{ joules}}{4.19 \text{ joules/cal}} \\ &= 7.8 \times 10^5 \text{ cal} \end{aligned}$$

and because the engine is only 50 percent efficient, we actually need

$$\begin{aligned} Q/50\% &= Q/(1/2) \\ Q_{\text{total}} &= 2Q = 1.6 \times 10^6 \text{ cal} \end{aligned}$$

Finally, since each gallon of gasoline produces 3.4×10^7 cal, we see that we need

$$\frac{1.6 \times 10^6 \text{ cal}}{3.4 \times 10^7 \text{ cal/gal}} = 0.05 \text{ gal}$$

How far up a hill can this car coast? We can determine this just by requiring that the increased potential energy be equal to the kinetic energy of the car at 100 km/hr (i.e., by applying conservation of energy).

The gravitational potential energy (see Optional Section 4.1) is

$$\text{P.E.} = mgh$$

where m = mass of the car, g = the acceleration of gravity (9.8 m/sec^2), and h = height up the hill. So we have

$$\begin{aligned} mgh &= \frac{1}{2} mv^2 = 7.8 \times 10^5 \text{ joules} \\ h &= \frac{7.8 \times 10^5 \text{ joules}}{(2000 \text{ kg})\ (9.8 \text{ m/sec}^2)} = 40 \frac{\text{joules}}{\text{kg (m/sec}^2)} \end{aligned}$$

Now 1 joule = 1 (kg) (m/sec^2), so

$$h = \frac{40 \text{ joules}}{(\text{kg})\ (\text{m/sec}^2)} \times \frac{\text{kg (m/sec)}^2}{1 \text{ joule}}$$
$$= 40 \text{ m } (131 \text{ ft})$$

Note that since we are ignoring friction, it does not matter how *long* the hill is, only that the height can increase by 40 meters. All of these calculations are examples of the kind of bookkeeping one can do using the first law of thermodynamics.

Appendix: Molecular Pressure and Speeds

It was stated in Section B2 that the average speed of air molecules at room temperature is 1000 miles per hour. Using the concepts developed in Section B2 and in Chapter 3, we can verify that statement. Let us consider a box full of air at normal temperature and pressure.

A molecule of mass m moving with velocity v, has momentum mv and kinetic energy $mv^2/2$. On the average, one-third of the kinetic energy may be associated with motion in any one direction (because space is three dimensional). Because kinetic energy is directly proportional to v^2, it must be that the average v in any one direction is $1/\sqrt{3}$ of the total v. Hence the momentum of a molecule, which is directly proportional to v, in that direction is $mv/\sqrt{3}$. When the molecule hits the wall, it will bounce off and move away from the wall with the same speed and momentum that it had before striking the wall, except that now the momentum is in the opposite direction. Hence the change in momentum, which is a vector quantity, is

$$\Delta p = + \frac{mv}{\sqrt{3}} - \left(\frac{-mv}{\sqrt{3}}\right) = \frac{2mv}{\sqrt{3}}$$

Now this molecule will move rapidly over to the other side of the box, bounce off that wall, and return again to the first wall. The time interval between collisions with the first wall is $\Delta t = 2L/(v/\sqrt{3})$ where L = length of one side of the box and $v/\sqrt{3}$ is the average velocity in one direction. The number of collisions by this one molecule per second is $1/\Delta t$. (If it takes ¼ sec between collisions, there are 4 collisions per second.)

Now according to Newton's second law (see Chapter 3), the applied force equals the change in momentum per unit time, that is,

$$F = \Delta p/\Delta t$$

In this case, the applied force is the force this one molecule exerts on the wall. This force is then

$$F = \left(\frac{2mv}{\sqrt{3}}\right)\left(\frac{v/\sqrt{3}}{2L}\right)$$
$$= \frac{mv^2}{3L}$$

If there are N molecules in the gas, then the total force is

$$F_{\text{total}} = NF = \frac{Nmv^2}{3L}$$

Now pressure is force per unit area, and the area of the wall is L^2. Hence

$$\text{pressure} = P = \frac{NF}{L^2} = \frac{Nmv^2}{3L^3}$$

But L^3 is the volume of the box, and Nm is the total mass of the gas, M. Hence we have,

$$P = \frac{Mv^2}{3V}$$

or

$$PV = \frac{Mv^2}{3}$$

From this expression we can estimate the speed of a molecule of air (air is mostly nitrogen and oxygen). We have

$$v^2 = \frac{3PV}{M} = \frac{3P}{(M/V)}$$

For air at room temperature, $P = 14.7\ \text{lb/in}^2 = 10^5\ \text{N/m}^2$, and $M/V = \text{density} \simeq 1.2\ \text{gm/liter} = 1.2\ \text{kg/m}^3$. Thus,

$$v^2 = \frac{(3)\ (10^5\ \text{N/m}^2)}{(1.2\ \text{kg/m}^3)}$$
$$= 25 \times 10^4\ \frac{\text{N}\cdot\text{m}}{\text{kg}}$$

Now $1\ \text{N} = 1\ \text{kg}\ (\text{m/sec}^2)$, so that

$$v^2 = 25 \times 10^4 \frac{\text{N}\cdot\text{m}}{\text{kg}} \left(\frac{1 \text{ kg}\cdot\text{m/sec}^2}{1 \text{ N}}\right)$$

or

$$v^2 = 25 \times 10^4 \frac{\text{m}^2}{\text{sec}^2}$$

and

$$v = 500 \frac{\text{m}}{\text{sec}} \approx 1100 \frac{\text{mi}}{\text{hr}}$$

So we see that air molecules are moving about 1000 mph!

STUDY QUESTIONS

1. What is a conservation law?
2. What is an "isolated" system?
3. What is conserved in an elastic collision?
4. Give an example of a (nearly) elastic collision, besides billiard-ball collisions. Give an example of a very inelastic collision.
5. What is the difference between momentum and kinetic energy?
6. What is the integrated effect of a force acting over a distance on an isolated object?
7. What is the integrated effect of a force acting during a time interval on an isolated object?
8. Define work.
9. Why is potential energy referred to as "potential" energy?
10. What is meant by a "conservative" system? Give an example of one.
11. Why is it understandable that scientists first believed heat to be a fluid?
12. What were believed to be the characteristics of caloric fluid?
13. Why do higher pressures make a steam engine work better?
14. What was Watt's main improvement for steam engines?
15. Define heat capacity and specific heat.

16. Why can steam scald one more than boiling water?
17. How did the calorists explain heat produced by friction?
18. Why did Rumford question the validity of the caloric theory?
19. What are the two important results of Joule's experiment with a water churn?
20. What is heat, in terms of the kinetic-molecular model?
21. What is the difference between heat and temperature?
22. Explain phase changes, including latent heats, in terms of the kinetic-molecular model.
23. State the law of conservation of energy in your own words.
24. How does recognizing heat as a form of energy "save" the conservation of energy principle for a pendulum?
25. How is total mechanical energy lost in a system consisting of the Sun and a planet?
26. Name the five basic forms of energy.
27. What kinds of energy are present in the following systems? (a) a car battery, (b) a bullet, (c) a rock on a mountaintop, (d) burning gas.
28. From a practical viewpoint, what is energy?

PROBLEMS

1. A rock of mass 1.0 kg is dropped from a height of 10 meters. What will be its velocity just before it strikes the ground? What if the height is 100 meters? (Neglect air friction.)
2. A pendulum bob at the bottom of its swing has a velocity of 10 m/sec. How high will it rise above the lowest point of its swing? (Acceleration of gravity $= g = 9.8\ \text{m/sec}^2$.)
3. A person throws a ball of mass 0.5 kg straight up in the air with a velocity of 10 m/sec. How high will the ball rise? What will be the velocity of the ball just before it strikes the ground? (Neglect air friction.)
4. How much heat is required to raise 10 liters of water from 0° to 50° C?
5. How much heat is required to melt 1 kg of ice? (Latent heat of fusion of water is 80 cal/gm.) How much heat is required to boil away 1 kg of water already at 100° C? (Latent heat of vaporization is 539 cal/gm.)
6. How many gallons of gasoline must be burned in order to raise the temperature of 10 liters of water from 25° C to 50° C, assuming 50 percent

efficiency in converting the heat released into heat in the water? (The heat of combustion of gasoline is 3.4×10^7 cal/gallon.)

7. Consider the example in the Appendix for the velocity of air molecules. If instead of air (70 percent nitrogen, 30 percent oxygen) the gas is hydrogen at normal temperature and pressure (density = M/V = 0.09 gm/liter), what is the velocity of the molecules?
8. If your body is 20 percent efficient in converting food intake to useful work, how many kcal of food must you eat in one day to do 25×10^5 joules of work without tapping any of your body's energy reserves?
9. If 60 percent of the potential energy of the water flowing over a dam can be converted to electricity, find the maximum power output, in kW, for a dam 50 m high where the water crosses at a rate of 300,000 kg/min (1 watt = 1 joule/sec, 1 kW = 1000 watts).
10. An automobile with mass 2000 kg is traveling at 20 m/sec. How much heat is developed in the brakes when the car is stopped?
11. A person of mass 70 kg walks up to the third floor of a building. This is a vertical height of 12 meters above the street. How many joules of work did the person do? By how much did the person's potential energy increase?
12. A bullet of mass 10 gm moving at 200 m/sec strikes and lodges in a block of mass 1 kg hanging from a string. It turns out that one percent of the kinetic energy of the bullet is transferred to the block. How high will the block rise as it swings on the string? What happened to the remaining 99 percent of the kinetic energy of the bullet?

REFERENCES

Sanborn C. Brown, *Benjamin Thompson, Count Rumford,* MIT Press, Cambridge, Mass., 1979.
A personal and scientific biography of one of the first American scientists.

Richard P. Feynman, Robert B. Leighton, and Matthew Sands, *The Feynman Lectures on Physics,* Addison-Wesley, Reading, Mass., 1963, Vol. 1, Chapter 4.
A more advanced undergraduate textbook with emphasis on understanding the physical principles involved. Uses elementary calculus.

Douglas C. Giancoli, *The Ideas of Physics,* Harcourt Brace Jovanovich, New York, 1974, Chapter 7.
An introductory text similar to this one with somewhat different emphases and topics.

Gerald Holton and Stephen G. Brush, *Introduction to Concepts and Theories in Physical Science,* 2nd ed., Addison-Wesley, Reading, Mass., 1973.
Chapters 15–17, 22. See comment in references for Chapter 2.

Morton Mott-Smith, *The Concept of Energy Simply Explained,* Dover Publications, New York, 1964.

A reprint of a very interesting book originally published in 1934, and written in a descriptive style for the lay public. Because it was written 60 years ago, some of its assertions are not acceptable today.

C. J. Schneer, *The Evolution of Physical Science,* Grove Press, New York, 1964 (a reprint of *The Search for Order,* Harper & Row, New York, 1960).
See comment in references for Chapter 2.

Sadi Carnot. (American Institute of Physics Niels Bohr Library.)

Chapter 5

ENTROPY AND PROBABILITY

Entropy tells it where to go

A. Introduction

The concept of entropy was originally developed from the study of possible limitations on the transformation of energy from one form to another. It is therefore particularly pertinent to concerns about energy shortages in modern industrial societies. Although it was once thought that boundless resources of energy are available for human use, some people fear that we may be running out of energy and that there is a great need for energy conservation. (This is not the same energy conservation expressed in the first law of thermodynamics.) Energy shortages are better described as shortages of *available* and useful energy, where the meaning of *useful* is at least partially determined by possible economic, political, and environmental consequences of the conversion of energy to a "useful" form.

The availability of energy of a system is related to entropy. The concept of entropy, however, is much more consequential and fundamental than its relationship to energy transformations indicates. It is particularly significant for chemistry and chemical processes, and is an important part of the studies of scientists and engineers. For example, the entropy of a system determines whether it will exist in a particular solid, liquid, or gas phase, and how difficult it is to change from one phase to another. The microscopic interpretation of the entropy concept leads to ideas of order–disorder, organization and disorganization, irreversibility, and probabilities within the kinetic-molecular model of matter. These ideas have been extended and broadened to become significant parts of information theory and communication theory and have been applied to living systems as well. They have also been applied by analogy to political and economic systems.

This chapter is primarily concerned with the development of the entropy concept and its applications to energy transformations, reversibility and irreversibility, and the overall direction of processes in physical systems. One of the major consequences of the entropy concept is the recognition of heat as a "degraded" form of energy. When energy is in the form of heat, a certain fraction of it is unavailable for use, and this fact must be taken into account in the design of energy transformation systems.

1. Energy as a Parameter of a System

The law of conservation of energy states that although energy may be changed from one form to another, the total amount of energy in an isolated system can be neither increased nor decreased, but rather must remain unchanged. As already stated in Chapter 4, if a system is not isolated, the energy conservation law is formulated in a different, but equivalent, manner. The state of a system can be characterized or described, partially at least, in terms of its total energy content. The state of the system can be further described in terms of other properties, such as its temperature, its size or volume, its mass, its internal pressure, its electrical condition, and so on. These properties, all of which can be measured or calculated in some way, are called **parameters** of the system. If the state of a system changes, at least some of the parameters of the system must change. In particular, it is possible that its total energy content may change.

The law of conservation of energy states that the change in energy content of the system must be equal to the amount of energy added to the system during the change minus any energy removed from the system during the change. In other words, the conservation law states that it is possible to do bookkeeping on the energy content of a system—any increases or decreases in the energy content must be accounted for in terms of energy (in any form) added to or taken away from the system. The law of conservation of energy used in this manner is called the **first law of thermodynamics.**

The concept of energy and its conservation, although making it possible to set up an equation, sets no limits on the quantities entering into the equation. It gives no idea about how much energy can be added or taken away or what rules, if any, should govern the transformation of energy from one form to another. Nor does the energy concept state what fraction of the energy of a system should be in the form of kinetic energy or heat or potential energy or electrical energy, and so on. In fact, it is not even possible to say how much energy a system has in an absolute sense because motion and position are relative. Therefore, kinetic energy and potential energy are relative, and it is only changes in energy that are significant, as discussed in the previous chapter.

2. Entropy as a System Parameter

There are rules that govern energy transformations. The concept of **entropy,** and related ideas, deals with these rules. In fact, the word *entropy* was coined by the German physicist Rudolf J. E. Clausius in 1865 (although he developed the concept in 1854) from Greek words meaning *transformation content.* Entropy, like energy, is also a parameter that can be used to describe the state of a system. Although the energy concept deals with an abstract intangible ''substance'' that exists only in relative amounts and is recognized only by its changes, by contrast the entropy concept deals with how much energy is subject to change or transformation from one form to another.

Historically, the entropy concept developed from the study of ''heat'' and temperature and the recognition that heat is a rather special form of energy that is *not* completely convertible to other forms. Other forms of energy, however, are completely convertible to heat. Moreover, it seems that in all transformations of energy from one form to another, some energy must be transformed into heat.

3. Statistical Nature of Entropy

Entropy (and temperature also) is ultimately explained in terms of statistical concepts because of the atomic nature of matter. (In fact, the full impact and power of the energy concept itself is not appreciated without statistical considerations and the resulting connection of energy with temperature and entropy.) The idea of entropy can be generalized, through its connection with statistics and probability, to be a useful means of describing relative amounts of organization (order versus disorder) and thus becomes useful for studying various states of matter (gas, liquid, solid, plasma, liquid crystal, etc.).

B. Heat and Temperature

1. Distinction between Heat and Temperature

To understand the relationship between the concepts of energy and entropy, it is first necessary to make a clear distinction between heat and temperature. One useful way to make this distinction is to use the caloric theory of heat. (Although the caloric theory is erroneous, the analogy it makes between heat and a fluid helps to introduce some otherwise very abstract ideas.) The relationship between heat flow into an object and its temperature is expressed by the heat capacity of the object. The heat capacity of an object or system is defined as the amount of heat it takes to raise the temperature of the object by one degree (see Chapter 4, Section B1).

Temperature is described in terms of *intensity* or *concentration* of internal molecular energy, that is, the internal molecular energy per unit amount of the substance. For example, a thimble filled with boiling hot water can have a higher temperature than a bathtub full of lukewarm water, yet it takes much less heat energy to raise the water in the thimble to its high temperature than the bathtub water to its lower temperature. The distinction between temperature and heat is explained in terms of the kinetic-molecular theory of matter, which describes temperature as related to the *average* energy of microscopic modes of disorganized motion per molecule of the water, and heat in terms of changes in the *total* energy of microscopic modes of disorganized motion of the molecules. The water molecules in the thimble have more energy on the average than the water molecules in the bathtub, but the thimble needed far less heat than the bathtub because there are far fewer molecules in the thimble than in the bathtub.

2. Properties of Temperature

Although temperature can be ''explained'' in terms of concentration of molecular energy, the usefulness of the temperature concept actually depends on two other fundamental aspects of heat. These aspects are recognized in common everyday experience: thermal equilibrium and heat flow.

2.1 Temperature and Thermal Equilibrium Thermometers are used to measure the temperature of an object or a system. Almost all thermometers depend on the existence of thermal equilibrium. This means that if two objects or systems are allowed to interact with each other, they will eventually come to have the same temperature. Depending on the circumstances, this may happen very quickly or very slowly, but eventually the temperature of one or both of the objects will change in such a way that they will both have the same final temperature.

For example, the measurement of the temperature of a feverish hospital patient depends on the interaction between one object, the patient, and a second object, the thermometer. The nurse puts the thermometer into the mouth of the patient and waits for several minutes. While the interaction is going on, the state of the thermometer is changing. For this example, we assume that the thermometer is the mercury in glass type. The length of the mercury column inside the thermometer is a parameter of the thermometer, and it increases as the temperature of the thermometer increases. When the temperature of the thermometer becomes the same as that of the patient, the length of the mercury no longer changes, and therefore the thermometer is in thermal equilibrium both internally and with the patient. This means two things: (1) the thermometer will not change any further, and (2) the temperature of the thermometer is the same as the temperature of the patient. Only after equilibrium is reached may the nurse remove the thermometer from the patient's mouth and read the scale to determine the temperature of the patient.

2.2 Temperature and Equations of State Actually the thermometer does not measure the temperature directly, but rather some other parameter is measured. In the case of the mercury in glass thermometer, the parameter measured is the length of the mercury in the glass tube. In other types of thermometers, other parameters are measured. In resistance thermometers, the electrical resistance of the thermometer is the parameter that is measured. In thermoelectric thermometers, the thermoelectric voltage is the parameter measured, whereas in a common oven thermometer, the shape of a coil to which a pointer is attached is the parameter that changes. Almost any kind of a system can be used as a thermometer. In all cases, the parameter that is actually measured must be related to the temperature parameter in a definite mathematical manner. This mathematical relationship is called an **equation of state** of the system. (In this case, the thermometer is the system.) This simply means that if some parameters of a system change, then the other parameters must also change, and the changes can be calculated using the equation of state. The dial reading or scale reading of the thermometer represents the results of solving the equation of state for the temperature. The equation of state of the system being used as a thermometer determines the temperature scale of the thermometer.

Every system has its own equation of state. For example, helium gas at moderate pressures and not too low temperatures has the following equation of state relating three of its parameters: $PV = RT$, where P stands for pressure, V for volume, T for temperature (using the thermodynamic or absolute temperature scale, which will be defined below), and R is a proportionality constant. This equation of state is often called the *general gas law* or the *perfect gas law* and is a mathematical relationship among the three parameters mentioned. At all times, no matter what state the helium gas is in, the three parameters—temperature, pressure, and volume—must satisfy the equation. (It is also possible to determine the energy content, U, and the entropy, S, of the system once the equation of state is known.) The equation of state of the system can be drawn as a graph of two of the parameters if the other one is held constant. Figure 5.1 shows graphs of pressure versus volume for helium gas at several different temperatures.

The equation of state and resulting graphs for helium are relatively simple because helium is a gas at all but the lowest temperatures. The equation of state and the resulting graphs of pressure versus volume for water are much more

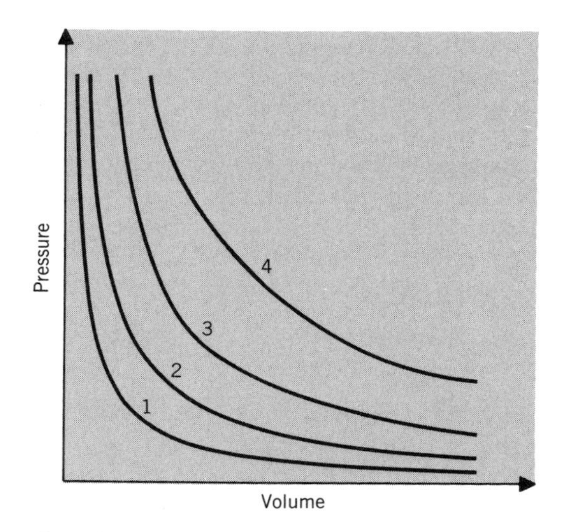

Figure 5.1. Equation of state for helium gas. Each curve is an "isotherm" showing how pressure and volume can change while the temperature remains constant. The curves 1, 2, 3, 4 are for successively higher temperatures.

complex because water can be solid, liquid, or gas, depending on the temperature and pressure. Some graphs for water are shown in Fig. 5.2.

Both helium and water could be used as thermometers, and in fact, helium is sometimes used as a thermometer for scientific measurements at very low tem-

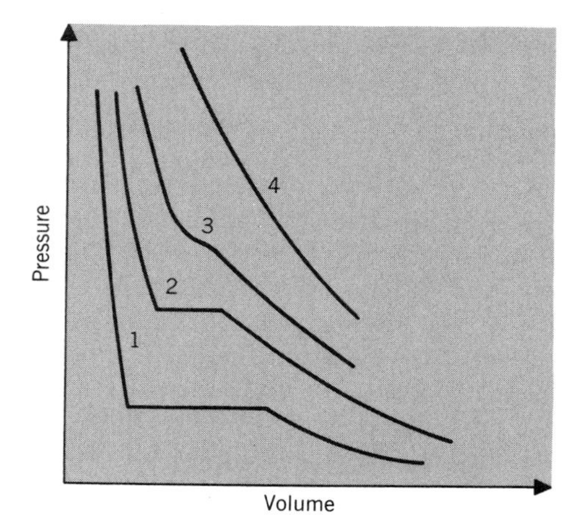

Figure 5.2. Equation of state for water. Each curve is an "isotherm" showing how pressure and volume can change while temperature remains constant. The curves 1, 2, 3, 4 are for successively higher temperatures. The flat parts of curves 1 and 2 cover a range of volumes for which part of the water is liquid and part gas. Water is completely liquid to the left of the flat parts, and completely a gas to the right. Curve 3 is at a temperature (374° Celsius) for which it is impossible to tell the difference between liquid and gas, whereas at all higher temperatures water can be only a gas.

peratures; except for such special purposes, however, helium is not a very practical thermometric material. Water is useful in establishing certain points on temperature scales, but otherwise is not a practical thermometric substance. When the barometric pressure is 760 mm (30 inches), water freezes at 0° Celsius (32° Fahrenheit), and boils at 100° Celsius (212° Fahrenheit). Temperature scales are discussed further in the following optional section and in Section F below.

Optional Section 5.1 *Temperature Scales*

The same concept of thermal equilibrium is used to define temperature scales in a very precise and reproducible manner; this is called "fixed point" definition of temperature scales. For example, the Celsius and Fahrenheit scales were at one time defined using the freezing points and boiling points of water. On the Celsius (sometimes called Centigrade) scale, the freezing point of water is by definition 0° and the boiling point of water by definition 100°. The water must be pure, and the measurement must be made at standard atmospheric pressure which is 760 mm or equivalently 14.5 lb/sq in. Thus for the common mercury in glass thermometer immersed in a mixture of ice and water, the equilibrium position of the "top" of the mercury column was marked on the glass tube (see Fig. 5.O.1) as 0°. Then with the thermometer immersed in boiling water, the mercury expanded along the length of the glass tube to a new equilibrium position, which was marked on the glass. Then the part of the

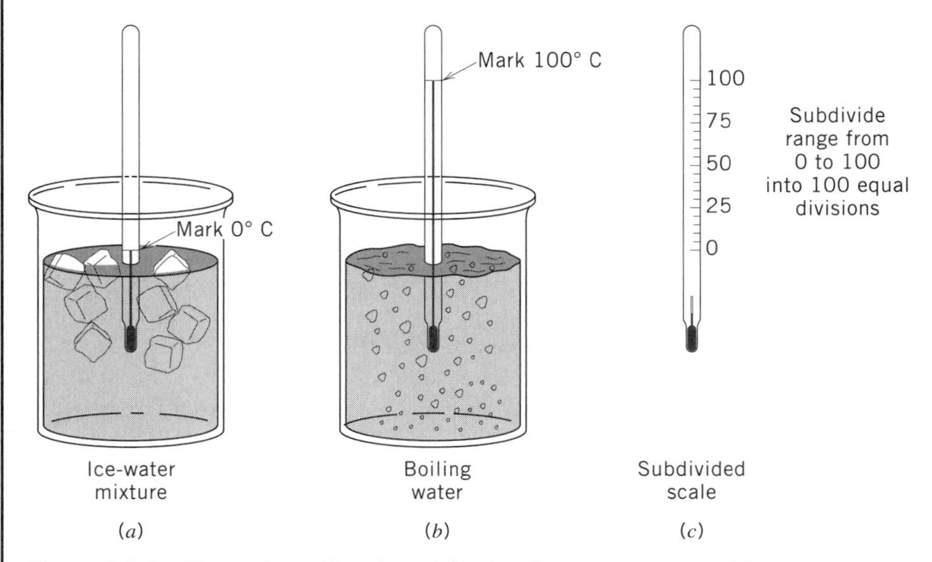

Figure 5.O.1. Two-point calibration of Celsius thermometers. (a) Thermometer immersed in ice-water mixture. Position of top of mercury column marked as 0° C. (b) Thermometer immersed in boiling water mixture. Position of top of mercury column marked as 100° C. (c) Space between 0° C and 100° C subdivided into 100 parts and markings inscribed on glass tube.

glass tube between the two equilibrium positions was divided into 100 equal parts, corresponding to 100 degrees.

For the Fahrenheit scale the position of the top of the mercury column at the freezing point of water was marked as 32° and the position of the top of the mercury column when immersed in boiling water was marked as 212° with the space between the two marks divided into 180 equal parts, corresponding to 180 degrees. Since 100/180 = 5/9, then with a little algebra, it can be shown that the following formulas should be used to convert from Fahrenheit temperature F to Celsius temperature C:

$$F = 32 + \frac{9}{5}C$$

$$C = \frac{5}{9}(F - 32)$$

Adding 273 to the Celsius scale gives the centigrade absolute scale, which like the Celsius scale is used in most scientific work, and in engineering practice other than in America. If 459 is added to the Fahrenheit scale, then the Rankine absolute scale, which is used in American engineering practice, is obtained. (The ratio of 273 to 459 + 32 = 273/491 = 5/9.) The reason for defining these absolute scales is given below.

OPTIONAL SECTION 5.2 *Significance of General Gas Law*

Helium gas is not the only substance that obeys the general gas law. Almost all normal gases (such as oxygen or nitrogen) at not too high pressures or not too low temperatures obey the same set of laws. Not too low temperatures means well above their boiling temperatures. (The boiling temperature for helium is −269° C and for nitrogen −196° C.) For all such gases there is a wide range of conditions for which it is found experimentally that when the product of their pressure P and volume V, that is, PV, is plotted against temperature, for example, as measured on the Celsius temperature scale, then a straight line is obtained. This is shown by the solid lines in Fig. 5.O.2. If these lines are extended toward the left of the figure (dashed lines in the figure) they all cross the horizontal temperature axis at −273° C (−459° F on the Fahrenheit scale). It does not matter whether the gas actually behaves this way at low temperatures (in fact, all real gases eventually deviate from the straight line relationship); but *by definition* the "ideal" or "perfect" gas follows the straight line behavior drawn in Fig. 5.O.2. The significant feature of the ideal gas behavior is that all the straight lines pass through the same point on the graph, with $P = 0$ and therefore $PV = 0$. At this point we remind ourselves that the choice of 0° C to be the freezing point of water was purely arbitrary. If we choose a new temperature scale, for which the freezing point of water is 273° and the boiling point of water is 373°, then the temperature at which $PV = 0$ will be 0°, and there will be no negative temperatures on this scale, which is

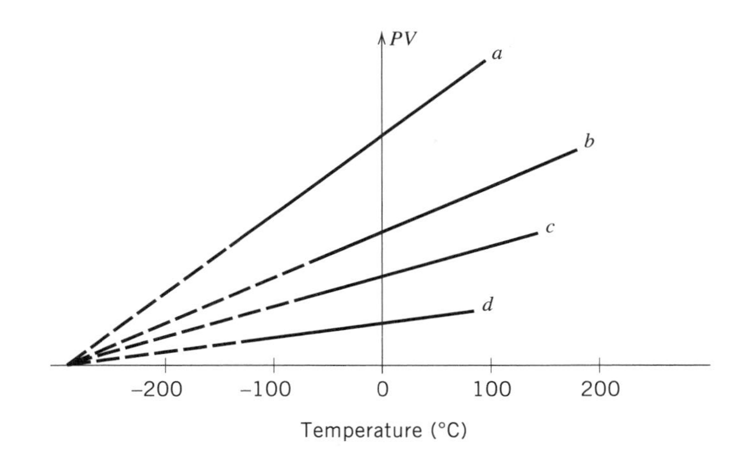

Figure 5.O.2. Use of general gas law to define "absolute" temperature scale. Solid lines *a*, *b*, *c*, *d* represent product of experimental measurements of *P* and *V*. Dashed lines are extrapolations of the straight portions of the plots. The extrapolations all converge at −273° C. Adding 273 to all Celsius readings is equivalent to shifting the zero of the scale to the convergence temperature, thereby making it 0° absolute.

commonly called an *absolute* temperature scale. It should be noted that if the temperature were less than 0° on the absolute scale, then either the pressure or the volume of the ideal gas would be negative. Both of these possibilities are physically impossible.

On this new absolute temperature scale, the straight lines representing the ideal gas all pass through the zero of the horizontal axis, and the equation describing them can be written

$$PV = CT$$

where C is a proportionality constant. Eventually it was recognized that the **constant** C depended in a simple way on the amount and kind of gas, so that $C = nR$ and the general gas law can be written as

$$PV = nRT$$

where n is the number of "moles" of gas, and R is called the general gas constant per mole. A mole of any pure substance is defined as an amount of the substance that has a mass in grams numerically equal to the sum of the atomic masses of the atoms making up one molecule of the substance. The quantity n is not required to be a whole number. A mole of any pure substance contains a specific number of molecules, called *Avogadro's number,* which is approximately equal to 6.02×10^{23}.

The general gas law leads to three other laws, which were actually recognized before the general gas law, and which were used to infer the general gas law. These laws can be derived from the general gas law by keeping one of the quantities P, V, or T constant and unchanging.

If the temperature is held constant, the result is Boyle's Law, which simply states that if the temperature of a gas does not change, then the product of pressure and volume, PV, must remain constant. Thus if P is multiplied by a certain factor, then V must be divided by that same factor in order to keep the product constant. Similarly if P is divided by a certain factor, then V must be multiplied by that same factor. Figure 5.1 shows plots of Boyle's Law at four different temperatures.

On the other hand, if P is held constant, increasing the temperature causes an increase in the volume of the gas that is exactly proportional to the increase of temperature, if the absolute temperature scale is used. This is called Charles' Law or Gay-Lussac's Law, V/T = constant. Basically this law says that gases expand significantly when heated. It is this law that is responsible for the flight of hot air balloons, since "expanded air" is lighter than ordinary air.

The third law derived from the general gas law has no particular name, but it says that if the volume is held fixed, then the "absolute" or true pressure of gas is proportional to the absolute temperature, that is, P/T = constant. This is the law that explains why it is dangerous to throw used spray containers into a fire.

2.3 Temperature Differences and the Flow of Heat When a thermometer is used to measure temperature, the thermometer must come into thermal equilibrium with the system whose temperature is to be measured. However, it is, in principle, possible to measure the temperature of the system even before the thermometer reaches thermal equilibrium. If two systems at different temperatures are interacting thermally with each other, heat will flow from the higher temperature (hotter) system to the lower temperature (colder) system. Thus a nurse can tell that the patient has a fever by feeling the patient's forehead. If the patient has a fever, the forehead will feel hot; that is, heat will flow from the patient's forehead to the nurse's hand. If, on the other hand, the patient's temperature is well below normal, the forehead will feel cool; that is, heat will flow from the nurse's hand to the patient's forehead. This very fact that temperature differences can cause heat to flow is very significant in understanding the concept of entropy and how heat energy can be transformed into other forms of energy.

C. THE NATURAL FLOW OF HEAT

Everyone knows that heat flows from hotter objects to colder objects. If an ice cube is dropped into a cup of hot coffee, the ice cube becomes warmer, melts into water, and the resulting water becomes warmer, whereas the coffee becomes cooler. The heat "flows" out of the hot coffee into the cold ice cube. As already discussed in Chapter 4, this was easily explained by the erroneous caloric theory, which also indicated that heat cannot by itself flow from colder to hotter objects, because that would be making the caloric fluid flow ''uphill.'' Of course, the caloric theory is wrong, and so it is not possible to explain so simply why heat flows from higher temperature bodies to lower temperature bodies, and *not* vice versa. In terms of the distinction between heat and temperature made in Section B1 above, energy flows from high concentrations (high temperature) to low concentrations (low temperature); that is, energy ''spreads out'' by itself. This fact is the essential content of the second law of thermodynamics. Like Newton's laws of motion or the law of conservation of energy, this law is accepted as a basic postulate. There are several different ways of stating the second law, all of which can be shown to be equivalent to each other.

In terms of the natural flow of heat, the **second law of thermodynamics** is stated as follows: **In an isolated system, there is no way to systematically reverse the flow of heat from higher to lower temperatures.** In other words, heat cannot flow ''uphill'' overall. The word *overall* is very important. It is, of course, possible to force heat ''uphill,'' using a heat pump, just as water can be forced uphill using a water pump. An ordinary refrigerator and an air conditioner are examples of heat pumps. However, if a refrigerator is working in one part of an isolated system to force heat ''uphill,'' in some other part of the system more heat will flow ''downhill'' or further ''downhill,'' so that *overall* the net effect in the system is of a ''downhill'' flow of heat. If one system is not isolated, a second system can interact with it in such a way as to make the heat flow ''uphill'' in the first system, but of course there will be at least a compensating ''downhill'' flow of heat in the second system.

Although the second law of thermodynamics as stated above seems obvious and not very important, it leads ultimately to the result that heat is a form of energy not completely convertible to other forms. Indeed the ideas involved in the second law of thermodynamics provide a deep insight into the nature of the physical universe. Some general applications of the second law and different ways of stating it are discussed further below.

D. Transformation of Heat Energy Into Other Forms of Energy

1. Heat and Motion

Long before heat was recognized to be a form of energy, it was known that heat and motion were somehow related to each other. The friction that is often associated with motion always results in the generation of heat. Steam engines, which require heat, are used to generate motion. By the middle of the nineteenth century it was recognized that a steam engine is an energy *converter,* that is, a system that can be used to convert or transform heat energy into some form of mechanical energy—work, potential energy, kinetic energy—or electrical energy or chemical energy, and so on.

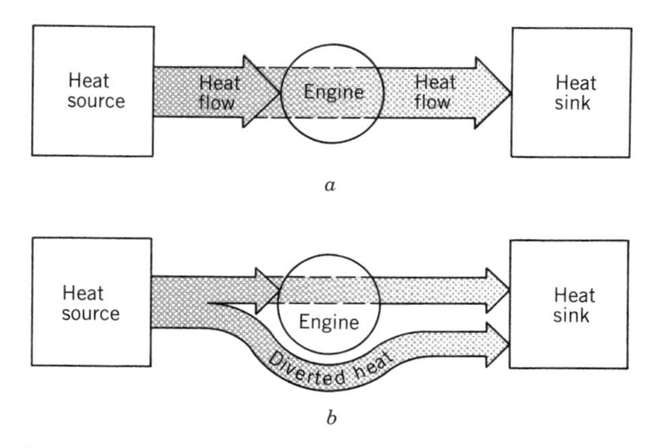

Figure 5.3. Schematic representation of heat flows for a heat engine. (a) All the heat from the source passes through the engine. (b) Some of the heat from the source bypasses the engine.

Even before the true nature of heat as a form of energy was understood, as discussed in the preceding chapter, Carnot realized that the steam engine was only one example of a *heat engine.* Another familiar example of a heat engine is the modern gasoline engine used in automobiles. Carnot realized that the essential, common operating principle of all heat engines was that heat flowed *through* the engine. He pointed out that the fundamental result of all of Watt's improvements of steam engines was more effective heat flow through the engine.

In the spirit of Plato, in an effort to get at the "true reality," Carnot visualized an "*ideal heat engine,*" whose operation would not be hampered by extraneous factors such as friction or unnecessary heat losses. He therefore considered that this ideal engine would be reversible (could move the fluid uphill) and would operate very slowly. The principle of the engine was that heat had to flow through the engine to make it operate, and it was necessary to ensure that all the available heat would flow through the engine. This is shown diagrammatically in Fig. 5.3. In Fig. 5.3a all the heat from the source flows through the engine. In Fig. 5.3b some of the heat from the source is diverted, bypassing the engine, going directly to the heat sink, and thus is "wasted."

OPTIONAL SECTION 5.3 *The Reversible Heat Engine*

Carnot proved that the best possible heat engine, what is called the ideal heat engine, is also a *reversible* heat engine. In the context used here reversible means more than moving heat from lower temperatures to higher temperatures (this is what refrigerators and air conditioners do). If a heat engine runs in the "forward" direction, a certain amount of "heat" moves from a higher temperature reservoir to a lower temperature reservoir (see Fig. 5.4) and work is done by the engine. If the engine runs in a "backward" direction, work is done on the engine and heat is moved from a lower temperature to a higher temperature reservoir. If the engine is truly reversible, the exact same amount of work as it puts out in the forward direction can be used to drive the engine "backward," and the exact same amount of heat moves from the low-temperature reservoir to the high-temperature reservoir as had previously moved from high temperature to low temperature. Everything is then exactly the same as before the engine operated in the forward direction.

This means of course that there is zero friction in all of the moving parts of the engine and that there is no air resistance to the motion of any parts. Friction and any other resistance to motion results in irreversibility of the heat engine because (1) it reduces the work output of the engine when running in the forward direction, and (2) it increases the amount of work needed to run in the backward direction. Note the analogy to Galileo's suppression of friction effects in order to get at the underlying nature of motion.

Carnot also inferred from Watt's introduction of the condenser and use of continued expansion after the steam valve was closed (see the discussion in Section B1 of Chapter 4 of Watt's improvements of the Newcomen engine) another cause of irreversibility—heat flow between engine and heat reservoirs had to be accomplished at as small a temperature difference as possible. For example, a temperature difference between heat reservoirs and engine of 1° is much more desirable than a temperature difference of 50° because it "wastes" less of the heat flow's capability to drive the process (compare the two parts of Fig. 5.4). In the ideal case the temperature difference between engine and reservoir when heat transfer takes place is zero degrees. This is the same kind of idealization as saying that friction and air resistance to motion can be totally eliminated only at a speed of zero miles per hour. Indeed, the reversible engine must operate very slowly.

Carnot then specified a particular type of reversible heat engine, now called the Carnot engine, which would operate between two particular temperature reservoirs, that is, it would take in heat at a "hot" temperature T_H and discharge heat at a "cold" temperature T_c and would satisfy these requirements. The operation of this engine proceeded in a cycle of four steps, called the Carnot cycle, as described further below.

There are other possible reversible engine designs that can operate over the same temperature range as the Carnot engine (and that have a different set of steps in their cycle), but they have the exact same efficiency as the Carnot engine provided all the steps in the cycle are reversible. The key element is reversibility.

Carnot then made an analogy between the ideal engine and a well-designed waterwheel or water mill through which water flows, causing rotation of a shaft, which can therefore operate any attached machinery. In fact, because the caloric theory considered that the heat fluid flowed from high temperatures to low temperatures just as water flows from high levels to low levels, the analogy was very close. The heat engine could be considered to be a *heat wheel,* which could be designed just like a waterwheel. (Carnot's father had made many suggestions for improving the design of waterwheels, and Carnot was influenced by these.)

Waterwheels operate by virtue of having water fall from a high-level reservoir to a low-level reservoir, and similarly heat wheels operate by virtue of having heat flowing from a high temperature or hot "heat reservoir" to a lower temperature or relatively cooler "heat reservoir." For most efficient operation, the waterwheel should be tall enough to take in water exactly at the level of the high reservoir (see Fig. 5.4). The water should stay in the "buckets" of the waterwheel and not spill out as the wheel turns (otherwise it would be wasted and its motive power lost), until the buckets are about to dip into the low-level reservoir, at which time they can discharge their water. The buckets must discharge all their water at the low-level reservoir, and not carry any water back up again or have any water fall into them until they get back up to the high-level reservoir.

Carnot reasoned further that the ideal or most efficient heat engine must be operated in exactly the same way as the most efficient waterwheel: Heat must be taken in only when the engine is at the temperature of the high-temperature reservoir. The engine must be thoroughly insulated while its temperature is changing from high to low, so that no heat will leak out (be diverted) and its motive force be lost. The heat must then be discharged at the temperature of the low-temperature reservoir, so that no heat will be carried back "uphill."

He concluded that the efficiency (i.e., the relative effectiveness of the engine in providing motion or work) of his ideal heat engine would depend only on the

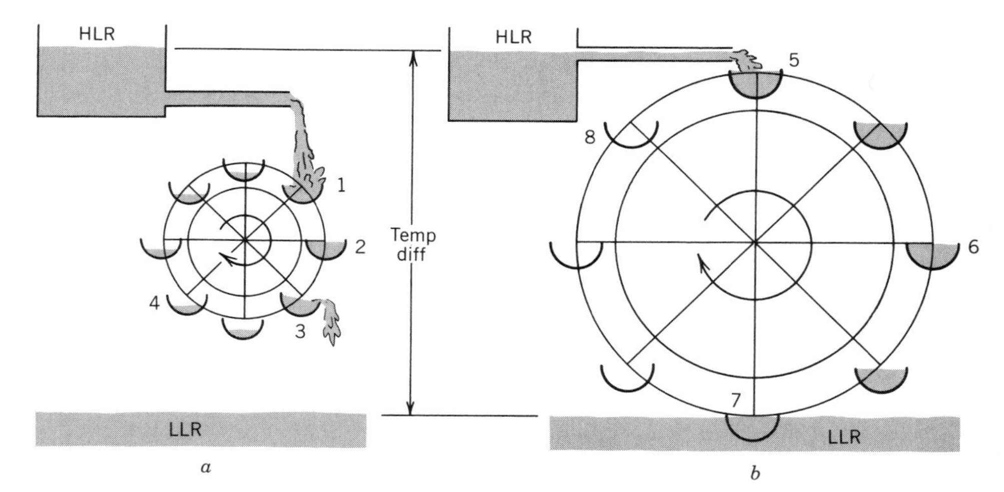

Figure 5.4. Carnot's water wheel analogy for a heat engine. HLR—high-level reservoir; LLR—low-level reservoir. (a) An inefficient design: 1—Caloric fluid being taken into the engine at a temperature far below that of the high-level reservoir. 2—Fluid being lost through "spillage" (leakage). 3—Fluid being discharged before buckets reach temperature of low-level reservoir. 4—Caloric fluid being carried back up to high-level reservoir. (b) An efficient design: 5—Fluid taken in only at temperature of high-level reservoir. 6—No loss of fluid as buckets move down to lower level. 7—Fluid discharged only when buckets reach temperature of low-level reservoir. 8—No fluid carried up as buckets move from low-level to high-level reservoir.

difference in temperature of the two reservoirs, and not on whether it was a steam engine using water as the working substance or some other kind of engine using some other working substance (such as helium gas or oil or diesel fuel or mercury vapor). He concluded, however, that for practical reasons the steam engine was probably the best one to build.

2. First Law of Thermodynamics: Conversion of Heat to Another Form of Energy

About 1850, the English physicist Lord Kelvin (as discussed in Chapter 4) and the German physicist Rudolf J. E. Clausius both recognized that Carnot's basic ideas were correct. However, Carnot's theory needed to be revised to take into account the first law of thermodynamics (i.e., the principle of conservation of energy), which means that heat is not an indestructible fluid, but simply a form of energy. Although heat must flow through the engine, the waterwheel analogy had to be discarded. The amount of heat flowing out of the engine into the low-temperature reservoir is less than the amount of heat flowing from the high-temperature reservoir, the difference being converted into work or mechanical energy. In other words, when a certain amount of heat energy flows into the engine from the high-temperature reservoir, part of it comes out as some other form of energy, and the rest of it flows as heat energy into the low-temperature reservoir. All the energy received must be discharged because after the engine goes through one complete cycle of operation, it must be in the same thermodynamic state as it was at the beginning of the cycle. Thus its internal energy parameter, U, must be the same as it was at the beginning. Therefore, according to the first law of thermodynamics, all the energy added to it (in the form of heat from the high-temperature reservoir) must be given up by it (as work or some other form of energy and as heat discharged to the low-temperature reservoir). This is shown schematically in the energy flow diagram of Fig. 5.5.

3. Alternative Forms of Second Law of Thermodynamics

The essential correctness of Carnot's ideas, according to Clausius, lay in the fact that at least part of the heat flowing into the engine must flow on through it and be discharged. This discharged heat cannot be converted to another form of

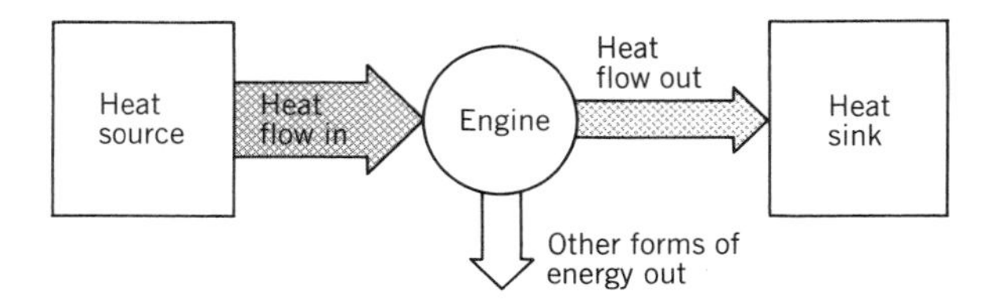

Figure 5.5. Energy flow for a heat engine. Energy input as heat; energy output as heat and work or other useful forms of energy.

energy by the engine. In other words, it is not possible to convert heat entirely into another form of energy by means of an engine operating in a cycle. To do so would violate the second law of thermodynamics (see section C above), which states that heat cannot flow ''uphill'' in an isolated system. This is shown schematically in Fig. 5.6, which illustrates what would happen if an engine could convert heat entirely into mechanical energy, for example.

Suppose such an engine were being driven by the heat drawn from a heat reservoir at a temperature of 30° Celsius (86° Fahrenheit). The assumption is that the engine draws heat from the reservoir and converts it completely into mechanical energy. This mechanical energy could then be completely converted into heat once again, as Joule was able to show by his paddle wheel experiment. In particular, the mechanical energy could be used to stir a container full of water at 50° with no other change, because the engine working in a cycle returns to its original state. The ''driving force'' for the entire process was the heat in the 30° reservoir. Heat would have ''flowed uphill'' by itself, which is impossible according to the second law of thermodynamics.

As will be discussed in Section E2, the flaw in this process is in the assumption of total transformation of heat into mechanical energy. Some heat must move ''downhill'' untransformed to allow for the possibility that the portion of the heat that is transformed into mechanical energy could be retransformed into heat energy at a higher temperature than the reservoir from which it was originally derived. Overall there must be a net ''downhill'' movement of heat. Section I below provides a detailed discussion of how this ''overall downhill movement'' of heat is calculated. Although it is possible in general to show that total conversion of heat to the other forms of energy is not possible, as has just been indicated, further insight into this principle can be obtained from a study of engine

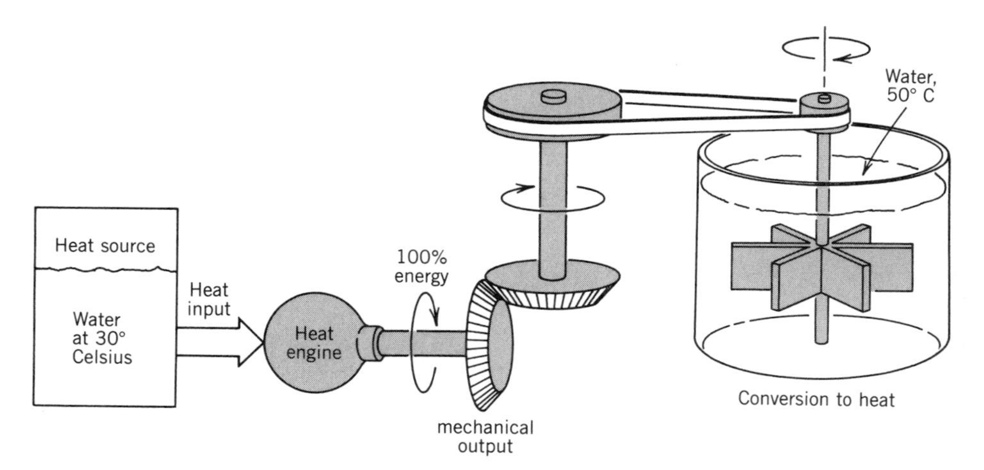

Figure 5.6. Schematic of an impossible machine. Heat taken from a large body of water at 30° Celsius is completely converted into mechanical energy. The mechanical energy is then reconverted into heat by stirring a container full of water at 50° Celsius with the resulting overall effect that heat flowed ''uphill.''

cycles, as is given in Section E below and in the appendix at the end of the chapter.

Sometimes the second law of thermodynamics is stated specifically in terms of these limitations on the convertibility of heat to another form of energy. As just shown, such statements are equivalent to the original statement in terms of irreversibility of heat flow:

1. **It is impossible to have a heat engine working in a cycle that will draw heat from a heat reservoir without having to discharge heat to a lower temperature reservoir.**
2. **It is impossible to have a heat engine working in a cycle that will result in complete (100%) conversion of heat into work.**

Whenever energy is released in the form of heat at some temperature (e.g., by burning some coal or oil or by nuclear reaction), some of the heat must eventually be discharged to a lower temperature reservoir.

4. Perpetual Motion of the Second Kind

Any violation of the second law of thermodynamics is called ''perpetual motion of the second kind.'' If it were possible to build a perpetual motion machine of the second kind, some beneficial results could be obtained.

For example, an oceangoing ship could take in water from the surface of the ocean, extract heat from the water, and use that heat to drive the ship's engines. The result would be that the water taken in would be turned into ice and the ship propelled forward. The ice would be cast overboard, the ship would draw in more ocean water, and the process would continue. The ship would not need any coal, fuel oil, or nuclear energy for its operation. The only immediate by-products of the operation of the ship would be the ice, plus various previously dissolved salts and minerals that would have crystallized out of the ocean water during the heat-extraction process. All of these could be returned to the ocean, with the only net result being that the ocean surface is very slightly, essentially insignificantly, colder.

An even more spectacular example can be conceived. Imagine electricity generating stations located on the banks of the various polluted rivers of the world: the Cuyahoga in Ohio, the Vistula in Poland, the Elbe in Germany and the Czech Republic, the Ganges in India, and so on. These stations would take in polluted river water and extract heat from the water for conversion into electrical energy, thereby lowering the temperature of the water to below its freezing point. The process of freezing the water results in a partial separation of the pollutants from the water. The now cleaner frozen water is allowed to melt and is returned to the river, and the separated pollutants can be either reclaimed for industrial or agricultural use or stored with proper precautions. The result of all this would be a significant improvement of the environment as a by-product of the generation of electrical energy.

Neither of these schemes will work, because although they would take heat from an existing heat reservoir, they do not discharge any heat to an even lower temperature heat reservoir to conform to the requirement that heat must flow "downhill." Unfortunately, the second law of thermodynamics complicates the solution to environmental and energy problems.

E. EFFICIENCY OF HEAT ENGINES

1. Efficiency of Any Heat Engine

The efficiency of any engine is simply the useful energy or work put out by the engine divided by the energy put into the engine to make it operate. In a heat engine, coal or oil or gas might be burned to generate the energy input in the form of heat. This is called Q_H. Suppose that in a particular case 50,000 kilocalories (a common unit of heat) were released by the combustion of fuel to be Q_H. Suppose that of this amount 10,000 kilocalories were converted into electricity. This is called W for work. (In the following discussion, the word *work* is used for any form of energy other than heat. As already pointed out, work is completely convertible to any other form of energy.) The fractional efficiency is W/Q_H = 10,000/50,000 = 1/5 = 20 percent. Because only 10,000 kilocalories were converted into electricity, 40,000 kilocalories (the remainder) were discharged to a lower temperature reservoir. This is called Q_C.

It can be shown that the fractional efficiency can also be calculated from $1 - Q_C/Q_H$. In the example at hand, $1 - Q_C/Q_H = 1 - 40{,}000/50{,}000 = 1 - 0.8 = 0.2$ = 20 percent. This relationship is true for any heat engine, whether it is ideal or not. The efficiency of the large Newcomen engines was only about ½ percent, and for Watt's improved engine was about 3 or 4 percent. The fractional efficiency of the heat engines in a modern electric generating station is of the order of 0.4 = 40 percent. Table 5.1 lists the efficiencies of several types of engines. The temperatures given in Table 5.1 are based on the Kelvin scale, which is discussed in Section F.

2. Carnot or Thermodynamic Efficiency

The second law of thermodynamics states that even the very best heat engine conceivable—the ideal heat engine—cannot be 100 percent efficient. It is natural to ask how efficient this best heat engine actually is. This is an important practical question, because the ideal heat engine sets a standard against which other practical engines can be measured. If a particular engine comes close to the ideal efficiency, then its builders will know that they cannot improve it much more; on the other hand, if the engine has only one third or less of the ideal efficiency, then perhaps it can be improved significantly. The ideal efficiency is sometimes called the Carnot or thermodynamic efficiency.

Table 5.1. Efficiencies of Some Heat Engines[a]

Type of Engine	Hot Reservoir Temperature (Kelvin)	Cold Reservoir Temperature (Kelvin)	Efficiency (percent)
Newcomen engine	373	300	½
Watt's engine	385	300	3–4
	1500	300	80
Ideal Carnot	1000	300	70
	811	311	62
Rankine	811	311	50
Actual steam turbine power plant	811	311	40
Binary vapor cycle	811	311	57
Gasoline engine with Carnot efficiency	1944	289	85
Ideal Otto gasoline engine	1944	289	58
Actual gasoline engine	—	—	30
Actual diesel engine	—	—	40

[a]See appendix and references at end of this chapter.

2.1 The Carnot Cycle Even though he used the erroneous caloric theory, Carnot did list the correct steps in the cycle of the ideal heat engine. Therefore, the ideal heat engine is called the Carnot engine and the cycle is called the Carnot cycle for an engine operating between two heat reservoirs. These steps are enumerated below and are shown schematically in Fig. 5.7 and given in more detail for a special case in the appendix at the end of this chapter.

STEP 1. Isothermal heat intake (**Isothermal** means constant temperature.) The engine is at the temperature T_H of the hot reservoir, and slowly takes in heat, while its temperature is held constant. In order to do this, some other parameter of the engine system must change, and as a result the engine may do some useful work. (For example, in a steam engine the piston is forced out, increasing the volume of the cylinder.)

STEP 2. Adiabatic performance of work by the engine (**Adiabatic** means no heat can enter or leave.) The engine is completely thermally insulated from the heat reservoirs and the rest of the universe, and does some useful work. The performance of work means that energy (but not in the form of heat) leaves the engine, and therefore the internal energy content of the engine is less at the end of this step than at the beginning of this step. Because of the equation of state of the working substance of the engine, the temperature of the engine decreases. This step is terminated when the engine reaches the same temperature T_C as the low temperature or cold reservoir. (In Fig. 5.7, the piston is still moving outward.)

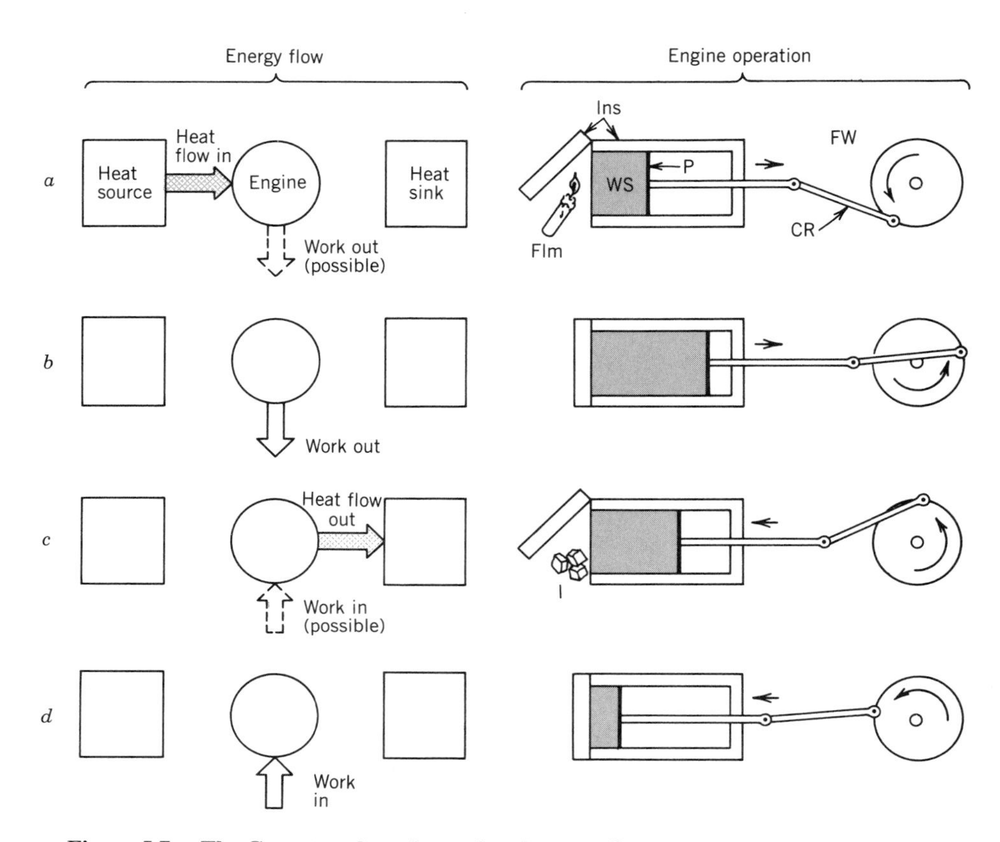

Figure 5.7. The Carnot cycle and associated energy flows. The left side of the figure shows the energy flow associated with diagrams on the right side of the figure. P, piston; WS, working substance; CR, connecting rod; FW, flywheel; Ins, insulation; Flm, flame; I, ice. (a) Isothermal heat intake from hot reservoir, piston moving outward, work done by engine. (b) Adiabatic work by engine, piston moving outward. (c) Isothermal heat discharge to cold reservoir, piston moving inward as work is done on the engine by the flywheel. (d) Adiabatic temperature increase, piston moving inward as more work is done on engine by the flywheel.

STEP 3. Isothermal heat discharge The engine is now at the temperature T_C of the cold reservoir, the thermal insulation is removed, and heat is slowly discharged to the low-temperature reservoir, while the temperature of the engine remains constant. Again, in order to do this, some other parameter of the engine must change, resulting in some work being done on the engine. It is at this step and the next step at which some of the work taken out of the engine must be returned to the engine. These two steps, which are *necessary* for an engine working in a cycle, can be regarded as the steps where nature reduces the efficiency from 100 percent to the thermodynamic values. (In Fig. 5.7, the piston is now driven inward by the flywheel and connecting rod, decreasing the cylinder volume.)

STEP 4. Adiabatic temperature increase of engine The engine is again completely thermally insulated from the heat reservoirs and the rest of the universe. Work is done on the engine, usually by forcing a change in one of its parameters, resulting in an increase of temperature until it returns to the starting temperature, T_H, of the hot reservoir. The internal energy content of the engine is now the same as at the beginning of Step 1, and the engine is ready to repeat the cycle. (In Fig. 5.7, the piston continues its inward motion.)

It should be noted that there may be portions of an engine cycle in which heat is converted 100 percent to work. Step 1 of the Carnot cycle, when applied to the engine described in the appendix at the end of this chapter, is such a portion. In order for the engine to continue working, however, it must go through a *complete* cycle. In the process of going through the complete cycle, the second law of thermodynamics requires that a portion of the work be returned to the engine and transformed back into heat, some of which is then discharged to the low-temperature reservoir, and the rest used to return the engine to the starting point of the cycle. The important point is that this is true for all engines, regardless of how they are operated.

2.2 Efficiency of the Carnot Cycle Carnot had originally reasoned that the efficiency of the Carnot cycle should be proportional to the temperature difference between the hot and cold reservoirs, $T_H - T_C$. Clausius corrected Carnot's analysis to show that, if the proper temperature scale is used (the so-called *absolute* temperature scale), the efficiency would be equal to the difference in temperature of the two reservoirs divided by the temperature of the hot reservoir; that is, the efficiency equals $(T_H - T_C)/T_H = 1 - T_C/T_H$.

Using Clausius's formula for the Carnot or thermodynamic efficiency, it is possible to calculate the efficiency that could be expected if all friction, irreversibilities, and heat losses were eliminated from a modern electricity generating plant, and if it were operated in a Carnot cycle. The results of such calculations are also shown in Table 5.1. It can be seen that actual modern generating stations do get reasonably close to the Carnot efficiency for the temperatures at which they operate.

3. Possibilities for Improving Heat Engine Efficiencies

The only way to improve the theoretically possible efficiencies of heat engines is by either increasing T_H or decreasing T_C. On the planet Earth, T_C cannot be lower than 273 kelvins (32° Fahrenheit) because that is the freezing temperature of water, and present-day power plants use cooling water from a river or ocean for their low-temperature reservoirs. Theoretically, it would be possible to use some other coolant, say *antifreeze* (ethylene glycol) as used in automobile engines, which remains liquid at temperatures much lower than the freezing point of water. In order to do this, however, a refrigeration system must be used to lower the temperature of the antifreeze, so that it can be a very low-temperature reservoir. Unfortunately, energy is required to operate the refrigeration system. It can be shown from the second law of thermodynamics that the amount of energy required to operate the refrigeration system, which would have to be taken from the power plant, would more than offset the energy gain from the increased efficiency resulting from the lower value of T_C. In fact, the net effect of installing the refrigerator would be to *decrease* the overall efficiency below the Carnot efficiency.

Of course, one can imagine building an electricity generating station on an orbiting satellite in outer space, where T_C is only a few degrees absolute. It would then be necessary to consider how the electricity generated at the station would be transmitted to Earth.

Alternatively, it is necessary to consider the possibility of raising the temperature T_H of the hot reservoir. This temperature cannot be higher than the melting temperature of the material out of which the heat engine is built. Actually T_H will have to be considerably less than the melting temperature, because most materials are rather weak near their melting points. If T_H were about 1500 kelvins (melting point of steel is 1700 kelvins), and T_C were 300 kelvins, the Carnot efficiency is calculated to be $1 - 300/1500 = 0.8 = 80$ percent. Because of the need for safety factors (to keep boilers from rupturing under high pressure associated with high temperatures), T_H would be more reasonably set at 1000 kelvins, giving a Carnot efficiency of 70 percent. An actual engine might not achieve much more than ¾ of the Carnot efficiency, yielding a maximum feasible efficiency in the range from 50 to 55 percent. Thus it can be expected that half or more of the fuel burned in a practical heat engine operating between two ''reasonable'' temperature reservoirs will be wasted so far as its convertibility into ''useful'' work is concerned. (See the data given in Table 5.1 for more specific examples.)

An idea has been proposed, which is in effect a three-reservoir heat engine to obtain more efficient transformation of the heat liberated at high temperatures. Some heat is added to the system at an intermediate temperature between T_H and T_C, for example, during step 4 outlined above for the Carnot cycle. This heat added at an intermediate temperature then helps to do the work *on* the engine to bring it back up to T_H. Step 4 would no longer be adiabatic. The result is that the heat entering the system at T_H is then transformed more effectively into some other form of energy; however, the heat added at the intermediate temperature is transformed less effectively. Overall the efficiency is less than Carnot efficiency if *all* sources of heat are taken into account. In effect, Q_C, the heat discharged to the low-temperature reservoir, comes mostly from the intermediate-temperature reservoir, rather than from the high-temperature reservoir. The advantage of the idea as proposed is that this intermediate-temperature reservoir would be a source of heat that requires that no additional fuel be burned; that is, the source would be either solar or geothermal heat, or some low-grade heat source that is already available for other reasons than energy conversion. (See the article by Powell et al. listed in the references at the end of this chapter.)

4. Types of Heat Engines

The ideas involved in heat engine theory are almost universal. There are many varieties of heat engines, and they can be classified according to several categories. For example, there are external combustion engines, such as reciprocating steam engines, steam turbines, and hot air engines; also internal combustion engines, such as the various types of gasoline and diesel engines, jet engines, and rocket engines. Any engine that involves the burning of some fuel or depends on temperature in order to operate, or in any way involves the liberation or extraction of heat and its conversion to some other form of energy, is a heat engine. Even the Earth's atmosphere is a giant engine that generates energy. (Wind energy is kinetic energy of systematic motion of air molecules, and thunderclouds contain electrical energy.) All heat engines are subject to the same rules concerning their maximum theoretical efficiency, which is completely governed by the theory developed by Carnot and corrected by Clausius. As already indicated, the ideal efficiency (the Carnot efficiency) for all these engines depends only on the temperatures of their two heat reservoirs and is independent of anything else or any details or features of engine construction.

A detailed discussion of a *helium gas* external combustion engine operating in a Carnot cycle is given in the appendix at the end of this chapter. It was the mathematical analysis of an engine of this type that permitted Clausius to derive the correct formula for engine efficiencies. Other than for theoretical purposes, an engine built specifically to conform as closely as possible to the Carnot cycle is not always practical. There is, in fact, a story about a ship's engine that was built to have an efficiency close to that of a Carnot engine. When it was installed in the ship for which it was designed, however, it was so heavy that the ship became overloaded and sank! (See Chapter XVII of Mott-Smith's book listed in the references.)

F. The Thermodynamic or Absolute Temperature Scale

In Section B2 above, thermometers and their scales were described as depending on temperature equilibrium and the equation of state of the thermometer system. It was also indicated that temperature scales could alternatively be defined in terms of the heat flow between two different temperatures. The Carnot cycle makes it possible to do just that. Instead of using the physical parameters of a thermometric substance such as mercury, it is possible to use the properties of heat and heat transformation.

Imagine two bodies at different temperatures. Imagine further a small Carnot engine that would use these two bodies as its heat reservoirs. According to the Carnot engine theory, the ratio of the heat flowing out of the engine to the heat flowing into the engine, Q_C/Q_H, should depend only on the temperatures of the two bodies. Lord Kelvin proposed that the temperature of the two bodies, T_C and

T_H, should be defined as proportional to their respective heat flows; that is, $T_C/T_H = Q_C/Q_H$. The temperature scale so defined is called a *thermodynamic temperature scale.* For example, suppose that a particular body is known to be at a temperature of 500 absolute, and that we wish to know the temperature of some cooler body. It might be imagined that a small Carnot engine is connected between the two bodies, and the heat flow in, Q_H, and the heat flow out, Q_C, might be determined. If the ratio of Q_C/Q_H were ½, then according to Kelvin, the ratio of T_C/T_H should be ½ also. Thus the temperature of the cooler body is ½ that of the hotter body or ½ of 500; that is, 250.

The thermodynamic temperature scale is defined earlier only in terms of ratios. To actually fix the scale it is necessary also to specify a size for the units of temperature (degrees). If there are to be 100 units between the freezing and boiling points of water (as in the Celsius temperature scale), then it turns out that on the thermodynamic scale the freezing temperature of water is 273 kelvins and the boiling temperature is 373 kelvins. This thermodynamic temperature scale is called the Kelvin scale.[1] If there are to be 180° between the freezing and boiling points of water (as in the Fahrenheit scale), then it turns out that the freezing temperature of water is 491° and the boiling point is 671°. This thermodynamic scale is called the Rankine scale.

G. The Third Law of Thermodynamics

Because of the connection between absolute temperature and engine efficiencies, it is of interest to consider what would happen if the low-temperature reservoir were at absolute zero. As discussed above, the Carnot efficiency of a heat engine is equal to $1 - T_C/T_H$. If $T_C = 0$, the Carnot efficiency would be 100 percent. Such reservoirs do not normally exist, however. As already pointed out, the energy expended in the attempt to make a cold reservoir at zero absolute temperature would be greater than the extra useful energy obtained by operating at zero absolute temperature. Nevertheless, it would be interesting to see whether the theory would be valid, even at that low temperature. Thus it is necessary to examine the possibilities of making such a reservoir.

It turns out that in order to make a low-temperature reservoir, the effectiveness of the refrigerator used to achieve the low temperature becomes less as the temperature goes down, and it takes longer and longer to reach lower temperatures as the temperature gets closer to absolute zero. In fact, this experience is summarized in the **third law of thermodynamics: It is not possible to reach the absolute zero of temperature in a finite number of steps.** This means that it is

[1]In common usage the units on a temperature scale are usually called *degrees;* by international agreement, however, the units on the Kelvin scale are called *kelvins.* The specification of the Kelvin scale is somewhat more complicated than described here, and involves a particular fixed point, the *triple point of water.* The Kelvin scale turns out to be identical with the absolute temperature scale, discussed in Optional Sections 5.1 and 5.2 above.

possible to get extremely close to absolute zero temperature, but it is always just beyond being attained. In other words, "absolute zero can be achieved, but it will take forever."[2]

The three laws of thermodynamics are sometimes summarized humorously in gambling terms. Comparing heat energy with the stakes in a gambling house, the laws then become:

1. "You can't win; you can only break even."
2. "You can only break even if you play long enough and are dealt the exactly correct set of cards."
3. "You should live so long as to get the right cards."

H. Energy Degradation and Unavailability

Even though energy is neither created nor destroyed, but only transformed, the concept of "lost" or "degraded" energy is useful. If a given heat engine is only 40 percent efficient, then the other 60 percent of the heat generated in operating the engine is not converted to "useful" forms, but is discharged to the low-temperature reservoir. Moreover, this heat cannot be recycled through the engine for another attempt to convert it into a useful form. Any heat that is discharged to a low-temperature reservoir is wasted, in that it is no longer *available* for transformation to other forms of energy (unless there is a still lower temperature reservoir accessible for use). The energy is said to be *degraded.* Indeed, a given amount of energy in the form of heat is degraded in comparison with the same amount of energy in some other form, because it is not completely convertible.

If energy in the form of heat flows out of a high-temperature reservoir directly to a lower temperature reservoir, it will become much more degraded than if it were to flow through a heat engine, where at least part of it is transformed to another form to be available for further transformations. As already mentioned, Clausius sought to describe the availability for transformation of the energy content of a system by coining the word *entropy* from Greek roots, meaning transformation content. In fact, he used the word to describe the *unavailability* of the system's energy for transformation. Thus a body or system or reservoir whose internal energy is *less* available for transformation has a *greater* entropy than a body of equal size whose internal energy is more available (i.e., available if a heat engine were set up between the body and another one at a lower temperature).

[2]Strictly speaking, it is preferable to say "zero Kelvin" rather than "zero absolute" and "thermodynamic temperature" rather than "absolute temperature," because the temperature scale is based on heat flow, which is a dynamic process. The use of words such as "absolute zero" can cause confusion, because in certain circumstances the question of "negative absolute temperature" arises. For further details, see Chapter 5 of the book by Zemansky, listed in the references at the end of this chapter.

There are two characteristics of the unavailability or degradation of energy: (1) that it must be in the form that is described as heat, and (2) that the lower the temperature of the body, the more unavailable its heat energy.

I. ENTROPY: MACROSCOPIC DEFINITION

1. Changes in Entropy

One of the best ways to describe a concept in physics is to tell how it is determined or calculated. Entropy, like energy, is a relative concept, and thus it is changes in entropy that are significant. In some cases, changes in entropy of a body or system are calculated in terms of the amount of heat added to or taken away from the body or system, per unit of absolute temperature. If a body that is initially at equilibrium at some temperature has some heat added to it, then by definition the entropy of the body increases by an amount equal to the amount of heat added, divided by the absolute temperature of the body.[3] For example, if a body initially at 300 kelvins (about room temperature) receives 5000 kilocalories of heat, then its entropy change is $+5000/300 = +\ 16.67$ kilocalories per kelvin. On the other hand, if heat is taken away from a body, its entropy decreases. Thus if a body at 600 kelvins loses 5000 kilocalories of heat, its entropy change is $-\ 5000/600 = -8.33$ kilocalories per kelvin. (If the temperature of the body is changing, then its entropy change can be calculated from the area underneath a graph of heat versus $1/T$.)

A thermodynamic system consisting of a hot reservoir at temperature T_H, a cold reservoir at temperature T_C, and a heat engine can be used as an example of how entropy calculations are employed. To be specific, T_H will be assumed to be 600 kelvins and T_C to be 300 kelvins. If 5000 kilocalories of heat were to flow directly from the hot reservoir to the cold reservoir, that amount of heat would become totally "degraded" or unavailable for transformation to some other form. Alternatively, the 5000 kilocalories could flow through a heat engine having an efficiency of 20 percent. Another alternative would be for the heat to flow through a Carnot engine, which in this case could have an efficiency of 50 percent, as can be verified using the concepts developed in Section E1.

If the heat flows directly from the hot to the cold reservoir, the entropy change of the hot reservoir is $-5000/600 = -8.33$ kilocalories per kelvin (a decrease) as calculated above. The entropy change (an increase) of the cold reservoir was also calculated above as $+5000/300 = 16.67$ kilocalories per kelvin. Combining the two entropy changes, there is a net entropy increase for the total system of $+16.67 - 8.33 = +8.34$ kilocalories per kelvin. This number is a measure of

[3]This definition of entropy change, just as, for example, the definition of kinetic energy as $\frac{1}{2}mv^2$, is presented without justification.

the fact that heat has been ''degraded'' by passing from the high-temperature reservoir to the low-temperature reservoir; that is, it is less available for transformation.

On the other hand, if the heat flows through the 20 percent efficient engine, 1000 kilocalories of heat are converted into some other form of energy and 4000 kilocalories of heat are discharged to the cold reservoir. The entropy increase of the cold reservoir is then $4000/300 = +13.33$ kilocalories per kelvin. The entropy change of the hot reservoir is still -8.33 kilocalories per kelvin, so the net entropy change of the total system is $+13.33 - 8.33 = 5$ kilocalories per kelvin, which is less than before; this means that some of the heat energy that could have been transformed is degraded.

Finally, if the heat flows through the 50 percent efficient Carnot engine, 2500 kilocalories are converted into some other form of energy and 2500 kilocalories of heat are discharged to the cold reservoir. The entropy increase of the cold reservoir is $+2500/300 = 8.33$ kilocalories per kelvin. The entropy change of the hot reservoir is still unchanged, and the net entropy change of the total system is $+8.33 - 8.33 = 0$. In other words, for the Carnot engine, all the energy available for transformation is transformed to some form other than heat and is still available for further transformation. Thus in this case the entropy of the total system is not changed, although the entropy of various parts of the system (the two reservoirs) did change. Even in the Carnot case, some energy was ''lost,'' because it was transferred to the low-temperature reservoir; but that energy was already understood to be unavailable for transformation because it had previously been converted into heat within the isolated system. Whenever energy is in the form of heat, we know that some of it is unavailable for transformation and is in a degraded form. Only if the heat is discharged through a Carnot engine is it kept from degrading further in subsequent processes.

2. Another Form of the Second Law of Thermodynamics

It should be noted that in the sample entropy calculations above, the entropy of the total system either increased or was unchanged. The question then arises as to whether there could be any changes in the total system such that there would be a net entropy decrease. In the particular case considered above, this would require that the entropy increase of the low-temperature reservoir be less than 8.33 kilocalories per kelvin. This would mean that less than 2500 kilocalories would be discharged from the engine, and therefore more heat energy would be converted into another form in the Carnot engine than is allowed by the second law of thermodynamics. But the second law cannot be violated. This is an example of yet another version of the second law: **The entropy of an isolated system can never decrease; it can only increase or remain unchanged.**

Whenever some other form of energy in a system is converted into heat, the entropy of the system increases. In Joule's paddle wheel experiment, the falling weights start with potential energy determined by their mass and their height above the ground. After they fall, their mechanical potential energy is converted into "heat" added to the water. The total energy of the system consisting of the water and the weights is conserved; however, the entropy of the water is increased (calculated as in the examples above), whereas the entropy of the weights is unchanged (no heat was added or taken from them). As a result, there has been a net entropy increase in the total system because of the transformation of mechanical potential energy into heat energy.

The examples of entropy calculations discussed before, and others like them, show that it is possible to use the numerical value of the entropy of various systems to characterize mathematically the ways in which the energy is **distributed** throughout the systems. The flow of heat from a hot part to a cold part of a system is a redistribution of the energy of the system, and as a result the entropy of the system as a whole increases. Of course, in time the entire system comes to thermal equilibrium at some uniform temperature. In that case, the entropy of the system, having increased as heat flowed, has reached a high point. Because all parts of the system are now at the same temperature, no more heat will flow, and thus the entropy of the system is said to be "at a maximum" when the system is in thermal equilibrium.

It is possible to disturb this equilibrium by converting part of the energy of the system into additional heat (e.g., some coal or oil in the system can be ignited and burned). This will increase the entropy of the system, marking the fact that the system is now different because some of the chemical energy of the coal or oil is now in the form of energy of random molecular motion and, therefore, less available for transformation to other forms. This "new" heat energy is now ready to flow to other parts of the system. As discussed above, if it flows through a Carnot engine, there will be no further entropy increase; if it flows through any other engine or directly to other parts of the system, however, the entropy of the system will increase to a new maximum value.

J. Entropy Increase and Irreversibility

If some process takes place that causes an increase in the entropy of an isolated system, the second law of thermodynamics states that the process can *never* be reversed so long as the system remains isolated, because such a reversal would decrease the entropy (from its new value), which is forbidden. Once the coal is burned, producing heat and ashes, the process cannot be reversed to make the heat flow back into the ashes and the ashes become coal again. A process can be reversed only if the accompanying entropy change is zero.

A Carnot engine is completely reversible because the total entropy change associated with its operation is zero. Any engine having any amount of friction (and, therefore, transforming mechanical energy into heat) carries out an irreversible process. Even a frictionless engine operating between two temperature reservoirs that does not operate in the Carnot cycle carries out an irreversible process because the total entropy change of the system of reservoirs plus engine is greater than zero. The engine itself may be capable of being reversed,[4] whether it has friction or not, but nevertheless there has been an irreversible change in the total system. As already stated, this irreversible change is associated with the approach to thermal equilibrium of the two reservoirs.

Thus in this sense, the principle of increase of entropy ''tells the isolated system which way to go.'' The system, by itself, can only go through processes that do not decrease its entropy. One British scientist, Arthur S. Eddington, referred to entropy as ''time's arrow'' because descriptions made at different times of an isolated system could be placed in the proper time sequence by arranging them in order of increasing entropy.

K. Entropy as a Parameter of a System

If a system is not isolated but can be acted on by other systems, it may gain or lose energy. If this energy is in the form of heat, it is possible to calculate the entropy change of the system directly from the heat received or lost, provided the temperature of the system can be calculated while the heat is being gained or lost. This can be done using the first law of thermodynamics as an equation, together with the equation of state of the system. The mathematical analysis shows that the entropy of the system can be treated as a physical parameter determining the state of the system, just as temperature is a physical parameter determining the state of the system. The value of the entropy can be calculated from the temperature and other parameters of the system such as volume, pressure, electrical voltage, or internal energy content. Every system has entropy. Although there are no ''entropy meters'' to measure the entropy of a system, the entropy can be calculated (or equivalently looked up in tables) if the other parameters (e.g., temperature, volume, mass) are known. Similarly, if there were no ''temperature meters'' (i.e., thermometers), it would be possible to calculate the temperature of an object from its other parameters.

Just as temperature and pressure and the other **macroscopic** parameters, which are parameters describing gross or bulk properties of the system, are ultimately ''explained'' in terms of a microscopic model (the kinetic-molecular theory), which considers matter to be made up of atoms and molecules in various states of motion and position, so too entropy can be ''explained'' in terms of the kinetic-molecular theory of matter.

[4]In principle, a heat engine can be reversed, or run ''backwards,'' by putting work into it and causing it to draw heat from a low-temperature reservoir and transferring that heat, plus the equivalent heat of the work put in, to a higher temperature reservoir. The reversed heat engine is called a *heat pump*. Examples of heat pumps are air conditioners and refrigerators, which are used to cool buildings and food, respectively. Heat pumps are also used to heat buildings. A Carnot engine run ''backwards'' is the ideal heat pump. According to the second law of thermodynamics, the operation of a heat pump in an isolated system can never result in a decrease of the entropy of the system as a whole.

L. Microscopic Interpretation of Entropy

1. Review of Kinetic-Molecular Model

As speculated even as early as the times of Francis Bacon, Robert Hooke, and Isaac Newton, the effect of transferring heat to a gas is to increase the microscopic random motions of molecules, but with no resulting mass motion. Even "still air" has all its molecules in motion, but with constantly changing directions (because otherwise the air would have an overall mass motion—i.e., there would be a wind or breeze). The absolute temperature of an ideal gas can be shown to be proportional to the average random translational kinetic energy per gas molecule.

It is important to make a clear distinction between *random* and *ordered* motion. If a bullet is moving through space at a velocity of several hundred miles per hour, the average speeds of its molecules are all the same and in the same direction; and their motion is said to be organized. When the bullet has an adiabatic (no heat lost) collision with another object that brings it to a halt, the molecules still have the same average kinetic energy as before, but the motion has now become totally microscopic and randomized. The molecules now are not all going in the same direction, nor are they traveling very far in any one direction, so that their net average velocity (as contrasted to their speed) is zero. The kinetic energy of the bullet, which formerly was calculated from the gross overall motion of the bullet, has been transformed into heat and added to the energy of the microscopic random motions of the molecules.

2. Energy Distribution among the Molecules

The use of terminology such as "the average random translational kinetic energy per molecule" implies that some molecules are moving faster and some moving slower than the average, and that there are other forms of kinetic energy, such as tumbling, twisting, and vibrations of the different parts of the molecule. In fact, any particular molecule may be moving faster than average at one time and slower than average at another time. One may ask what proportion of the molecules is moving only a little faster (or slower) than the average, what proportion is moving much faster than the average, and so on. A graph of the answers to these and similar questions is called a *distribution function* or a *partition function,* because it shows how the total kinetic energy of the gas is distributed or shared among the various molecules.

It should be, in principle, possible to calculate this distribution function from the basic principles of mechanics developed from the ideas of Isaac Newton and his contemporaries and successors. But this would be a very difficult and complicated calculation because the large number of molecules in just one cubic inch of gas (about 200 million million million or 2×10^{20} using "scientific notation") at normal temperature and pressure would require that perhaps an equally large number of equations would have to be solved. Moreover, it would be extremely difficult to make the measurements to determine the starting conditions of position and velocity of each molecule, which are required to make the calculations. The next best thing to do is to try to make a statistical calculation; that is, make assumptions as to where "typical" molecules are and with what velocities they are moving. It is necessary to use ideas of probability and chance in these assumptions. Therefore, the result of such assumptions is that there will be *random* deviations of particular individual molecules from the "typical" speeds and directions.

The essential meaning of *random* is summed up in words such as *unpredictable* or *unknown* or *according to chance.* Nevertheless, although the motions of particular individual molecules may be unpredictable, the average motion of the "typical" molecule is predictable. One can, in fact, even predict what a particular individual molecule will be doing (but not with certainty, only with probability) so that a large enough statistical sample of similar molecules will behave on the average according to predictions.

But this poses a dilemma. How can there be unpredictable results if everything is based on Newtonian mechanics, which assumes certainty? It is necessary to make a further hypothesis or assumption that the *averages* obtained using probabilities are the same as the values that would be obtained if calculations were performed with Newton's laws and then averaged. This assumption, called the ergodic hypothesis, coupled with the laws of Newtonian mechanics, then logically leads to the result that the distribution function over a period of time will develop in such a way as to be identical to a random probability distribution about an average kinetic energy.

3. The Equilibrium Distribution

The microscopic model and the idea of distribution functions can be used to "explain" how mixing a hot gas with a cold gas results in a transfer of heat from the hot gas to the cold gas, and in obtaining an equilibrium temperature for the mixture of the two gases. Figure 5.8 shows a schematic representation of a container containing hot (i.e., high temperature) gas molecules on the right and cold (i.e., low temperature) gas molecules on the left. Initially, the boundary between the two halves of the container was made by an adiabatic (perfectly insulating) wall, but this is now completely removed, so that the molecules are free to cross over between the two halves of the container. Even though the "hot" molecules will not travel very far between collisions, they will interact with the "cold" molecules, and "share" the energy they gained with the other molecules. In the meantime, some of the cold molecules will "diffuse" into the right half of the container, increasing their average energy as they collide or interact with the hot molecules. Similarly, some of the hot molecules will diffuse into the left half of the container, losing energy as they collide or interact with the cold molecules. In time, all the hot molecules may interact directly or through a chain of collisions with cold molecules, and similarly all the cold molecules will interact with hot molecules.

As this process continues, the two previously distinct energy distribution functions will look more and more like each other, as shown in Fig. 5.8. Eventually

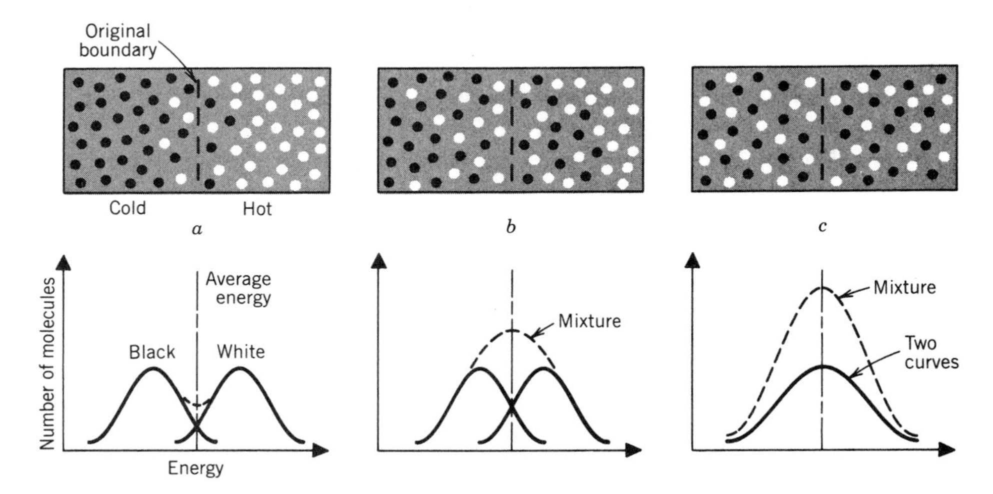

Figure 5.8. Mixing of hot and cold gases. Black molecules originally all on left side and at lower temperature than white molecules, which originally were all on the right side. (a) Original distributions shortly after removing adiabatic wall. (b) Mixing process about half completed. Energy distributions becoming more alike. (c) Mixing process complete. The energy distributions of the two sets of molecules are now identical and merged into one overall equilibrium distribution. The solid curves in the graphs represent the energy distributions for the black and white molecules; the dotted curves, which are the sum of the solid curves, represent the energy distribution for all the molecules.

they become essentially identical; that is, the two separate distribution functions will become one. The average kinetic energy of translational motion per molecule (corresponding to the absolute temperature) will be the same as the average (weighted according to relative numbers of originally hot and cold molecules) of the two original averages, but now there will be a random distribution of all the molecules about the new average. This new distribution function is the equilibrium distribution function; that is, if the system is not further disturbed, the distribution will not change.

Moreover, the equilibrium distribution is the *most probable* distribution function in the following sense: It is possible to imagine many ways in which a given total energy content of a system can be shared among the various molecules of the system. For example, after all the molecules are mixed, half the molecules could have exactly 10 percent more energy than the average, and the other half could have exactly 10 percent less energy than the average. Such a distribution function is shown in Fig. 5.9. The probability that such a distribution function would occur is very small. It is more likely that the distribution function will be random about the average value, because there are many more ways of being random (unpredictable) than of being either 10 percent above or 10 percent below the average.[5]

[5]There are cases in which the average value of a random distribution is not the same as the most probable value (see for example, Fig. 7.15b in Chapter 7), but these can be regarded as refinements of the foregoing discussion.

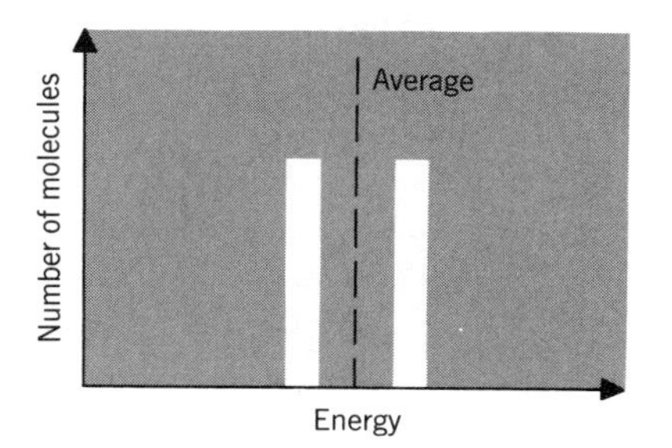

Figure 5.9. An improbable energy distribution.

This is exactly analogous to gambling with two dice. If the dice are thrown a great many times, then it turns out that the average value of the sum of the two dice is equal to 7, and the most probable value is also equal to 7. The next most probable values are 8 and 6, whereas the least probable values are 2 and 12. This occurs because, as shown in Table 5.2, there are six different ways to make 7, five different ways to make 8 or 6, but only one way to make 2 or 12. Successful gamblers using "honest" dice are well aware of this.

4. Entropy and Probability

Because the equilibrium distribution is one of maximum probability in the microscopic picture, and the second law of thermodynamics leads to the idea that entropy is maximum at equilibrium, it is reasonable to assume that there is a connection between the entropy of a system and the probability that its particular energy distribution will occur. (It is possible to show mathematically that the logarithm of the probability of occurrence of a particular energy distribution for

Table 5.2. Ways in Which Two Dice Can Be Thrown to Give a Particular Sum

Sum of Two Dice	Combinations Giving the Sum	Number of Combinations
2	1,1	1
3	1,2; 2,1	2
4	1,3; 2,2; 3,1	3
5	1,4; 2,3; 3,2; 4,1	4
6	1,5; 2,4; 3,3; 4,2; 5,1	5
7	1,6; 2,5; 3,4; 4,3; 5,2; 6,1	6
8	2,6; 3,5; 4,4; 5,3; 6,2	5
9	3,6; 4,5; 5,4; 6,3	4
10	4,6; 5,5; 6,4	3
11	5,6; 6,5	2
12	6,6	1

a given system behaves in the same way as the entropy of the system.) In fact, the microscopic picture coupled with the introduction of concepts of probability makes entropy a very powerful idea. We can explain, for example, why all the molecules in a container do not gather on one side of the container: It is simply more probable for molecules to spread throughout the container, because there are more ways in which they can be placed throughout the container than ways in which they can be placed alongside one wall. Therefore, the entropy is greater for the situation in which the molecules are spread out, and this is the equilibrium state. Similarly, if two types of gas molecules are present in a container, it is more likely they will be randomly mixed than segregated into layers (this assumes that the force of gravity is not so strong that differences in the weights of the gas molecules are significant).

M. Entropy and Order: Maxwell's Demon

1. Order and Probability

If several truckloads of bricks were dumped onto one location, they would probably fall into a jumbled pile. It is highly unlikely that they would fall into place to make an orderly structure such as a building. This is because there are more ways to make a jumbled pile of bricks than a building. In other words, a disorderly arrangement is more probable. (Similarly, it is more likely that a person's living quarters will be disorganized because there are many more ways to be "sloppy" than to be neat.)

Returning to the kinetic-molecular theory of matter, when a substance such as water is in a solid state (ice), its molecules are arranged in a very definite regular geometric pattern. As heat is added to the solid, the entropy increases. The molecules, although maintaining almost the same average pattern as before, are moving about much more at any instant of time. The entropy increase is therefore described as an increase of disorder on the molecular level. When the ice melts, the water molecules are moving about even more, and in fact the previous regular geometric pattern has been almost entirely broken, making the system much more disorganized and its entropy, therefore, much larger by an amount equal to the latent heat of melting (see Chapter 4) divided by the thermodynamic temperature. Still, on the average, the molecules are about as close together as before. As more heat is added and the temperature rises well above the melting temperature, the average distance between molecules increases, and the range of the momentum values of the molecules increases. When the water evaporates, the molecules become quite separated, and their range of momentum values increases even further. The molecules are now in a very disorganized arrangement, and as a result the entropy of the system is much larger.

The ideas of order and disorder (or organization or lack of it) include not only arrangements of molecules in space, but also the energy distribution functions as well. The mixing of hot and cold molecules discussed earlier in connection with the approach to equilibrium (Section K3) can also be described as a loss of organization in their energy distributions. Initially, half the molecules were clustered in a distribution with a low average temperature (less energy per molecule) and the other half in a distribution with a high average temperature. This represents a discernible grouping or ordering of the molecules—those with low energy and those with high energy. After they have mixed and reached equilibrium, there is only one distribution, and there is no longer organization or spatial segregation according to energy.

Finally, the principle that the entropy of an isolated system cannot decrease simply means that a system by itself will not separate its molecules into groups of different average energy (or spacing or both), because such a separation would represent the acquisition of some higher degree of orderliness or organization—which, statistically, is quite improbable. Furthermore, microscopically, entropy is a measure of the disorder of a system. When the entropy of a system has increased, it has become less disorderly.

2. Maxwell's Demon

Many schemes have been proposed to circumvent the principle of no entropy decrease for an isolated system. One of the most famous and fanciful of these schemes was considered by the great Scottish theoretical physicist, James Clerk Maxwell (1831–1879). He imagined a box containing a gas of molecules in thermal equilibrium. A wall was mounted across the middle of the box, dividing it into two parts. In the wall there was a trapdoor, which was initially open, so that molecules from both halves of the container could pass through. Maxwell then proposed that there might be a tiny elf sitting by the door who could close or open the door very quickly. This elf had the demonic desire to frustrate the second law of thermodynamics, and was therefore called "Maxwell's Demon."

The demon wanted to separate the molecules into two groups—"hot" or fast molecules on the right side of the container, and "cold" or slow molecules on the left side of the container. The elf would watch all the molecules, paying particularly close attention to those molecules that seemed to be approaching the door. If a fast molecule approached from the right, the demon could quickly "slam the door in its face" so that the molecule would bounce off the door and head back toward the right side of the box. On the other hand, if a slow molecule approached from the right, the demon would open the door and let the slow molecule pass through to the left side of the box. Fast molecules approaching the door from the left side of the box were allowed to pass through to the right side, whereas slow molecules approaching the door from the left bumped into the closed door and were kept on the left side. Thus after a while, the demon would separate the molecules into two groups, one of higher average translational kinetic energy per molecule and therefore of higher temperature on the right, and the other group having lower temperature on the left.

The second law of thermodynamics states that elves and demons exist only in fairy tales, and Maxwell's Demon was only a figment of his imagination, as he himself well knew. It says moreover that if, instead of a demon, any sort of mechanical or electronic device that could sense molecular speeds were installed to control the operation of the door, such a device will not work unless it receives energy from outside the system. (This conclusion is not obvious from the information presented above, but a detailed analysis shows it to be so.) If energy can be transferred to or from outside the system, however, then the system is no longer isolated and its entropy can be decreased.

Because the microscopic picture deals with probabilities, it cannot be said with absolute certainty that a group of molecules cannot spontaneously increase its order by the significant amounts proposed by Maxwell's Demon. But such an occurrence is so rare, unless the number of molecules in the group is very very small, that it would scarcely be credible if it happened; in any case a second (or third or fourth) occurrence of the event, which would be needed to verify for skeptics that it actually happened, would be so unlikely as to be beyond experience. One should not ''hold one's breath'' waiting for it to recur.

N. ''Heat Death'' of the Universe

1. ''Ultimate'' Energy Sources in the Universe

Julius Robert Mayer considered that the Earth's energy resources originate in the Sun. For example, the energy stored in coal was derived from the decay of vegetation that had grown because it previously received energy in the form of sunlight. Even the heat energy in the Earth's interior, which is thought to be due to confined radioactivity in the Earth's core, originally came from the material from which the Earth was formed. But this material, according to one current idea, had its origin in the supernova explosion of some ancient star, of which the entire solar system is a remnant. Generalizing from this postulated history of the Earth and solar system, it is now considered that the overall sources of energy in the universe are the stars, which are, of course, quite hot. Depending on their temperature, some of the nuclear energy of the stars is released in the form of heat.

2. The Universe as a Closed System

If it is assumed that the universe is a closed system, and there is nothing else, then it might be further assumed that there is a fixed amount of energy in this system. This energy is distributed throughout the universe in some manner. The various stars represent very high concentrations of energy, and thus there is order in the universe. It might be reasonably expected that ultimately any particular star will cool and lose its internal energy. It may go through a rather complicated cycle in doing so. The Sun, for example, might use up all its presently available energy in five billion years, according to some estimates. As it cools off, its internal pressure will decrease (according to its equation of state), and as a result of the cohesive force of gravitation, the Sun would then "collapse," thereby igniting a new series of nuclear reactions, converting some of its nuclear energy into heat and thus raising its temperature again. In time, however, the Sun will radiate away large amounts of energy while cooling to some stable low temperature. Presumably, similar processes should occur in all stars, so that ultimately all the stars in the universe will have cooled and redistributed their previously available energy.

This process might take a length of time ranging from a thousand million to a million million years (10^9 to 10^{12} years); but if the second law of thermodynamics holds in the closed universe, it must happen, and the universe will reach some equilibrium temperature, which is estimated to be well below 10 kelvins. All temperature differences will have disappeared; all heat engines will be inoperable; even life itself will be impossible because the energy needed to sustain life will be unavailable. The universe as a whole will have become essentially disorganized, because the equilibrium condition is one of maximum entropy and hence maximum disorder. Moreover, anything that increases the order in any one part of the universe leads to a greater decrease of order in the rest of the universe.

This ultimate end of the universe is called the "heat death of the universe," although the universe will be very cold indeed if this happens. The concept is implicit in the writings of Clausius and of the Austrian physicist Ludwig Boltzmann, but it was first explicitly discussed by Lord Kelvin in 1852.

3. Philosophical Implications and Criticisms of the "Heat Death"

If "heat death" represents the ultimate state of the universe, then such human ideals as the inevitability of progress and the perfectibility of the human race are delusions. If the ultimate end is chaos, what is the point of scientific endeavor, which seeks order? Thus the idea of "heat death" has been attacked on philosophical grounds. There are other grounds for skepticism, however. The assumption that the universe is a closed system has on occasion been questioned. Furthermore, the assumption that the laws of thermodynamics apply throughout the universe in the same way they apply in our small portion of the universe is an extrapolation from experience based on relatively small systems. Too often, extrapolations of known principles beyond the domain in which they were verified have been failures. We do know, of course, that some physical phenomena in the far reaches of the universe seem to be similar to the same phenomena in the vicinity of the Earth—the analysis of the light from stars both near and far is quite consistent. But not much is known about the universe. As yet, there is no direct evidence that the extensions of known theories are not valid; if there were such evidence efforts would be made to modify the theories appropriately. There is as yet simply not much evidence—observational astrophysics and cosmology are relatively young sciences.

Even if statistical arguments are attempted, however, the extension of various probability assumptions from finite to essentially infinite systems has never actually been justified.

Nevertheless, the ideas of entropy increase in isolated systems, of energy degradation, and of the heat death of the universe have had a significant impact on cultural and philosophic thought since the latter half of the nineteenth century. The increasing degradation of personalities of isolated individuals or groups or societies are major themes in modern thought and literature. The writings of such people as Charles Baudelaire, Sigmund Freud, Henry Adams, Herman Melville, Thomas Pynchon, and John Updike among others show evidence of this. Some economists have also included thermodynamic analogies in their studies, as have individuals concerned with ecology. (See for example the books by Zbigniew Lewicki, Rudolf Arnheim, and Nicholas Georgescu-Roegen as listed in the bibliography at the end of this chapter.)

APPENDIX: HEAT ENGINE CYCLES

A. DEFINITION OF HEAT ENGINE CYCLES

1. Basic Principles of Heat Engines

A heat engine is a device for converting heat energy into another form of energy. Any substance or system whose physical properties change with temperature can be used as a heat engine. If the temperature of the substance can be made to vary cyclically, then the associated physical properties will vary cyclically. By suitable mechanical, electrical, or other linkages to some one particular physical property, the cyclical changes in temperature of the system can be used to convert heat energy into kinetic or potential energy, for example.

2. Examples of Heat Engines

Among the more common examples of heat engines are steam engines in ships, trains, and electric power generating stations, gasoline engines, diesel engines, jet engines, and rocket engines. All of these have mechanical moving parts. On the other hand, fuel cells and thermoelectric cells, which have no moving parts, and generate electrical energy, are also heat engines.

3. Specification of a Heat Engine

The essential working parts of a heat engine are not always obvious from the external appearances of the overall assembly of the heat engine and its auxiliary apparatus. For example, a complete steam engine such as might be used in a railway locomotive is shown in Fig. 5.10. In terms of size the major components

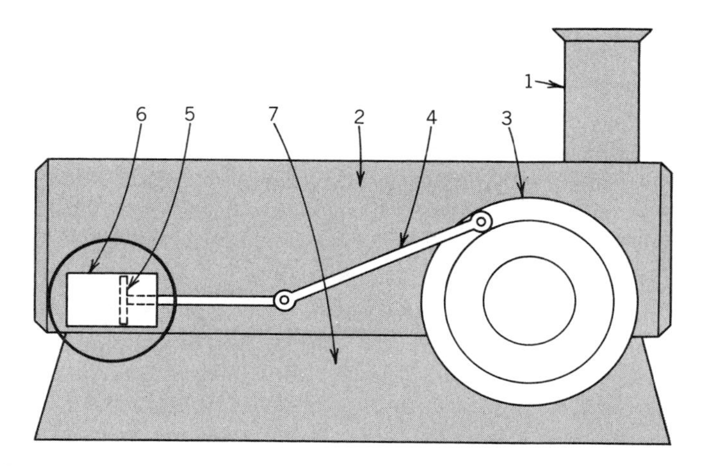

Figure 5.10. Major components of a steam engine. 1, smokestack; 2, boiler; 3, flywheel; 4, connecting rod; 5, movable piston; 6, steam cylinder; 7, mounting.

are the boiler, the smokestack, and the driving wheel or flywheel. The essential heat engine, however, is the steam cylinder, which is circled in the diagram. The properties that change with temperature are the pressure and volume of the steam in the cylinder. The linkages for converting these property changes into mechanical energy are the movable piston, connecting rod, and flywheel. Not shown in the figure are the condenser and valves for admitting and removing steam.

Figure 5.11 focuses on the basic system involved in the steam engine, the cylinder filled with steam (the working substance) having a variable volume as determined by the movable piston. In oversimplified terms, when the gas (e.g., the steam) in the cylinder is hot and at high pressure, the gas expands, increasing its volume as it pushes the piston outward. When the gas is cool and at lower pressure, it contracts, decreasing its volume as the piston moves inward. It is the expansion and contraction of the volume of the gas that results in the motion of the piston. In fact, the analysis of the heat engine behavior is determined from an analysis of the pressure and volume of the gas in the cylinder as the temperature is varied.

The analysis is done by means of a graph of pressure versus volume. Such a graph is called a *P-V* graph or alternatively an indicator diagram.

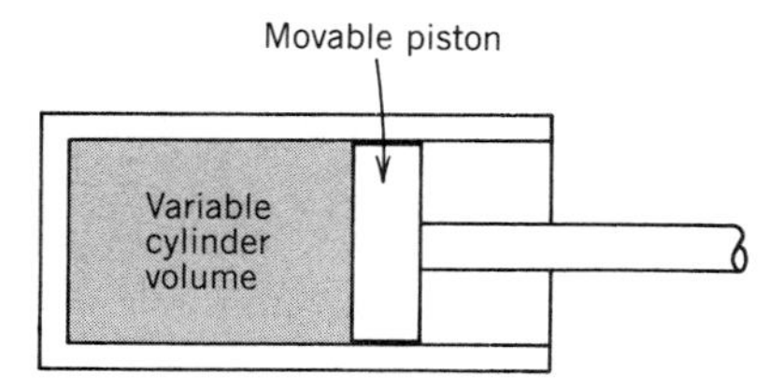

Figure 5.11. The steam cylinder.

The P-V graph does two things: (1) It makes it possible to specify the thermodynamic state of the working substance (i.e., the value of its parameters P and V—see Section A1) and through the equation of state of the substance, its temperature T; and (2) it makes it possible to calculate the work done by or on the substance as its state changes. In the case of the cylinder, the force exerted on the piston by the gas is proportional to the pressure of the gas. The distance that the piston moves under the influence of the force is proportional to the change in volume of the gas. Thus a P-V graph is equivalent to a graph of force versus distance; and, as discussed at the beginning of Chapter 4, the area under such a graph is proportional to the work done as the piston moves under the influence of the force.

One additional statement must be made in interpreting the area underneath a P-V diagram. If the pressure is such as to cause the piston to move outward, the work is done *by* the engine on the piston, and is counted as a positive energy output. If the piston is moving inward, so that the internal pressure is resisting the motion, then the work is done *on* the engine by the piston, and is counted as a negative energy output, or alternatively an energy input to the system. When a complete cycle is executed, the negative energy outputs must be subtracted from the positive energy outputs, in order to determine the net work done by the engine during the cycle.

B. Perfect Gas Heat Engine: The Helium Gas Engine

Analysis of the operation of the steam engine is not easy. However, according to Carnot (as discussed in Chapter 4 and in Section E2 of this chapter), the properties of the ideal or best possible heat engine are independent of the working substance of the heat engine; rather, they depend only on the temperatures at which heat is taken into and discharged from the heat engine. Therefore, the study of any one ideal heat engine will provide information about all other kinds of ideal heat engines.

A particularly easy engine to study is one that uses helium gas rather than steam as its working substance, because the equation of state of helium gas is fairly simple. Helium is one of the family of so-called perfect gases for which the equation of state $PV = RT$ is approximately correct. In this equation, P stands for pressure, V for volume, T for absolute temperature, and R is a constant number called the *general gas constant.*

It is also true that the internal energy of perfect gases depends only on temperature T (or equivalently on PV).

To simplify matters further, the helium gas engine is considered to be totally enclosed in a perfectly insulating jacket having a movable flap at one end as shown in Fig. 5.12. When it is desired to admit heat to the engine, this is done by opening the flap and placing a flame at the uncovered end. When heat is to be removed from the engine, the flap is also opened and a cooling substance (e.g.,

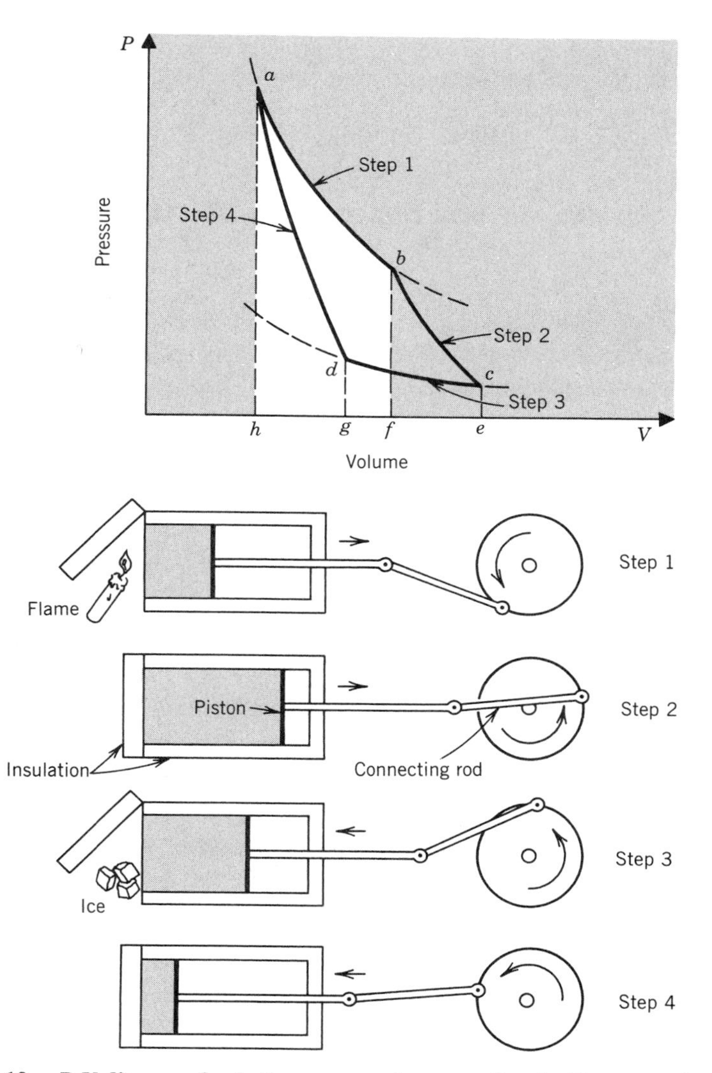

Figure 5.12. *P-V* diagram for helium gas engine operating in Carnot cycle. Step 1—Isothermal heat intake at temperature T_H. Step 2—Adiabatic temperature decrease to T_C. Step 3—Isothermal heat discharge at temperature T_C. Step 4—Adiabatic temperature increase to T_H.

some ice) is placed at the uncovered end. Thus the flame and the ice serve as the high-temperature and low-temperature reservoirs, respectively. For adiabatic operation, the flap is closed, thereby completely thermally insulating the engine.

The function of the flywheel is to serve as the output of converted energy from the engine, in this case mechanical rotational kinetic energy. Thus when work is done by the engine, as the gas expands the work appears as an increase in the rotational kinetic energy of the flywheel. Similarly, some of the rotational kinetic energy of the flywheel can be used to push the piston back and compress the gas, thereby doing work on the engine.

C. Carnot Cycle for Helium Gas Engine

As shown in Fig. 5.12, the Carnot cycle for a helium gas engine can be analyzed in detail. It is assumed that the working substance is initially in the state indicated by the point *a* on the *P-V* diagram. This means that the piston is at its maximum inward position, and engine pressure is maximum and volume minimum.

1. Isothermal Heat Intake

In this process, the insulating flap is opened, a flame is brought to the closed end of the cylinder, and an isothermal heat intake occurs. The pressure and volume of the gas change continuously along the line from *a* to *b*, while the temperature of the gas is maintained constant at the temperature T_H of the flame. The pressure drops while the volume increases. The exact shape of the line is determined by the equation of state with $T = T_H$, the temperature of the flame. As mentioned in Section B of this Appendix, the internal energy of helium gas depends only on temperature, and therefore is constant during this isothermal heat intake. This means that none of the heat taken in during this step can be accumulated within the engine; it must be discharged as work and thus is converted to another form of energy. This energy is in the form of additional kinetic energy of the flywheel.

The amount of this output energy or output work is calculated from the area underneath the curve, which is the area enclosed by the solid line from *a* to *b*, the dotted vertical line from *b* to *f*, the solid horizontal line along the *V* axis from *f* to *h*, and the dotted vertical line from *h* to *a*. This area is shown light in Fig. 5.12. All the heat energy taken in at this time, Q_H, is converted to another form. This is not in violation of the second law of thermodynamics because the cycle is not yet complete.

It should be noted also that if T_H were larger, the general shape of the line from *a* to *b* would be similar but displaced further upward on the graph.

2. Adiabatic Decrease of Temperature

The flame is now removed, the insulating flap is closed, and because the gas is still at high pressure, it will continue to expand while the system is completely thermally insulated. Work is still being done by the engine; that is, energy is being removed from the gas and transferred to the flywheel. During this process, the pressure decreases precipitously as the volume increases. Because energy is being removed from the gas, its internal energy *U* decreases, and therefore, for this perfect gas, the absolute temperature drops proportionately and continuously as the process continues along the line from *b* to *c*. At point *c* the temperature of the gas is now T_C, the temperature of the low-temperature reservoir.

The exact shape of the line can be calculated from the equation of state and the heat capacity (or specific heat) of the gas for this process. The work done by the engine (or energy removed) is calculated from the area enclosed by the solid line from *b* to *c*, the dotted vertical line from *c* to *e*, the solid horizontal line from *e* to *f*, and the dotted vertical line from *f* to *b*.

3. Isothermal Heat Discharge

When the point *c* is reached, the insulating flap is opened and the ice at the temperature T_C is brought into contact with the closed end of the cylinder. The ice ensures that the gas will remain at the temperature T_C. The piston is now driven inward by the flywheel, which continues to rotate. The kinetic energy of the flywheel is used to do work against the internal pressure of the gas; that is, work is done on the engine. But because the temperature and internal energy of the gas are maintained constant by the ice, this work cannot be accumulated internally within the gas and must be discharged as heat to the ice, causing some of the ice to melt.

Note that at this step some of the original engine input, which had been in the form of heat Q_H, and had actually been totally converted to another form of energy, is now returned to the engine, converted to microscopic motions of the molecules of gas, and then discharged as heat Q_C to the cold reservoir at temperature T_C. The amount of energy involved in this step is again the area underneath the *P-V* curve; that is, the area enclosed by the vertical dotted line from *c* to *e*, the horizontal line from *e* to *g*, the vertical dotted line from *g* to *d*, and the solid curve from *d* to *c*.

Here it should be noted that if T_C were lower, the shape of the line from *c* to *d* would be similar but further down on the graph.

4. Adiabatic Increase of Temperature

When the process reaches point *d* on the *P-V* graph, the ice is removed, the flap closed, and the system is once again thermally insulated. The flywheel continues to do work on the engine, resulting in an increase in its internal energy and hence of its temperature along the line from *d* to *a*, thereby completing the cycle. The energy added is determined from the area enclosed by the solid line from *d* to *a*, the vertical dotted line from *a* to *h*, the horizontal line from *h* to *g*, and the vertical dotted line from *g* to *d*.

5. Net Work Output: Efficiency of the Carnot Cycle

In Step 1 of the Carnot cycle, all of the heat input to the helium gas engine is converted to mechanical energy. In Step 2, additional mechanical energy is derived from the internal energy of the system. In Steps 3 and 4, which are necessary to complete the cycle, the second law of thermodynamics plays an active role. In Step 3, mechanical energy is returned to the engine and converted into heat for discharge to the low-temperature reservoir. In Step 4, the additional mechanical energy that has been derived from the internal energy of the system is returned, restoring the system to its original state at the beginning of the cycle.

The net work done, W, is calculated from the area enclosed within the solid lines traced out on the P-V graph as the engine went through a complete cycle, that is, the area enclosed in the solid lines from a to b to c to d and back to a again. This is simply an expression of the first law of thermodynamics: $W = Q_H - Q_C$.

The efficiency of the engine is simply the ratio of W to Q_H and is determined from the ratio of the area *abcd* enclosed by the cycle to the light area *abfh*. Knowing the equation of state of the helium gas and, therefore, the mathematical shapes of all the curves in Fig. 5.12, Clausius was able to calculate that the efficiency is

$$\frac{W}{Q_H} = \frac{(Q_H - Q_C)}{Q_H} = \frac{(T_H - T_C)}{T_H}$$

This is the efficiency given in Section E2.2 of this chapter.

By inspection of Fig. 5.12, it is easy to see that the closer the portion from c to d of the cycle is to the V axis, the greater the area enclosed within the cycle, and in particular the closer the enclosed area is in size to the light area underneath the line from a to b. This condition is obtained by making T_C smaller. Similarly, if the line from a to b were made higher by increasing the temperature T_H, the relative areas *abcd* and *abfh* become more nearly equal (even though their shapes are different), and the efficiency is higher.

D. Steam Engine Cycles

As already pointed out, the usefulness of the helium gas heat engine is in the relative ease with which it can be mathematically analyzed. In actual applications, the steam engine is far more useful. Nevertheless the information derived from the helium gas engine is very important and universally applicable.

1. Carnot Cycle for a Steam Engine

Figure 5.13 shows a P-V diagram for a steam engine operating in a Carnot cycle. The points marked *a, b, c, d* correspond to the same points in Fig. 5.12 for the helium gas engine. It is quite clear that the cycle for water-steam shown in Fig. 5.13 is much different in appearance than that for helium in Fig. 5.12. Actually, the difference between the two cycles is even greater than it appears from a comparison of the figures, because it is necessary to draw the volume axis for Fig. 5.13 on a logarithmic scale (each unit along the scale does not just represent

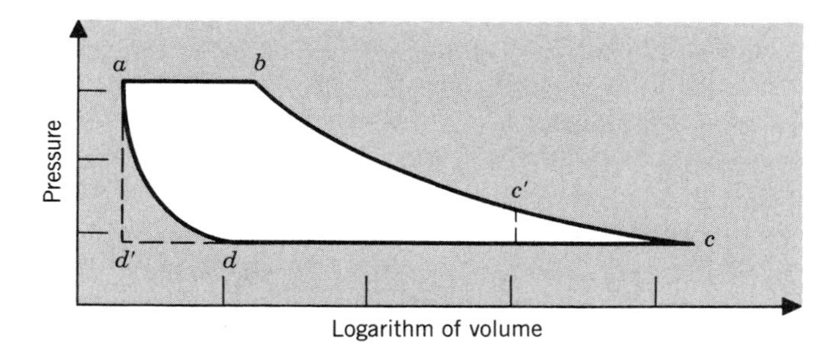

Figure 5.13. P-V diagram for steam engine. Carnot cycle along solid path, *abcda;* Rankine cycle along dotted path *abcd'a.*

equal additional units of volume, but rather multiplicative factors) in order to keep the volume range within a reasonable size. Moreover, the Carnot cycle for water-steam can be reasonably drawn over only a restricted temperature range. Carnot himself felt that it would be essentially futile to build a steam engine operating in a Carnot cycle, with particular difficulty associated with achieving Step 4, the adiabatic compression step from *d* to *a*.

2. Rankine Cycle

Instead of evaluating actual steam engines against the Carnot cycle as a comparison standard, another ideal cycle, which is less efficient than the Carnot cycle, is used. This is shown also in Fig. 5.13 by the dotted extension to the solid cycle. Step 3, the isothermal discharge of heat to the low-temperature reservoir at temperature T_C, is prolonged to the point d' to its starting state *a*. The efficiency of the engine is less than Carnot efficiency because of the extra heat discharged between the points d' and *d* on the graph, and because of the extra heat added between the points d' and *a*.

3. Actual Steam Engine Cycles

An actual steam engine cycle is modified even further from the Rankine cycle. The adiabatic Step 2 from *b* to *c* in Fig. 5.13 would require a great change in the volume of the cylinder, making it excessively long. (Because volume is plotted on a logarithmic scale, each unit along the scale represents a factor of 10 change in volume.) Therefore, in order to keep the cylinder to a convenient size, Step 2 is terminated at the point c', and the pressure abruptly dropped by discharging heat at a temperature higher than T_C, thereby decreasing the efficiency further but not very much.

The *P-V* diagram for a typical steam engine cycle is shown in Fig. 5.14. In addition to the previously mentioned changes from Fig. 5.13, the "corners" of the cycle are "rounded" because of various heat leaks inherent in design. Even this cycle is an "idealized" cycle in that it does not take into account such factors

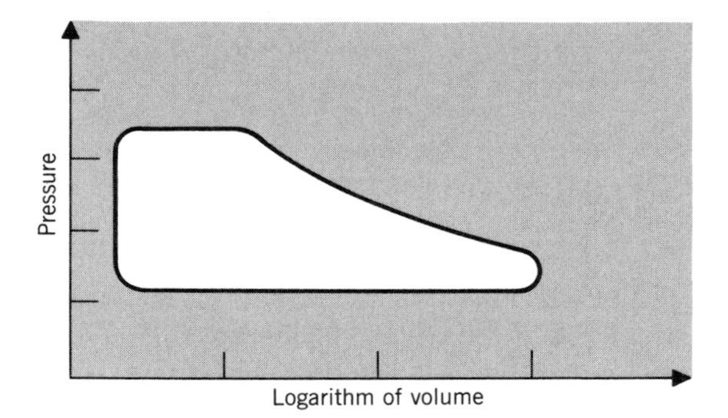

Figure 5.14. Idealized representation of actual steam-powered piston engine cycle.

as friction, turbulence in steam flow, substantial temperature differences between reservoirs and engine, and the normal rapid rate of engine operation.

E. OTHER ENGINE CYCLES

All heat engines operate according to a characteristic cycle. In the process of analyzing them, it is usually necessary to describe the engine in somewhat idealized terms.

For example, in the analysis of an internal combustion gasoline engine, the fact that the working substance (mixture of gasoline and air) is continuously being removed from the engine is ignored and the working substance is assumed to be conserved—only its temperature being changed. This is done to simplify the analysis. The cycle, of course, is neither Carnot nor Rankine. The following steps occur: (1) a heating step in which the pressure increases very rapidly while the volume remains constant; (2) an adiabatic expansion step in which work is done by the engine and the gas cools; (3) a further cooling step in which the pressure drops very quickly while the volume remains unchanged; and (4) an adiabatic compression step in which work is done on the engine, and the gas is returned to its original thermodynamic state. This particular cycle is called the *Otto cycle,* after one of the early German pioneers of gasoline engine development. Similarly, there is a *Diesel cycle.* There is also the *Stirling cycle,* which is used for external combustion engines with air as the working substance. The Stirling cycle is a very efficient cycle which, although proposed over a century ago, is currently the subject of much intensive research and development for possible automotive use. There are, of course, many other cycles that have been studied, but not all of them are based on gases in cylinders. Where other working substances are used, instead of graphing the cycle on a *P-V* diagram, other parameters are used to study the cycle. All such cycles can be compared on a similar basis by using temperature, *T*, and entropy, *S*, of the working substance as the parameters to be studied. Such a graph, called a *T-S* diagram, makes it possible to estimate engine efficiencies very simply.

Study Questions

1. What is meant by a parameter of a system?
2. One example of a system with its parameters would be an automobile tire inflated to a particular volume and pressure at a certain temperature. Can you give some other examples?
3. How is temperature distinguished from heat?
4. People sometimes talk about putting ''cold'' into a system. Can this be explained by the caloric theory or any other theory? Justify your answer.
5. What is meant by an equation of state? How is an equation of state used to determine the temperature of a system?
6. What is the first law of thermodynamics? How does it apply to systems that are not isolated?
7. Do all systems have the same equation of state?
8. What basic ideas are involved in the measurement of temperature?
9. In which direction is the natural flow of heat?
10. State several alternative forms of the second law of thermodynamics.
11. What is meant by perpetual motion of the second kind? Give some examples.
12. If you were told that some company had invented a new kind of electric power plant that was, in theory, three times more efficient than the best present-day power plants, would you invest money in that company?
13. How did Clausius modify Carnot's original analysis of the ideal heat engine?
14. Why is the Carnot engine important?
15. Define the terms *isothermal* and *adiabatic*.
16. How much heat can be transferred into a system during an adiabatic process?
17. Which is more efficient, a Carnot engine using steam or a Carnot engine using gasoline?
18. How much is the efficiency of an ideal heat engine?
19. How is temperature defined on the thermodynamic temperature scale?
20. It is sometimes said that absolute zero is the temperature at which everything is ''frozen solid,'' that is, all motion ceases. As will be pointed out in Chapter 7, this statement is incorrect. What might be a better definition of absolute zero?
21. State the third law of thermodynamics.

22. Why is heat called a degraded form of energy?
23. Heat is allowed to flow from a high-temperature reservoir to a low-temperature reservoir. How can the resulting entropy increase be minimized?
24. What is the connection between entropy change and irreversibility?
25. When hot food items are placed into a working refrigerator, their entropy decreases. Is this a violation of the second law of thermodynamics? Justify your answer.
26. Why is entropy referred to as "time's arrow"?
27. A bullet has an adiabatic collision with a large block of wood and is stopped by the wood. What happens to the entropy of the bullet and why? What happens to the temperature of the bullet and why?
28. What is meant by "the energy distribution for the molecules of a system"?
29. Why should the equilibrium distribution be the most probable distribution?
30. What is the relationship between entropy, probability, and disorder?
31. Who is Maxwell's Demon and what does he do?
32. What is meant by the "Heat Death" of the universe and how is this related to the second law of thermodynamics?

PROBLEMS

1. Temperatures in desert areas often reach 40° C. Calculate the temperature on the Fahrenheit scale.
2. Pressure cookers are usually used to cook food at temperatures higher than the normal boiling point of water without burning the food. They can also be used to verify the third special law derived from the general gas law, i.e., P/T = constant, assuming that V is constant, provided no water is present in the cooker (water does not obey the general gas law). A six-quart pressure cooker is sealed at normal atmospheric pressure (14.7 lb/sq in.) at a temperature of 20° C. Calculate the Celsius temperature to which the pressure cooker must be heated in order to double the internal pressure. (*Hint:* Remember that the general gas laws require the use of absolute temperatures.)
3. Verify the efficiencies for the four Carnot engines specified in Table 5.1.
4. The steam turbine power plant specified in Table 5.1 can generate electrical energy at a rate of 5 million kilowatts. (a) Calculate the rate of deposit of heat into the power plant's cooling system during the course of normal operation. [*Hint:* Efficiency can be expressed as useful energy output (work) per heat energy input or alternatively as useful power output per rate of heat input.] (b) Calculate the rate of entropy increase of the universe during the normal operation of this power plant.

REFERENCES

Henry Adams, *The Degradation of the Democratic Dogma,* Macmillan, New York, 1920.
This book typifies the impact of the ''Heat Death'' on social thinking.

Rudolf Arnheim, *Entropy and Art: An Essay on Disorder and Order.* University of California Press, Berkeley, 1971.
Contrasts what should be the artist's view of these subjects with the physicist's view.

George Gamow, *Mr. Tompkins in Paperback,* Cambridge University Press, New York, 1971, Chapter 9.
See footnote 11 in Chapter 6, Section D5.

Nicholas Georgescu-Roegen, *The Entropy Law and the Economic Process,* Harvard University Press, Cambridge, Mass., 1971.
Attempts a mathematical analysis of economic theory, employing general physics concepts from thermodynamics and quantum theory.

Charles C. Gillispie, *The Edge of Objectivity,* Princeton University Press, Princeton, N.J., 1960, pp. 400–405.
See comment in references for Chapter 1.

Gerald Holton and Stephen G. Brush, *Introduction to Concepts and Theories in Physical Science,* 2nd ed., Addison-Wesley, Reading, Mass., 1973, Chapter 18.
See comment in references for Chapter 1.

Zbigniew Lewicki, *The Bang and the Whimper: Apocalypse and Entropy in American Literature.* Westport, Conn.: Greenwood Press, 1984.
The latter half of this book discusses the entropy concept as employed by various American writers, including Herman Melville, Nathaniel West, Thomas Pynchon, William Gaddis, Susan Sontag, and John Updike.

Morton Mott-Smith, *The Concept of Energy Simply Explained,* Dover Publications, New York, 1964.
See comment in references for Chapter 4.

J. R. Powell, F. J. Salzano, Wen-Shi Yu, and J. S. Milau, ''A High Efficiency Power Cycle in Which Hydrogen is Compressed by Absorption in Metal Hydrides,'' *Science,* Vol. 193, 23 July 1976, pp. 314–316.
Describes a three-reservoir heat engine.

John F. Sandfort, *Heat Engines,* Anchor Books, Garden City, N.Y., 1962.
An introductory essentially nonmathematical discussion of heat engines, written for the lay public.

Mark W. Zemansky, *Temperatures Very Low and Very High,* D. Van Nostrand, Princeton, N.J., 1964.
An introductory discussion on the college level, which assumes some knowledge of physics.

Albert Einstein. (American Institute of Physics Niels Bohr Library.)

Chapter 6

ELECTRO-MAGNETISM AND RELATIVITY

The facts are relative, but the law is absolute

The theory of relativity is commonly associated with Albert Einstein, with the atomic bomb and nuclear energy, and with the notion that everything is relative. The idea of relativity did not originate with Einstein, however; nuclear energy is primarily a by-product, and in Einstein's original work he was as concerned with determining what is *absolute* as with what is relative. The true significance of Einstein's theory comes from its reexamination of certain metaphysical assumptions and its reassertion of other assumptions, and from its recognition of how physical facts and laws are ascertained. The theory of relativity is sometimes described as a theory about theories, in that other theories must be consistent with the theory of relativity. At the same time, in its applications it demonstrates how abstract and esoteric concepts may have very concrete consequences for everyday life. Among the specific consequences, the recognition of the equivalence of mass and energy and their mutual convertibility led directly to the idea of nuclear energy. Other consequences include insight into the structure of matter, solution of many astronomical problems, revision of cosmological concepts, and deeper insights into the nature of electromagnetism and gravitation. The roots of this theory go back to the Copernican revolution and its accompanying questions about absolute motion, absolute rest, and absolute acceleration. These questions were given additional pertinence as a result of nineteenth-century investigations of electric and magnetic phenomena and their connections to each other.

A. Galilean–Newtonian Relativity

1. Absolute Space and Time According to Newton

As discussed in Chapter 3, Newton laid down certain postulates in his *Principia,* which were stated without proof. The only requirement was that they be consistent with each other. These postulates were necessary in order to have a defined terminology that would serve as a basis for later developments. Among the things postulated were the nature of space and time. Because Newton planned to deduce from the motions of objects the forces acting between them, he had to have a basis for describing motion. Instinctively one thinks of motion through space and time, and thus it becomes necessary to be precise about what is meant by the concepts of space and time.

Newton's definition of space was as follows: "Absolute space, in its own nature, without relation to anything external, remains always similar and immovable . . . by another name, space is called extension." Newton also stated that the parts of space cannot be seen or distinguished from each other, so that instead of using absolute space and motion, it is necessary to use relative space and motion (this will be discussed further below).

He defined time as follows: "Absolute, true, and mathematical time, of itself, and from its own nature, flows equably without relation to anything external, and by another name is called duration." There is a difference between objective time and psychological time. We all know that time flies on some occasions and drags on other occasions, depending on circumstances. Focusing on objective time, Newton recognized that, as a practical matter, objective time is measured by motion. For example, during the day the passage of time can be measured by the apparent movement of the Sun in the sky. Even though Newton recognized it is relative space and time that are measured, he emphasized that in his system there were underlying concepts as given above. There are, of course, ambiguities in Newton's definitions. For example, to say that time flows equably or smoothly means that there has to be an independent way of determining what is *equable* or *smooth*. But these words themselves involve a knowledge of time.

2. Frames of Reference

Because Newton recognized that relative position, relative motion, and relative time are usually measured, it is necessary to discuss what is meant by relative measurements. If it is stated that a certain classroom in Kent, Ohio, is 40 miles due south of Cleveland, Ohio, then the position of the classroom is given relative to Cleveland. That same classroom can be described as being 10 miles east of Akron, Ohio. Similarly, an official proclamation might be dated 5754 years after the Creation, or 1994 years after the birth of Christ, or in the two-hundred eighteenth year of the independence of the United States, depending on the reference date from which time is measured.

The location of the classroom was not complete as given above; it is necessary to state on what floor of a particular building the classroom is to be found. In fact, more commonly, locations on Earth are given by latitude, longitude, and altitude above sea level. There are any number of different ways in which location can be specified, but in all cases, three *coordinates* are necessary to locate a particular point in space because space is three-dimensional. In effect, a *point in space* is a location on a three-dimensional map. Such a three-dimensional map is called a *frame of reference*. The actual values of the coordinates will depend on the choice of the origin (the point relative to which all others are measured) of the frame of reference; it can be Cleveland, Ohio, or Topeka, Kansas, or Greenwich, England, or Moscow, Russia, or Quito, Ecuador. For that matter, the origin does not have to be on Earth. It can be the center of the Sun or even in the middle of our galaxy. The actual numbers used for the coordinates depend on the origin of the frame of reference, and also whether distances are measured in miles or feet or kilometers or degrees. Once a frame of reference is established, motion is measured with respect to the frame of reference by specifying how the coordinates of a particular object change as time goes on. The velocity of an object is specified relative to a particular frame of reference, and the actual values of velocity may depend on the choice of the frame of reference, because the frame itself may be in motion. For example, a student sitting in a chair in a lecture room has zero velocity relative to any fixed point on the Earth, but relative to the Sun the student is moving with a velocity of 67,000 miles per hour, because the Earth (on which the student sits) is traveling at a speed of 67,000 miles per hour relative to the Sun. A rocket traveling at 1000 mph relative to the surface of the Earth might be traveling at 68,000 mph relative to the Sun, or 66,000 mph or some intermediate velocity, depending on its relative direction of motion. Because velocity is a vector quantity (see Chapter 3), the direction of the velocity relative to the Sun will also depend on the position of the Earth in its orbit, and the relative rotation of the Earth about its axis. Because the Sun itself is in motion relative to the center of the galaxy, the velocity of the rocket relative to the galaxy has yet another value. The galaxy itself is in motion relative to other galaxies.

Thus both coordinates and velocities are *relative* quantities. All quantities derived from coordinates and velocities (see Chapter 3, Section D2) must also be relative, for example, momentum, kinetic energy, and potential energy.

In view of Newton's definition of absolute space and time, it is reasonable to ask what the true velocity of the Earth is as it travels through space—not its velocity relative to the Sun or the galaxy, but relative to absolute space. How would that velocity be measured? It is necessary to have a reference point that is fixed in absolute space. One suggestion is to take the *midpoint* of the average position of the fixed stars. But what proof is there that the stars are fixed in space? These questions are clearly related to the problem that concerned Ptolemy, Copernicus, Galileo, Kepler, and their contemporaries: Is it the Earth that is moving, or the stars, or the Sun, or all of them? Galileo asserted that it is impossible to tell from an experiment done on the Earth whether the Earth is moving in a true absolute sense.

As he pointed out, if a ship is moving in a harbor, an object dropped from the mast of the ship falls straight down and hits the deck at the foot of the mast of the ship, as seen on the ship. But as seen from the shore, the object cannot fall straight down—it must undergo projectile motion. It maintains its forward motion because the ship is carrying it along, and simultaneously it falls toward the Earth (Chapter 3, Section B3). This results in a parabolic path, as seen from the shore. Nevertheless, the object will still hit the deck at the foot of the ship's mast, because the ship was moving along and keeping pace with the forward motion of the object. The sailors on the ship cannot tell whether the ship is moving by watching the falling object, because it hits in the same place on the ship regardless of the ship's motion.

It is only by looking at the shore that they can tell that there is relative motion between the ship and the shore. Of course, they say that it is the ship that is moving; it is hard to believe that the ship is standing still and the shore is moving. Similarly, Galileo (and Copernicus before him) found it hard to believe that the Earth was standing still and all the stars were moving. But he recognized that he had not proven which is actually moving, or how fast.

In fact, the law of inertia, that is, Newton's first law of motion, states that so far as forces are concerned, all reference frames which move at constant velocities with respect to each other are equivalent. It is impossible to determine which one of these reference frames is actually at rest in absolute space, or which one is going faster than the others. An object moving at constant velocity in one reference frame is also moving at a (different) constant velocity in another reference frame that moves at constant velocity with respect to the original reference frame. The object is also at rest in yet another reference frame.

For example, a man throws a ball at a speed of 60 miles per hour from the rear toward the front of the ship, which is itself traveling eastward at 30 mph. Relative to the ship, the ball is moving at a velocity of 60 mph eastward; relative to the ocean the ball is moving at a velocity of 90 mph eastward. In the absence of external forces (gravity and wind resistance), the ball will continue to move at those same velocities. Moreover, if a helicopter is also flying eastward at a speed of 90 mph relative to the ocean, the ball has a velocity of zero relative to the helicopter—it is at rest. We cannot say that the ball is actually moving at 90 mph because the Earth itself is in motion. All reference frames in which Newton's laws of motion, in particular his first law, are valid are called **inertial reference frames.** These reference frames may be in motion relative to each other, but their relative motion is at constant velocity.

It is worthwhile to contrast an inertial reference frame with a **noninertial reference frame.** An automobile traveling at a steady 50 mph in the same direction can be used as an inertial reference frame. If the driver of the automobile suddenly applies the brakes, then while the automobile is decelerating it becomes a noninertial or **accelerated reference frame.** Objects within the automobile suddenly begin accelerating (with respect to the automobile) even though no force is applied. A passenger in the front seat who is not wearing a seat belt is accelerated from rest and crashes through the windshield of the car. The law of inertia (Newton's first law) has been violated in this reference frame (because the passenger accelerated with no force acting upon him), and so it is called a noninertial reference frame. A rotating reference frame is also a noninertial reference frame. In making calculations, noninertial reference frames are often rejected or discarded, if at all possible, because Newton's laws of motion are not valid in such frames.[1]

Because all dynamical quantities are measured in terms of space and time, the measurements that are made may depend on the reference frame used in carrying out the measurements. The velocity of the ball discussed above depends on the reference frame from which the ball is examined. Similarly, the kinetic energy and the potential energy of the ball will depend on the reference frame. Of course, it should be possible to determine the physical properties of the ball in any reference frame. Given the velocity of the ball in one reference frame (e.g., with respect to the ship) it should be possible to calculate its velocity, or any other dynamical quantity, relative to the helicopter.

The equations that make it possible to carry out such calculations are called **transformation equations.** In effect, transformation equations are like a dictionary that permits "words" that are spoken in one reference frame to be "translated" into another reference frame. In using transformation equations, the coor-

[1]Strictly speaking, the Earth is a noninertial reference frame because it does not travel in a straight line and also because it spins on its axis. The effect of the resulting acceleration is very small compared with the effect of the force of gravity at the surface of the Earth, however; so for many purposes the error made by assuming the Earth to be an inertial reference frame is not too large. Nevertheless, it is the noninertial effects of the Earth's spin that cause cyclones and tornadoes.

dinates and velocities measured in one frame of reference are used to calculate the coordinates and velocities that would be measured in another reference frame. (Because frames of reference can be in motion relative to each other, the transformation equations also involve time. This will be discussed further in Section D below.)

3. Relative Quantities in Newtonian Physics

As already mentioned, many quantities or measurements will be different in different frames of reference: position, velocity, momentum, kinetic energy, potential energy. On the other hand, some quantities do not change, regardless of which inertial frame of reference is used. Such quantities are called **invariant** or **constant.** In Newtonian physics, examples of such invariant quantities are mass, acceleration, force, electric charge, time, the distance between two points (length), temperature, energy differences, and voltage differences. When these quantities are measured in one frame of reference, and their values are calculated in another frame of reference, using the appropriate transformation equations, the quantities have exactly the same values as in the first frame of reference. If the quantities in the second frame of reference are measured rather than calculated, they will again be exactly the same as measured in the first frame of reference.

But more than quantities are invariant. Newton's laws of motion are invariant from one inertial frame of reference to another one. This means that the mathematical form of Newton's laws of motion is the same in all inertial frames of reference. The law of conservation of energy is the same in all inertial frames of reference, even though the total amount of energy will be different in the different inertial frames. Similarly, the law of conservation of momentum is the same in all inertial frames of reference, even though the total momentum is different in different inertial frames.

Galilean–Newtonian relativity is the name given to the aspects of relative and invariant quantities and relationships and transformations between inertial reference frames, as discussed so far. These relationships and transformation equations depend in a very fundamental way on the concepts of time and space.

B. Electromagnetism and Relative Motion

Two of the great triumphs of nineteenth-century physics were the development of the concept of energy with all of its implications and the development of a unified theory of electricity and magnetism. The technological fruits of these two theories have shaped our material culture, and their scientific consequences have been equally pervasive.

As the theory of electromagnetism developed, it raised once again the possibility of determining the velocity of the Earth in absolute space. Whereas according to Newtonian mechanics there was no experimental way of determining whether the Earth was ''really'' moving in an absolute sense, at least from a mechanical experiment, electromagnetic theory suggested a number of experiments that could be observed to determine the ''true'' motion of the Earth.

Electromagnetic theory is also important for understanding the structure of matter, a subject of importance in the next two chapters, so it is appropriate to discuss certain aspects of electricity and magnetism.

1. Electric Charges and Electrical Fields

It is particularly easy to generate static electricity at times of low humidity, as occur in the winter in cold climates, by means of a phenomenon known as the *triboelectric effect.* If one removes an article of clothing such as a pullover sweater, or brushes long dry hair, or strokes a cat, or walks across an untreated wool carpet, one feels a tingle, hears crackling sounds, sees flashes of light, or observes *static cling* of items of clothing. All these effects, and many more, are due to the generation of static electricity by rubbing two or more objects together. This effect was known over 2600 years ago by Thales, the Ionian Greek philosopher who first recognized the value of studying mathematics (Chapter 2, Section A2).

The very word electricity is derived from the Greek for amber, a material that when rubbed exhibits triboelectricity. In the eighteenth and nineteenth centuries a number of investigations of electrical phenomena were carried out. Particularly notable was the recognition by Benjamin Franklin, among others, of the electrical nature of lightning. In his day all electrical effects were considered to be due to some sort of electric fluid, just as thermal effects were ascribed to a heat fluid. Franklin considered that triboelectricity was due to the transfer of this fluid from one object to another. Thus the object that gained fluid could be said to be positively (+) charged because it had more fluid than normal, and the object that had lost fluid became negatively (−) charged because it had lost fluid. This led to the principle of conservation of charge because the gains and losses naturally balanced. In time it was suggested that there had to be two types of electricity, which were called positive and negative. If an object had some positive fluid deposited on it, it was said to be positively charged, and similarly, if it had a small amount of negative fluid deposited on it, the object was said to be negatively charged. Thus Franklin's fluid excess became positive charge and his fluid deficiency became negative charge. It was also recognized that positive and negative charges exerted attractive forces on each other, but positive charges would repel other positive charges and negative charges would repel other negative charges. The principle of conservation of charge then became ''the total amount of electric charge in a closed system, taking into account the sign of the charge, must be constant.''

In 1785 Charles Augustin de Coulomb was able to show that the magnitude of the force between two electric charges was proportional to the product of the quantity of each charge and inversely proportional to the square of the distance between the charges and directed along the line joining the two charges. The mathematical formula for calculating the strength of the forces between two charges is called Coulomb's law and is similar to the universal law of gravitation

(Chapter 3, Section D3): The force is proportional to the product of the magnitude of the charges and inversely proportional to the square of the distance between them. The magnitude or amount of an electric charge is designated by the symbol q. The algebraic formula for the force between two charges Q and q is $F = kQq/r^2$, where k is a proportionality constant and r is the distance between the charges. Experimentally, it is found that charges come in multiples of one definite size, called the *charge of an electron,* which is designated by the letter e.[2]

Coulomb's law is mathematically identical to Newton's Universal Law of Gravitation with the notable differences that the force is not due to mass, but to some other property of matter, the electrical property, and that the force may be attractive or repulsive depending on whether the charges have opposite signs or the same signs.

Another apparent similarity between electric charge and mass is that both properties obey a conservation principle, in the same way that energy obeys a conservation principle. We will see later in this chapter that mass can be regarded as a form of energy, so that the principle of conservation of mass is not immutable, but the principle of conservation of electric charge is indeed always applicable.

Once the mathematical similarity between Coulomb's law and the Universal Law of Gravitation was recognized, some of the same concepts and questions applicable to gravity were equally applicable to electricity. Although the mathematical formula makes it possible to calculate the force between two charges, it does not explain how that force is transmitted from one charge to another. This is the same question that was noted in Chapter 3 in connection with Newton's law of universal gravitation—how do objects manage to "reach" across empty space and exert forces on each other? This question will be addressed further later in this chapter and in somewhat more detail in Chapter 8. One way of dealing with questions such as this, about which little is known, is to give the answer a name. In this case, the means by which one electric charge exerts a force on another electric charge is called the **electric field.**

Indeed the concept of the electric field was adapted from the concept of gravitational field. Thus every electric charge, q, by virtue of its very existence, has associated with it an electric field $\vec{E}$ which describes how it exerts a force on other charges. This is the only way electric charge is "seen" or manifests itself, as contrasted with mass, which can be "seen" or "felt" or "weighed." In a more fundamental sense, this is also the only way in which mass is detected. For example, as already noted in Chapter 3, the planets Neptune and Pluto were discovered only because of the forces they exerted on Uranus.

Electric fields (like gravitational fields) may extend throughout space. Electric fields differ from gravitational fields since electric fields from negative charges have effects opposite in direction from similarly located positive charges, and as a result it is possible to nullify or cancel out electric fields, whereas gravitational fields cannot be nullified.

[2]The most recent theory of the structure of subnuclear particles proposes that charges of two-thirds and one-third of e also exist, but such charges have not been detected in a free state (see Chapter 8).

The exact definition of electric field is as follows: If an electric field is present, then it will exert a force on a "test" charge q. (This charge is other than the charge or charges responsible for the electric field.) If the force is divided by the amount of the test charge, taking into account the sign of the charge, then the resulting vector quantity is by definition the electric field $\vec{E}$. (Strictly speaking, the size of the "test" charge should be as small as possible so that it will not disturb the charges that are responsible for the electric field.)

In sum, electric fields are said to be inherent in the very nature of electric charges. An electric charge q is said to have an electric field, designated by the symbol $\vec{E}$, associated with it. $\vec{E}$ is a vector quantity (Chapter 3, Section C5) and has the property that it extends over a fairly large region of space, although its magnitude becomes smaller at greater and greater distances from the electric charge q. Fig. 6.1a gives a schematic representation of the electric field from a concentrated electric charge (called a point charge). The arrows indicate the direction of the field, and the magnitude of the field depends on how close together

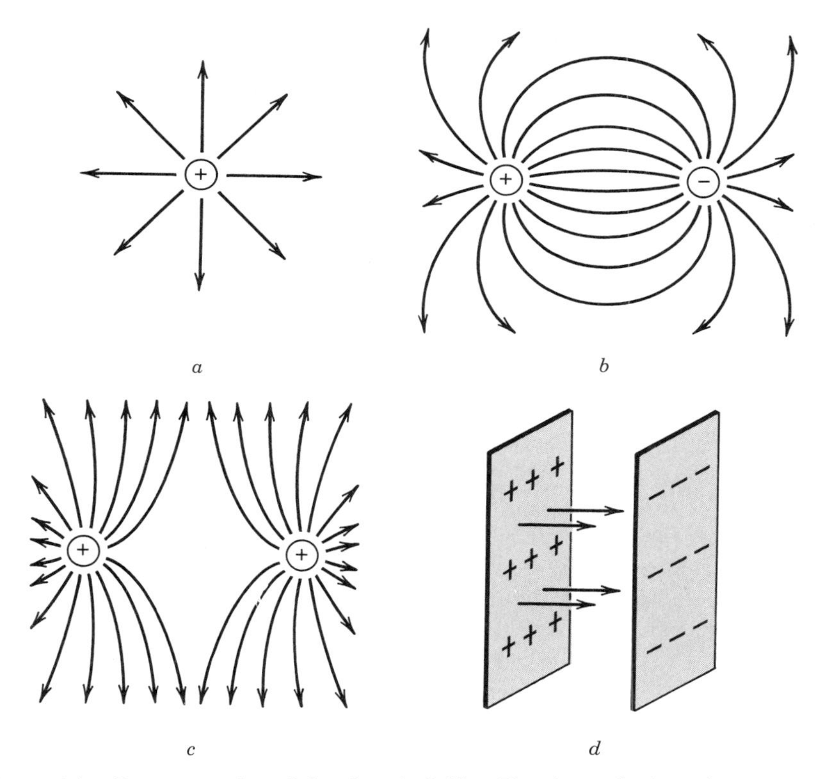

Figure 6.1. Representation of the electric field. The plus and minus signs represent positive and negative charges, the arrowed lines represent the electric field. (a) Isolated positive charge. (b) Equal + and − charges. (c) Equal + charges. (d) Sheets of + and − charges.

the arrows are. Because the arrows are directed radially, as shown in the figure, they are closer together near the charge and further apart far from the charge, showing that the magnitude of the field is greater close to the charge and lesser far from the charge.

If there are several electric charges present in a given region of space, the combined effect of their electric fields is calculated by adding (vectorially) their individual electric fields. Fig. 6.1b shows the resulting electric field from a pair of positive and negative charges of equal magnitude, and Fig. 6.1c shows the electric field resulting from a pair of positive charges. Fig. 6.1d shows the resulting electric field when sheets of positive and negative charges are placed opposite each other. If still another electric charge is brought into the region of space where an electric field has been established, then that other charge will be subjected to a force proportional to the combined electric fields of the charges already there, and in the direction of the combined fields (for a positive charge, and in the opposite direction for a negative charge). In one sense, the electric field can be considered a useful way of calculating the forces exerted by a collection of electric charges on another charge. As will be indicated below, however, an electric field has a more profound significance.

Magnetic effects have also been known from ancient times. In many ways magnetic effects are similar to electric effects. Magnetic charges are given a different name than electric charges—they are called **magnetic poles.** There are **north poles** (called north-seeking poles) and **south poles** (south-seeking poles). There are also **magnetic fields.** Fig. 6.2c shows a schematic diagram of the magnetic field associated with a bar magnet with field lines between the north (N) poles and the south (S) poles. Note the similarity to Fig. 6.1b, for field lines between + and − charges. However, it is impossible (at least to date) to separate north poles and south poles as distinct entities in the same way that it is possible to separate positive and negative charges from each other. Moreover, all presently known magnetic effects can be attributed to the motion of electric charges.

2. Electric Charges in Motion

When one or more electric charges are in motion, the moving charges constitute an **electric current.** (Electric currents are measured in units called *amperes.*) Associated with an electric current is a magnetic field designated by the symbol $\vec{B}$. The relationship between $\vec{B}$ and its associated electric current is quite different from that between $\vec{E}$ and q, as shown in Fig. 6.2. If, for example, the electric current is along a wire (because of an electric field directed along the wire), the associated magnetic field is directed in a ringlike pattern around the wire (Fig. 6.2a). The magnitude of the magnetic field decreases with increasing distance from the wire. On the other hand, if the wire is given a tight helical shape (Fig. 6.2b), then the magnetic field inside the helix is directed along the axis of the helix, flaring out at the ends of the helix. The pattern of the magnetic field exterior to the helix is identical to that of a bar magnet (Fig. 6.2c), including the location of north and south poles.

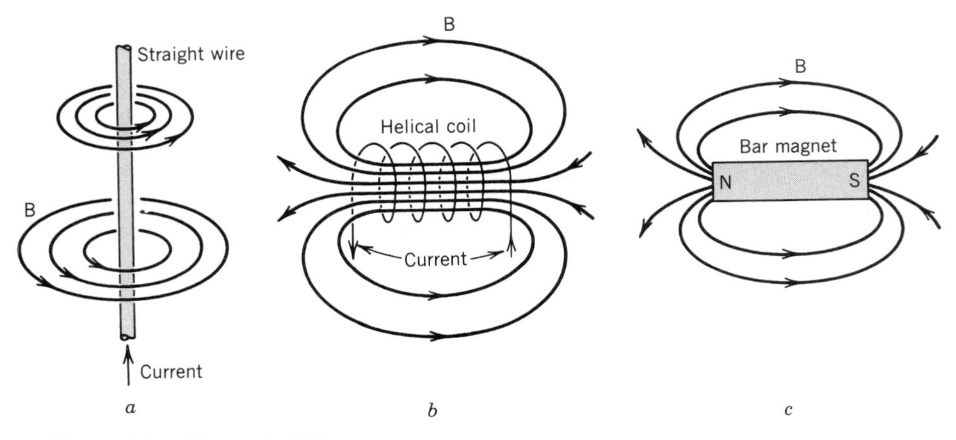

Figure 6.2. Magnetic fields. (a) From upward current in long straight wire. (b) From current in a helical coil or wire. (c) From a bar magnet.

OPTIONAL SECTION 6.1 *Conversion of Electrical Energy to Heat: Current, Voltage, and Power*

Electric currents are used in a very large fraction of the applications of electromagnetism to modern society. Prominent among these are electric heaters. In fact, James Joule's interest in the mechanical equivalent of heat (see Section B1 of Chapter 4) was initially aroused by his discovery of the fact that electric currents result in the generation of heat. The definition of many of the terms used to describe the properties of electric currents and electric circuits are based on energy concepts and mechanical concepts. The force acting on an electric charge q is calculated by multiplying the electric field $\vec{E}$ by q, that is, $\vec{F} = q\vec{E}$. For example, an object carrying a charge q equal to $+2$ coulombs when placed in an electric field of 4 newtons/coulomb will experience a force of $4 \times 2 = 8$ newtons in the direction of the field. If the charge were -2 coulombs, the direction of the force would be opposite to the direction of the field because of the minus sign.

Having given the name electric field to force per unit charge, a special name can also be given to potential energy per unit charge. This is called the *potential* or *voltage,* V. Thus the potential energy change Δ(P.E.) experienced by a charge q when it moves from one point to another is calculated by multiplying the voltage difference ΔV between the two points by q, that is, $\Delta(\text{P.E.}) = q\,\Delta V$. Like potential energy, voltage is a scalar quantity. The voltage difference ΔV is given in units of joules/coulomb, which are usually called volts. For example, a charge of 2 coulombs that moves between two points for which $\Delta V = 5$ volts will have a change of potential energy of 10 joules.

In the discussion of work and energy in Chapter 4, the primary emphasis was placed on potential energy P.E. and kinetic energy K.E., and the fact that in a conservative mechanical system, the sum of K.E. and P.E., the total mechanical energy, is always constant. Thus a decrease in the potential energy results in an equal increase in kinetic energy, assuming that the energy is not converted to some other form.

If there is some conversion of the energy to a form of nonmechanical energy, then the system is not a conservative mechanical system. In that case, when the potential energy decreases, the kinetic energy does not increase by the same amount, but rather by a lesser amount, and the total mechanical energy is not constant. The missing energy leaves the system. If the energy leaving the system is in the form of heat, then it is said that Joule heating has taken place.

The conversion of potential energy to heat instead of to kinetic energy is a prominent feature of electric currents flowing in wires. In fact once an electric current is established in a wire, the electric potential energy change of a charge moving between two points along the wire is completely converted into another form of energy, often heat energy, and none of it is converted into increased kinetic energy of the already moving electric charges. This heat is called Joule heat because Joule was the first to study the effect systematically.

The amount of electric energy converted to another form when an electric current flows in a wire is thus calculated as $q\ \Delta V$, where q is the total amount of charge that has moved through the potential difference ΔV. The electric charge, q, in coulombs equals $I \times t$, where I is the current and t is the time interval during which the current flows. I is expressed in amperes and t in seconds.

$$\text{energy} = (I \times t) \times \Delta V$$

To illustrate these relations we examine the operation of an electric heater. Such a heater might operate with current $I = 10$ amperes, $\Delta V = 110$ volts for a time $t = 1$ hour $= 3600$ seconds. Then the energy converted to heat $= 10$ amperes $\times$ 3600 seconds $\times$ 110 volts $= 3{,}960{,}000 = 3.96$ million joules.

It is customary to express the conversion of electrical energy in terms of a rate, that is, in terms of energy/time. Dividing $It\ \Delta V$ by t, gives power, P.

$$P = I\ \Delta V$$

Using the same numbers as before, that is, $I = 10$ amperes and $\Delta V = 110$ volts, we have $P = I\ \Delta V = 10 \times 110 = 1100$ watts $= 1.1$ kilowatts, where a kilowatt is 1000 watts. Then the total energy converted during a period of time t is $P \times t = 1.1$ kilowatts $\times$ 1 hour $= 1.1$ kWh, where kWh stands for kilowatt hours, an amount of energy equal to 3.6 million joules. Normally electric energy usage is billed per kilowatt hour (kWh). A typical rate is 12 cents per kWh. Thus the cost of operating a 1100 watt heater for one hour is 1.1 kWh $\times$ 12 cents $= 13.2$ cents.

One of the results of a systematic study of the generation of heat by an electric current is that the rate of conversion of electrical energy to heat is in many cases proportional to the square of the current in the circuit, that is, $P = RI^2$, where the proportionality constant R is called the resistance of the circuit. Since the relation $P = I\Delta V$ is also true, then $RI^2 = I\Delta V$ and therefore $RI = \Delta V$.

This last relationship is called Ohm's Law. (This law assumes that R is a constant for all values of the current. There are circuits for which R is not constant.) For our 1100-watt heater, we see that $P = 1100$ watts $= R \times (10$ amperes$)^2$. Thus $R = 1100/100 = 11$ ohms. Alternatively, $\Delta V = 110$ volts $= R \times 10$ amperes gives the same result, $R = 11$ ohms.

The mathematical relationships given above between power, P, current, I, voltage difference, ΔV, and electrical resistance, R, are extremely useful for analyzing electric circuits and appliances, not only in a technical sense, but also in an economic sense, because so much of modern life involves the use of electricity, and the conversion of electrical energy to or from other forms of energy.

Chapter 5 contains a broad outline discussion of how heat engines convert heat energy into other forms of energy, in particular into mechanical energy. This was done in terms of the expansion and compression of a gas, causing a piston to move back and forth in a reciprocating systematic manner. In a later section we will see how systematic mechanical motion of wires, perhaps as driven by a suitably linked reciprocating piston, results in the conversion of mechanical energy to electrical energy. In this section with the discussion of Joule heating due to electric currents, we have given some broad relationships involved in the conversion of electrical energy into heat energy.

Optional Section 6.2 *Free Electron Model of Electrical Conduction*

We wish now to discuss in somewhat more detail the specific heat conversion mechanism involved when an electric current is present in a wire. We want to do this in terms of a microscopic model based on the kinetic-molecular theory of matter. This model is called the "free-electron" model of electrical conductivity. The free-electron model like most modern models of matter assumes that bulk matter consists of a large number of atoms that on the whole apparently have no net electric charge, but that in fact have equal amounts of positive and negative charge. The electric charge is carried by tiny bits of matter. The bits of matter that are negatively charged are called electrons and are much smaller (less massive) than the carriers of positive charge. For good electrical conductors, the negative charge carriers (the electrons) are relatively free to move when subjected to an electric field. The positive charge carriers, on the other hand, are bound to the wire and are not free to move. In the free-electron model it is further assumed that the net effects of the electric repulsion between the free electrons are small enough that they can be totally ignored and that the effects of the electric attraction force between the mobile electrons and the stationary positive charge carriers can also be ignored most of the time. The "free electrons" are therefore assumed to behave like the molecules of gas discussed in Chapter 4, Section B-2, that is, they have kinetic energy proportional to the average temperature on the thermodynamic scale. This electron gas is in temperature equilibrium with the positive charge carriers because of occasional collisions with the stationary positive charge carriers.

When there is no applied electric field (or potential difference ΔV along a segment of wire), the motion of the free electrons is entirely random, that is, although they are in motion they have no net overall change of position, and hence there is no electric current. If, however, there is an applied electric field, there will be a resulting acceleration of the electrons along the wire. This acceleration results in a slight change of the velocity of the electrons in the direction of the acceleration. Thus the previously randomly moving electrons now have an overall "drift" along the wire corresponding to this slight change of velocity. The drift velocity depends on the magnitude of the electric field (which depends on ΔV and the length of the wire segment) and on the time between collisions of the electrons with the positive charge carriers.

Because of the "drift," just prior to a collision the electrons have average K.E. slightly greater than the original thermal K.E. However, just after the collision, this extra energy is shared with the positive charge carriers, as expected from the principle of entropy increase (see Chapter 5). Therefore the overall temperature of the system increases. This is the microscopic explanation of the I^2R or Joule heating of the wire.

A device such as an electric heater can be kept operating continuously only if there is a complete "electric circuit" available so that the electrons that have drifted along the wire segment and have "fallen" to a lower P.E. can be replaced by other electrons that have been raised to the higher P.E. at the beginning of the wire segment. Since all the electrons are moving along the complete circuit, one might visualize that each electron goes repeatedly around the circuit, having its P.E. increased each time it goes around so that it can once again drift through the heater. The necessary increase of P.E. for the electrons comes from an energy conversion device that converts some other form of energy into the electrical energy of the circuit. This increase of P.E. divided by the charge of the electron is called the EMF of the circuit and is expressed in volts. (The letters EMF are an abbreviation for electromotive force, which is a misnomer, but the letters are used and the word is discarded.) For the example of the electric heater, the EMF is exactly numerically equal to the potential difference of the heater.

In general, magnetic effects in matter are due to the presence of electrical currents associated with the dynamics of atomic and molecular structure. In other words, magnetic effects are due to the motion of electric charges, and thus the very existence of magnetic effects is an aspect of relativity. Magnetic effects require only relative motion of electric charges.

The discovery that magnetic fields are associated with electric currents was made by a Danish scientist, Hans Christian Oersted, who noticed when he moved a wire in which a current was flowing that the needle of a nearby magnetic compass changed its orientation. This meant that magnetic fields could be generated by electric currents. Depending on the configuration of the wire (how it is shaped), the magnetic field pattern can assume various forms as shown in Fig. 6.2. Note in particular that it is possible to simulate the field of a bar magnet by using a helical coil of wire to carry the current. In fact it is possible to use a small compass to map out the magnetic field associated with an electric current. Figure 6.3b shows a "photograph" of the magnetic field associated with a long straight

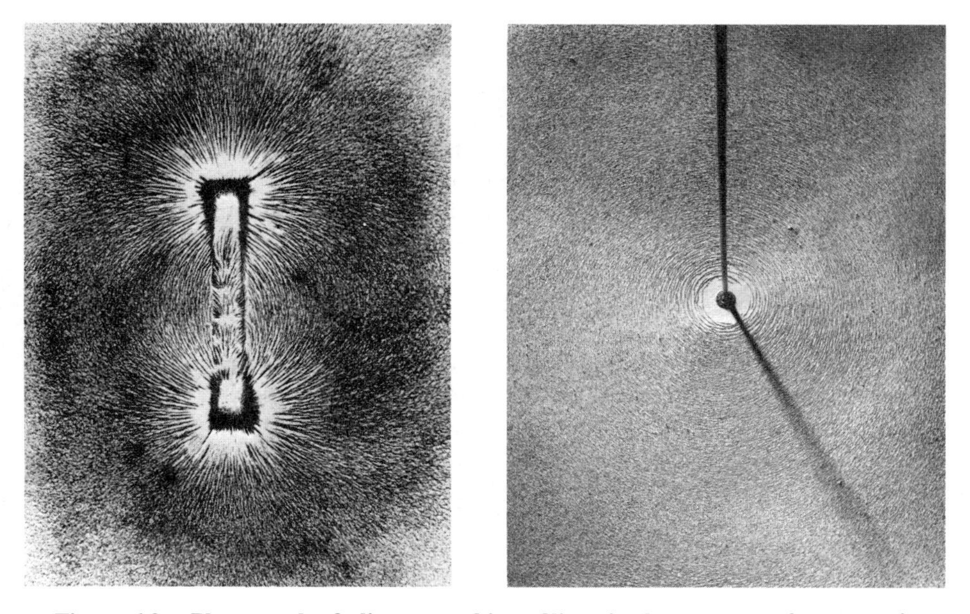

Figure 6.3. Photograph of alignment of iron filings in the presence of a magnetic field. (a) In the field of a bar magnet. (Kathy Bendo). (b) In the field of a current-carrying wire. (Educational Services, Inc.)

current-carrying wire. In the figure tiny bits of iron act as compass needles to show the orientation of the magnetic field. (Figure 6.3a shows a similar ''photograph'' of the magnetic field of a bar magnet.) Soon after Oersted's discovery it was demonstrated that current-carrying wires exert forces on each other. From these facts it was inferred that moving electric charges have a magnetic field associated with them, and that a magnetic field that is already in existence (due to the presence of one or more moving charges) will exert a force on any other moving electric charge. This force exerted by a magnetic field on a moving electric charge is called the Lorentz force.

Magnetic fields are similar in many ways to electric fields. Magnetic fields due to different currents or moving charges have combined effects that are calculated by adding (vectorially) the magnetic fields of the individual currents or moving charges. Also, a magnetic field will exert a force on any other current or moving charge that may be present.

Magnetic fields raise the same problem as electric fields. The magnetic field is the mechanism by which a moving electric charge (or an electric current) exerts a force on another moving electric charge. Formulas were developed for calculating the force, but without a clear explanation about how one moving charge ''reaches'' across empty space to exert a force on another moving charge. This question will also be addressed in Chapter 8. As in the case of the electric field, however, the magnetic field is more than just a useful way of calculating the force. In fact, magnetic fields, like electric fields, can even be made ''visible,'' as is shown in Fig. 6.3, which illustrates the use of iron filings to reveal the magnetic fields of wires and bar magnets.

Calculating the force is somewhat more complicated than calculating the force involved with an electric field. Not only does the force depend on the magnitude of the magnetic field and of the moving charge, but also on the velocity of the motion, including both speed and direction. The force is also perpendicular both to the original magnetic field and to the original direction of motion of the charge.

Because the force exerted by the magnetic field depends on velocity, relativistic effects are also present. This allows some possibilities for detecting absolute motion, and also for leading physics into a paradox: If the velocity depends on the frame of reference, and force depends on velocity, then force should depend on the frame of reference. But according to Galilean–Newtonian relativity (Section A3 above), force should be an invariant and should not depend on the frame of reference!

Optional Section 6.3 *Interconversion of Electrical and Mechanical Energy*

It was these electromagnetic effects that led to the development of new types of energy conversion devices in the later stages of the Industrial Revolution. Electric motors, for example, which are based on the forces between current-carrying wires, convert electrical (really electromagnetic) energy into mechanical energy. The reverse process, the conversion of mechanical energy into electrical energy, was accomplished by Faraday's invention of the dynamo, in which a wire is moved relative to a magnetic field. As a result the electric charges present in the wire have a force exerted on them. If these charges are free to move, then there will be within the wire an electric current; but in any case an electric field is established within the wire. If there is a current in the wires of the dynamo, then there will be forces between the wires which will oppose the motion that established the electric field within the wire.

The key to understanding the conversion of electrical energy into mechanical energy and vice versa is the recognition that a dynamo is an electric motor running in the reverse direction, just as a heat pump (or refrigerator) is a heat engine running in the reverse direction. In a heat pump a source of mechanical energy exerts a "force" which causes heat energy to flow in the "wrong direction." In the dynamo the source of mechanical energy exerts a "force" which causes the dynamo to try to turn in the "wrong direction." While the motor runs it is simultaneously a dynamo whose dynamic electric field opposes the electric field that caused the electric current through the motor. The dynamic electric field, or force per unit charge, results in a voltage or potential energy per unit charge, sometimes called a "back EMF" to distinguish it from the "applied EMF" needed to keep the motor running. (The back EMF is necessarily smaller in magnitude than the applied EMF; otherwise there could be no current.) The back EMF is expressed as a voltage difference ΔV which when multiplied by the current I in the motor gives the rate of conversion of electrical energy (power P) to mechanical energy:

$$P = \Delta V \times I$$

To illustrate voltage and current relationships in an electric motor, we consider the case of an electric motor that is operated with an applied EMF of 110 volts, a current $I = 10$ amperes, and a back EMF $\Delta V = 100$ volts for 1 hour. (These numbers are slightly different from numbers used in Optional Section 6.1 to illustrate the operation of an electric heater.) The total energy converted in one hour is power $\times$ time $= P \times t = I\Delta V \times t = 10$ amperes $\times$ 100 volts $\times$ 3600 seconds $=$ 3.6 million joules $=$ 1 kilowatt hour $=$ 1 kWh.

In the previous example given in Optional Section 6.1, 3.96 million joules were converted from electrical energy into heat energy; in this example 3.6 million joules are converted from electrical energy into mechanical energy (work done by the motor). The situation is a little more complex than in the case of the electric heater, because not all the energy converted by the device supplying the applied EMF of 110 volts has been accounted for. The electric motor has a "back EMF" ΔV of only 100 volts; there is another 10 volts (the difference between 110 and 100) that leads to additional energy conversion $=$ power $\times$ time $= P \times t = I\,\Delta V \times t = 10$ amps $\times$ 10 volts $\times$ 3600 seconds $=$ .36 million joules $=$ .1 kWh. This energy is heat energy dissipated by the electrical resistance of the motor, which in this case is equal to 1 ohm. To summarize, in this case, the 10 amperes of the motor current results in 3.96 million joules converted into electrical energy by the applied EMF that "drives" the circuit, 3.6 million joules of which is then converted into mechanical energy by the motor, with the remaining .36 million joules converted to heat. This example is a beautiful illustration of the workings of the first and second laws of thermodynamics.

3. Electrical and Magnetic Interactions

Figure 6.4 shows a symbolic scheme for displaying the relationship between electric charges and electric fields, and between magnetic charges and magnetic fields. At the apex of the triangular cluster of symbols is q, at the lower right

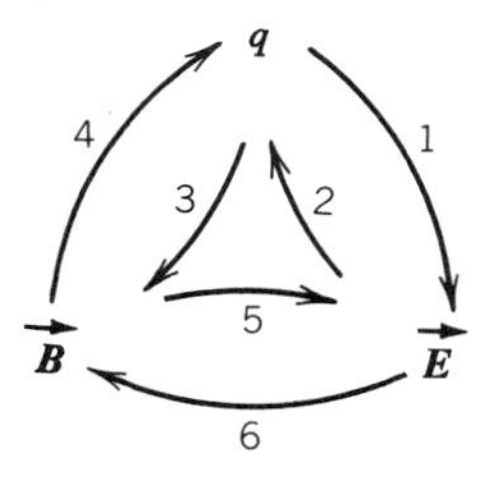

Figure 6.4. Triangle diagram showing relationships between q, $\vec{E}$ and $\vec{B}$. The arrows shown are not vectors, but represent relationships discussed in the text. (By permission from Kenneth W. Ford, *Basic Physics,* John Wiley & Sons, New York, 1968.)

corner is $\vec{E}$ and at the lower left corner is $\vec{B}$. The arrow labeled 1 indicates that associated with (or created by) an electric charge is an electric field. The arrow labeled 2 indicates that an electric field will exert a force on any other electric charge that may be present. In fact, if an electric charge has a force exerted on it, then it must be true that an electric field is present. Arrow 3 indicates that associated with a moving electric charge (an electric current) is a magnetic field $\vec{B}$. Arrow 4 indicates that a magnetic field will exert a force on any other moving electric charge that may be present. Because motion is relative, a moving magnetic field will exert a force on a stationary charge.

Even a *changing* magnetic field exerts a force on a stationary charge. However, as already discussed in connection with arrow 2, whenever an electric charge feels a force, then there must be an electric field present. Thus in effect, a changing magnetic field **creates** (or is equivalent to) an electric field. This is shown in Fig. 6.4 by arrow 5 between $\vec{B}$ and $\vec{E}$. The reciprocal relationship is also true: A changing electric field will create a magnetic field.

The six relationships represented by the arrows in Fig. 6.4 are involved in the technological applications of electromagnetism. For example, electric fields are used to create the beam of electrons that light up the picture tube of a television receiver. Magnetic fields due to electric currents are used in relays and electromagnets. Magnetic forces on electric currents are the basic mechanism at work in electric motors and generators, and magnetic forces on moving electric charges are used to deflect the electron beam to make *rasters* on television picture tubes. The creation of electric fields by changing magnetic fields represents the essential action of electric transformers.

All these applications were made possible by the explosive development of science and technology in the nineteenth century, but the crowning achievement, from the standpoint of pure science at least, was the realization by James Clerk Maxwell (the creator of Maxwell's Demon, Chapter 5, Section M2), that the phenomena represented by arrows 5 and 6 in Fig. 6.4 are significant for understanding the nature of light. Arrow 5 indicates that a changing magnetic field creates an electric field. The magnitude of the resulting electric field depends on the rate at which the magnetic field changes. If that rate of change itself is steady, then the resulting electric field will be steady. If, on the other hand, the rate of change of the magnetic field is not steady, then the resulting electric field will not be steady, that is, it will be *changing*. But arrow 6 represents the fact that a changing electric field will in turn create a magnetic field. Furthermore, a changing electric field that is changing at a nonsteady rate will create a nonsteady (i.e., a changing) magnetic field. A continuous chain of creation is possible: A nonsteadily changing magnetic field will create a nonsteadily changing electric field, which creates a nonsteadily changing magnetic field, which creates a nonsteadily changing electric field, and so on.

Maxwell was able to show that these changing electric and magnetic fields, which are constantly recreating each other, are also propagating (spreading out) through space at a definite speed, which he was able to calculate. This calculated speed is 186,000 miles per second—the speed of light! Maxwell thus concluded (in 1864) that light could be explained as arising from rapidly changing and propagating electric and magnetic fields, closely coupled with each other. These are called **electromagnetic fields.** In fact, light represents only a very tiny part of the general phenomena of propagating electromagnetic fields, as will be discussed in more detail below.

In his study of electric and magnetic forces, Maxwell, like most of his contemporaries, was concerned with the nagging fundamental question: "Just what is the mechanism by which one electrical charge (or one mass) reaches across empty space and exerts a force on another electrical charge (or mass)?" Newton said, as noted in Chapter 3, "I make no hypotheses." One possible answer is "through its electric (or gravitational) field." Such an answer was not very satisfactory to nineteenth-century scientists, because it does not reveal what an electric field is.

Michael Faraday (see Chapter 1, Section D4) proposed an analogy between the arrows depicting the electric fields represented in Fig. 6.1 and rubber bands. The arrows start on a positive charge and end on a negative charge. They are under tension like stretched rubber bands and therefore exert a force of attraction along their lengths between the positive and negative charges (6.1b). In addition to contracting along their length, Faraday visualized an additional property of the rubber bands: they push apart from each other, thereby explaining why two charges of the same sign repel each other (Fig. 6.1c), and why the pattern of spreading of the bands is as shown in the various parts of Fig. 6.1. These bands are called *field lines,* or somewhat more loosely, *lines of force.* Similarly, the magnetic field can be described in terms of magnetic field lines connecting magnetic poles, and the gravitational field in terms of gravitational field lines connecting masses.

It is possible to use the idea of field lines to actually work out the mathematical formulas for the forces between electric charges, but it is rather difficult to believe that they have objective reality. Nevertheless, in the spirit of Plato's Allegory of the Cave, one might attempt to justify them as representing the "true reality" behind the appearances. In the same way, the followers of the Ptolemaic theory conceived of the heavenly spheres, as discussed in Chapter 2. Of course, it is rather difficult to see how various objects could move about without getting all their electric, magnetic, and gravitational field lines tangled.

There was suggested, however, a more sophisticated and elegant way of approaching the problem of "action at a distance." It might be asserted (as Aristotle had done) that there is no such thing as empty space, but rather all space is filled with a substance called the **ether,** a special sort of substance having no mass whatsoever.[3] If a material body, or an electric charge, or a magnetic pole is inserted in the ether, then the ether has to be distorted or squeezed in some fashion to make room for whatever is inserted. The electric field lines represent the distortions of the ether resulting from the insertion of an electric charge. The propagating electromagnetic field that is light represents a fluctuation of the distortions present in the ether.

[3]In Chapters 2 and 3, it was noted that Aristotle's ether was present only in the regions above the lunar sphere. By the nineteenth century the ether was considered to be everywhere.

In the early part of the nineteenth century, the mathematical theory of elasticity had been thoroughly worked out and verified. Maxwell was able to adapt the theory of elastic solids to this hypothetical electric ether, and to represent both electric and magnetic fields as ether distortions. Light was described by Maxwell as a wave of distortions propagating through the ether. In this sense, electromagnetic waves are just like sound waves in air, or in water, or in any other medium. The mathematical theory of waves had been worked out in terms of the properties of the medium. Electromagnetic waves have special properties because the medium for these waves, the ether, is different from air or water.[4] One of the significant results of Maxwell's work was that he was able to show that it is not necessary to have one ether for electrical effects, and another ether for magnetic effects, and yet a third ether for light. One ether is sufficient—the electromagnetic ether—and different kinds of electrical, magnetic, and electromagnetic (light) effects are simply different types of distortions and combinations of distortions in the same ether. (Of course, he had not included gravitational effects, so conceivably there might be a separate ether for gravity.)

4. Electromagnetic Theory of Radiation; Light

As already mentioned, Maxwell, in 1864, showed that it is possible to develop waves in the electromagnetic field, and that these waves should travel with the speed of light. Actually, less than 50 years earlier, there had been a lively controversy about the nature of light. There had even been some thought in antiquity that light originated in the eyes of a person looking at an object. Indeed the expression "cast your eyes upon this" carries with it the implication that the eyes send out a beam that searches out the object to be seen. Even Superman sees in this way, except that his vision is particularly penetrating because he sends out X-rays, so that he can see through walls.

Isaac Newton did not consider that the eyes sent out a beam of light, but he did believe that light consists of a stream of corpuscles (i.e., little bullets) that makes objects visible by bouncing off the objects into the eyes of the beholder. One reason for his belief was the fact that light seems to travel in straight lines, as evidenced by the sharpness of shadows. Some of Newton's contemporaries, however, pointed to evidence that light is some sort of disturbance transmitted by a wave motion, just as sound is an air disturbance transmitted by a wave motion, or as a disturbance or distortion of the surface of a body of water is transmitted by a wave motion. In time, the experimental evidence in favor of the wave theory of light became so overwhelming that it was universally accepted.

One of the characteristics by which a wave is described is the distance between successive crests of the wave, which is called the **wavelength.** Wavelengths for visible water waves may vary from a few inches to tens or hundreds of feet. The wavelengths of audible sound waves in air range from a few inches to a few feet. The wavelengths of the light to which our eyes are sensitive, on the other hand,

[4]One of the differences, as mentioned below, is that waves in air or water or other fluids are longitudinal, whereas waves in a solid can be transverse as well as longitudinal.

are quite short—roughly 20,000 wavelengths would fit into one inch—as was shown by the beginning of the nineteenth century. It is because of these extremely short wavelengths that light can be formed into "beams," which seem to travel in straight lines and give sharp shadows. Figure 6.5 demonstrates the effect of wavelength on "shadow" formation for water waves. In Fig. 6.5b, where the wavelength is shorter, a sharper "shadow" is apparent.

The wave nature of light is also responsible for the appearance of colors in the light reflected from soap bubbles or from oil films on streets just after a rain. The colors result from an effect called **interference,** which will be discussed further in Chapter 7 (Section E3). The wave nature of light is also manifested in an effect called **diffraction,** which is responsible for the colors seen from "diffraction jewelry" or when light is reflected at a glancing angle from video discs or phonograph discs, for example those used for 33⅓ or 16⅔ rpm record players.

Another characteristic property of waves is their **frequency,** which is the number of wave crests per second that will pass by a given point as the waves spread out. The light that we see has a frequency of roughly 500 million million wave crests per second. Multiplying the frequency of a wave by its wavelength gives its speed—186,000 miles per second for light.

Until Maxwell's work, it was thought that light waves were transmitted through a special ether. A wave must travel in some medium; without any water there cannot be a water wave, and without air there cannot be a sound wave, as is demonstrated by the fact that sound waves do not exist in vacuum. For light waves, therefore, it was assumed that there is a special ether, called the *luminiferous ether,* which fills all space. As already discussed, Maxwell was able to show that the luminiferous ether, the electric ether, and the magnetic ether should be all one and the same, and that light waves represent a disturbance of this ether.

One of the important results of Maxwell's electromagnetic theory was the realization that light waves represent only a very small portion of the electromagnetic waves that can be generated. This led to the idea that radio waves can

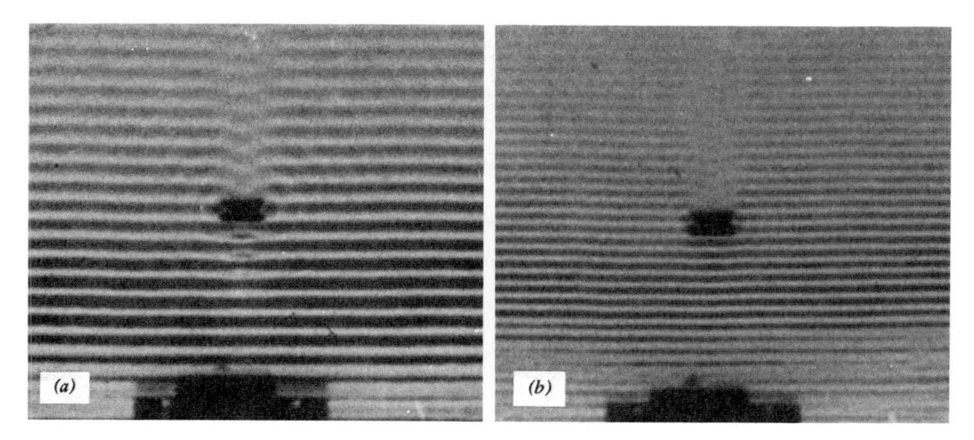

Figure 6.5. Photograph of water waves in a ripple tank. (Reproduced by permission of Educational Development Center, Newton, Mass.)

Table 6.1. The Electromagnetic Spectrum

Kind of Wave	Frequency in Hertz[a]	Wavelength in Meters[a]
Ordinary household alternating current	60	5×10^6 (3100 miles)
A.M. Radio	10^6	300
Television	10^9	0.3
Microwaves/radar	10^{11}	0.003
Infrared	10^{13}	0.00003
Visible	5×10^{14}	6×10^{-7}
X rays	3×10^{18}	10^{-10}

[a]Hertz means cycles or waves/second; 1 meter equals 39.37 inches; 10^6 means a one followed by six zeros, that is, one million; Similarly, 10^{11} means a one followed by 11 zeros; 10^{-7} means one divided by 10^7. The product of frequency by wavelength is 3×10^8, which is the speed in empty space of electromagnetic waves in meters/second.

be generated, as was demonstrated in 1886 by Heinrich Hertz, a German physicist. Electromagnetic waves are generated when an electric charge is caused to vibrate (oscillate). In effect, the electric charge ''waves'' its electric field, in a similar manner as a hand waving a long whip. Of course, the ''waving electric field'' is more like a bundle of ''rubber bands,'' to use Faraday's analogy; but this gives the nonsteadily changing electric field needed to create the electromagnetic wave, as described above. Hertz was able to generate oscillating electric currents at frequencies of a few hundred million vibrations per second. The wavelength of these waves was two to three feet, depending on the frequency. Subsequently, the Italian engineer, Guglielmo Marconi, invented the wireless telegraph to send messages by radio waves. The first successful trans-Atlantic radio messages were carried by waves having a frequency of 200,000 vibrations per second and a wavelength of almost a mile!

The entire range of electromagnetic waves is called the *electromagnetic spectrum,* and its extent is indicated in Table 6.1. All the electromagnetic waves listed in the table travel in empty space at the speed of light, 186,000 miles per second.[5] The differences among them are due entirely to their wavelengths or frequencies. Waves with short wavelengths have high frequencies and vice versa. When wavelength and frequency are multiplied together, the result is constant and is equal to the speed of the waves. Note that in Table 6.1, wavelengths are given in meters, as is usually done for radio waves.

[5]In a material medium, such as a glass of water, the speed of light is less than in empty space. The ratio of the speed of light in empty space to its speed in the material medium is called the index of refraction of the medium. Most often in a material medium the speed of the electromagnetic waves (and hence the index of refraction) is no longer constant, but is different for various wavelengths. This phenomenon is called dispersion.

Optional Section 6.4 *Wavelengths and Antennas*

The mathematical expression for the relationship between the frequency and wavelength of a wave is $f\lambda = v$, where f is the frequency of the wave, λ is the wavelength, and v is the speed of propagation of the wave, sometimes called the *phase velocity* of the wave. For light traveling in empty space, the phase velocity is often given the special symbol c, which is equal to 186,000 miles per second or 300 million meters per second. All of the values given in Table 6.1 satisfy this relationship.

It is usually true that the devices that transmit or receive or interact most strongly with electromagnetic radiation are comparable in size with the wavelength or a substantial fraction of a wavelength of the radiation involved. Thus transmission towers for A.M. radio stations are typically 75 to 150 meters tall, receiving antennas for F.M. radios (wavelength typically 3 meters) are .75 to 1.5 meters wide. Similarly CD records reflect white light as a band of colors because the spacing between the grooves is roughly .00025 inch (a millionth of a meter). The structure of crystalline matter is very effectively studied with X rays because the distance between atoms is comparable to the wavelength of X rays.

An electromagnetic wave can be described as a combination of pulsating electric and magnetic fields that are coupled to each other. Figure 6.6 shows how the electric and magnetic fields vary in an electromagnetic wave traveling in the direction of the heavy arrow. In the figure, the electric field vectors $\vec{E}$ are shown as varying in magnitude along the direction of travel, but being confined to the vertical plane. The magnetic field vectors $\vec{B}$ also vary along the direction of travel and are confined to the horizontal plane. When $\vec{E}$ and $\vec{B}$ vectors are confined as shown, the electromagnetic wave is said to be *plane polarized.* It should be noted that the pulsating $\vec{E}$ and $\vec{B}$ vectors are always perpendicular

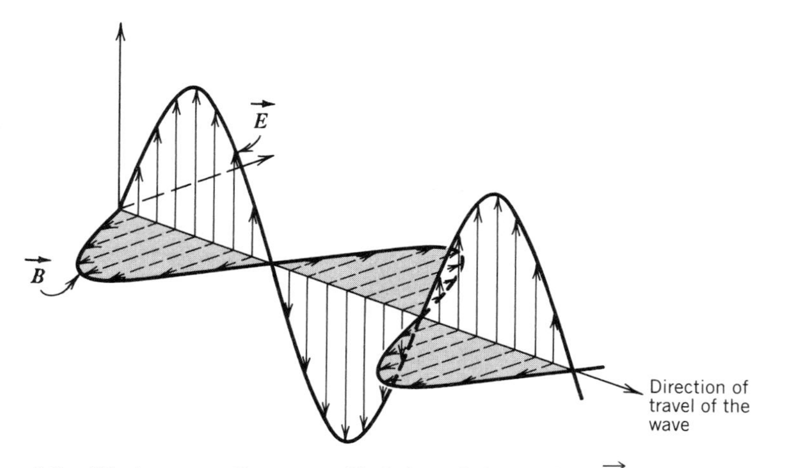

Figure 6.6. Electromagnetic wave. Variation of electric field $\vec{E}$ and magnetic field $\vec{B}$ along direction of travel for a plane electromagnetic wave.

to each other and to the direction of travel; this perpendicularity is true of all electromagnetic waves. When waves are perpendicular to the direction of travel, they are called **transverse waves.**

By contrast, the disturbance carried by sound waves traveling in air is in the form of a pulsating density of the air. This pulsation takes place in the direction of travel of the wave, and the wave is therefore called a longitudinal wave. When a sound wave travels in a solid medium, the resulting waves can be transverse as well as longitudinal. There are no longitudinal pulsations for electromagnetic waves.

In any wave there is energy associated with the pulsations. Transmission of the pulsations by the wave results in the "transport" of energy. An electromagnetic wave transports electromagnetic energy. The rate at which it transports energy is related to the product obtained by multiplying $\vec{E}$ and $\vec{B}$ (using special rules for multiplying vectors together). The immediate source of the energy transported by the electromagnetic wave is the kinetic energy of the vibrating electric charge that is "waving" its electric field. The generation of electromagnetic energy as described earlier is a process for converting kinetic energy into radiant energy (see Chapter 4, Section C2).

5. Electromagnetic Theory and Frames of Reference

With the development of electromagnetic theory and its concomitant electromagnetic ether, the question arose as to what happens to the ether when an object (e.g., the Earth) moves through it. Does the Earth move along without disturbing the ether, or does the Earth drag the ether along with it? If the Earth does not disturb the ether as it moves along, then it might be possible to use the ether as a frame of reference. It would then be possible to calculate the "true" force exerted by a magnetic field on a moving charge (see the end of Section B2 above). Additionally, it might be possible to settle once and for all the question that had plagued astronomers about which of the heavenly objects are actually moving and which are at rest. In order to use the ether as a frame of reference, however, it is necessary to have some knowledge of the physical nature of ether.

6. Properties of the Imponderable Ether

The ether was considered to fill all space, even empty space, because light travels across empty space. It was also considered to be massless. For this reason, it was called the *imponderable ether* (ponderous means heavy). It was also considered to offer no resistance to motion of objects through it, because otherwise the various heavenly objects would have gradually lost speed and come to rest, or at the very least showed some similar systematic changes of their orbits with time. Because light waves are transverse waves, as discussed in Section B4 above, the ether had to have the properties of an elastic solid. The mathematical theory of elasticity then required that the ether should be very "stiff," because waves travel faster in stiff media than in soft media. Because the speed of electromagnetic waves is very high compared with other waves (by a factor of almost a million), the stiffness had to be assumed to be quite large. The ether had to be considered to be at rest with respect to absolute space, at least in the regions of space between various astronomical bodies. It was also thought that a large, massive body might ''drag'' some ether along with it in its motion, although this would have certain consequences for the propagation of light in the vicinity of various massive objects.

In view of all these properties, the ether clearly had to be regarded as a wondrous substance—much superior to anything created for science fiction! But most important for the concerns of this chapter, the ether, as pointed out before and originally by Maxwell, offered the possibility of defining an absolute frame of reference, if only it could be detected.

C. Attempts to Detect the Ether

Although the existence of the ether may be inferred from the logical necessity to have a medium to support electromagnetic waves, it is clearly desirable to have some independent evidence or experiment to verify the properties of the ether. There is a rather long history of attempts to detect the ether.

1. Star Aberration

Starting about 1725, the English astronomer James Bradley attempted to measure the distance from the Earth to various stars, using a standard surveyor's technique. By sighting on a particular star from different points in the Earth's orbit, he expected to tilt his telescope at different angles. Knowing the angles of sight of the star, and the diameter of the Earth's orbit, by the use of simple trigonometric calculations he could then calculate the distance to the star (Fig. 6.7a). These telescope measurements also made it possible for him to verify the phenomenon of star parallax; that is, the sight angle between two different stars depended on the position of the Earth in its orbit (Chapter 2, Section C4). When traced by the telescope, however, all the stars seemed to move in small elliptical orbits, regardless of their distances from the Earth. Bradley found it necessary to tilt the telescope an additional amount in the direction of the Earth's motion.

A possible explanation of the additional tilt can be seen with the help of Figs. 6.7b and 6.7c. If the Earth were not moving, then the starlight would travel straight down the middle of the barrel of the telescope from the middle of the opening at the top to the eyepiece at the bottom. Because the Earth is moving, however, by the time the light gets to the eyepiece, the eyepiece will have moved in the direction of the motion (to the right in the figure), and the starlight would no longer pass through the middle of the eyepiece. In Fig. 6.7b, the solid line shows the position of the light and telescope just as the light from the star reaches the top of the telescope; the dotted lines show the positions when the light reaches

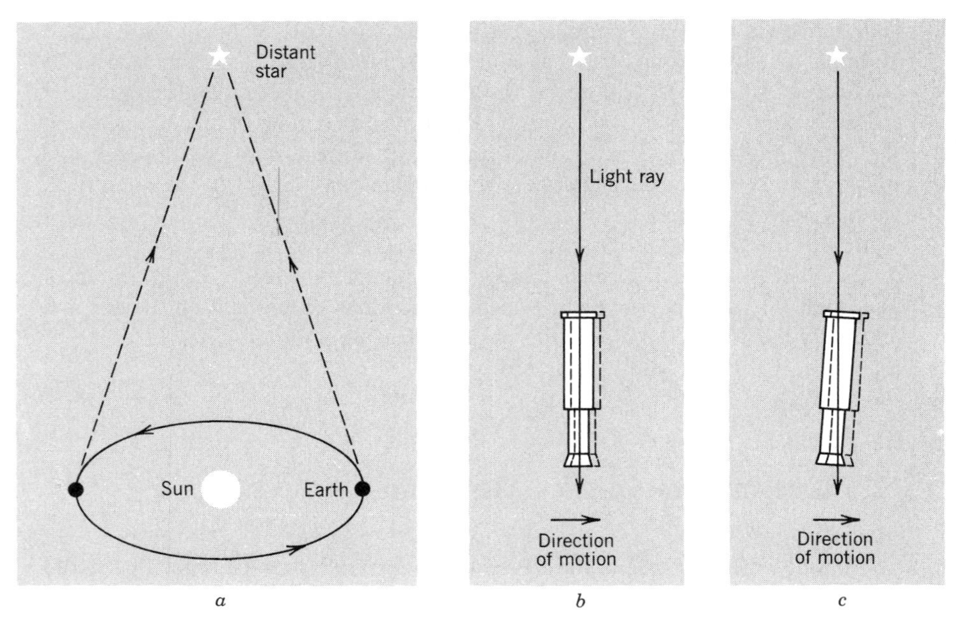

Figure 6.7. Star parallax and aberration. (a) Varying angle of view of distant star as seen from different points of Earth's orbit. (b) Motion of telescope across incoming ray of starlight. (c) Tilt of telescope to compensate for motion.

the bottom of the telescope. However, if the telescope were tilted so that the eyepiece is somewhat "behind" the opening at the top of the telescope, then the middle of the eyepiece will arrive at the light beam just in time for the light to pass through, as shown in the dotted portion of Fig. 6.7c.[6] The angle of tilt is proportional to the ratio of the speed of the telescope (i.e., the Earth) through the ether to the speed of light, and thus this tilt angle could be used to calculate the speed of the Earth with respect to the ether.

Some 150 years later a further check of this explanation was carried out. If the tube of the telescope were filled with water, then the angle of tilt should be expected to change substantially because light travels in water at three-fourths of the speed of light in air. Therefore it was expected that the tilt angle should increase by a factor 4/3, but in fact, the tilt angle did not change at all. This experiment led to the conclusion that somehow the water "dragged" the ether along with it enough to compensate for the expected effect on the tilt angle. A special experiment had been done some 20 years earlier, which implied that flowing water would indeed drag ether along with it, but not as much as needed to explain the failure of the experiment.

[6]Contrast the behavior of the beam of light with that of the object dropped from the mast of a moving ship, as described in Section A2 of this chapter.

2. Michelson-Morley Ether Drift Experiments

Starting in 1881, Albert A. Michelson, a young American, began a series of experiments that were intended to apply an extremely sensitive technique for measuring the motion of the Earth through the ether. For the purposes of the experiments, the important thing was to measure the relative motion of the Earth and ether; that is, one could imagine that the Earth was at rest and a stream of ether was slowly drifting past the Earth. For this reason, such experiments and others like them are called "ether-drift" experiments. The principles of Michelson's experimental technique can be understood from Fig. 6.8.

The basic idea is to compare two trains of light waves: one that has been transmitted in a direction parallel to the ether drift, and one that has been transmitted in a direction perpendicular to the ether drift. The first wave-train is sent

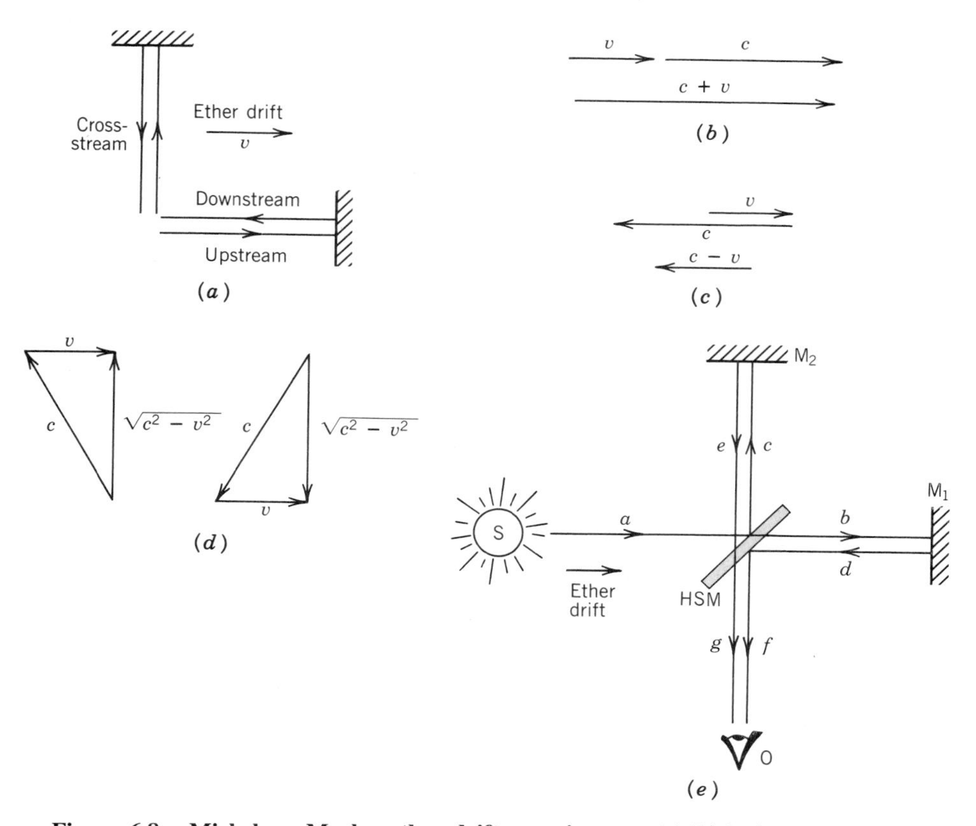

Figure 6.8. Michelson–Morley ether drift experiment. (a) Light beams traveling parallel and perpendicular to ether drift. (b) Vector addition of "downstream" velocities. (c) Vector addition of "upstream" velocities. (d) Vector additions of "cross-stream" velocities. (e) Schematic experimental arrangement. HSM, half-silvered mirror; M_1, fully silvered mirror 1; M_2 fully silvered mirror 2; S, light source; O, observer; *a, b, c, d, e, f, g,* light rays.

downstream to a mirror, which then reflects it back upstream. The second wave-train is sent cross-stream to a mirror as far away as the downstream mirror, and back again. The downstream portion of the first wave is expected to travel at the speed of light with respect to the ether *plus* the speed of the ether drift, with resulting absolute speed equal to $c + v$ (c is the speed of light in empty space and v is the speed of the ether drift). The upstream portion of the first wave-train is expected to travel at the speed of light *minus* the speed of ether drift, with resulting absolute speed equal to $c - v$. For the cross-stream wave the vector representing the velocity of the light relative to the medium is directed slightly upstream, so that its vector sum with the drift velocity will be directed straight across to the mirror. The magnitude of the resulting vector is expected to be smaller than c and is equal to $\sqrt{c^2 - v^2}$.

When the actual times for the two wave-trains to reach their respective mirrors and be reflected back to their starting points are compared, the overall result is that the downstream–upstream wave-train takes longer to make the round trip than the cross-stream wave-train (assuming the two mirrors are equidistant from their common starting point).

A schematic drawing of Michelson's apparatus is shown in Fig. 6.8e. A beam of light a from a light source falls on a "half-silvered" mirror oriented at an angle of 45 degrees. Half the beam continues along the path b to fully silvered mirror M_1, which reflects it along path d until it encounters the "half-silvered" mirror again, and part of it is reflected in direction f. The other half of the beam a is reflected at right angles along the path c to fully silvered mirror M_2, which then reflects it along path e back to the half-silvered mirror, where part of it continues along path g. The half-silvered mirror is particularly important for the experiment, because it serves as a starting point for the downstream–upstream wave-train and for the cross-stream wave-train. It also serves as a means of redirecting those two waves along a common path once they have completed their trip, so that they can be compared with each other by means of the phenomenon of interference (see Chapter 7, Section E3, for further discussion of interference). The results of such a comparison permit determining the difference in round-trip times for the two wave-trains.[7]

Michelson first performed the experiment in 1881 while he was serving as U.S. naval attaché in Germany. He repeated the experiment in 1887 in Cleveland, Ohio (at which time he was a professor at Case Institute of Technology), with the collaboration of Edward W. Morley, a professor at Western Reserve University. They used apparatus that was 20 times as sensitive as the 1881 apparatus. Morely continued to repeat the experiment over a period of years at different positions of the Earth's orbit. The experiment is now called the **Michelson–Morley experiment.**

[7]A number of other details of Michelson's experiment have been omitted. He had to reduce the effect of vibrations; he had to take account of the impossibility of actually making the distances between the mirror and half-silvered mirror exactly equal; the apparatus had to be of reasonable size, and so on.

The sensitivity of the 1887 apparatus was so great that it could have detected an effect smaller than expected by a factor of 40; however, the experimental result showed no difference in the round-trip times for the two wave-trains! Since then, the experiment has been repeated by many other investigators, with sensitivities up to 25 times greater than that of the 1887 apparatus, but with the same result.

One possible conclusion to be drawn from the result, called a *null result* because no difference was found between the two wave-trains, was that the Earth drags some of the ether along with it in the same way that an airplane drags some air along with it in the course of its flight. As mentioned above, the measurements of star aberration with a water-filled telescope seemed to suggest that the ether was being "dragged" along close to the surface of the Earth.

3. Ether Drag Experiments

Because the null result of the Michelson–Morley experiment implied that the ether was being dragged along by the moving Earth, it was apparent that a direct test of the ether drag hypothesis was necessary. In 1893, a British scientist, Oliver Lodge, reported the result of an experiment designed to simulate the Earth's drag on the ether. If a very massive concrete disk is set into rotation, it should drag the ether in its immediate vicinity into rotation as well. Another less massive stationary disk placed just below the rotating disk would then have an artificially created circulating ether drift just above its surface. Lodge then arranged a set of mirrors on the lower disk to produce two light wave-trains, one traveling in a counterclockwise path around the lower disk and the other in a clockwise path. One of the wave-trains would be traveling in the same direction as the artificial ether drift, and thus its velocity relative to the mirrors would be increased, whereas the wave-train traveling in the opposite direction would have its velocity decreased. The two wave-trains could then be compared through the interference phenomenon, and the drag-induced artificial ether drift could be measured.

The arrangement of the mirrors is shown in Fig. 6.9. A beam of light a from the light source is split at the half-silvered mirror. Half the beam follows the directions b, c, d, e, being reflected from the three fully silvered mirrors, back to the half-silvered mirror, with part going on through along the path f. The other half of the beam a is reflected by the half-silvered mirror in the directions of g, h, i, j, back to the half-silvered mirror; and finally part of it is reflected in the direction k. The two beams f and k are then compared by means of the interference phenomenon to determine a difference in the times required for them to complete the trip around the mirrors. Contrary to what was expected as a result of the Michelson–Morley experiment, a null result was again found! There was no detectible difference between the two oppositely circulating light waves.

4. Failure of All Ether Experiments

The Michelson–Morley experiment showed that the drift of the ether (i.e., the motion of the Earth through the ether) could not be detected. The Lodge experiment showed that a direct measurement of the drag of the ether by a massive

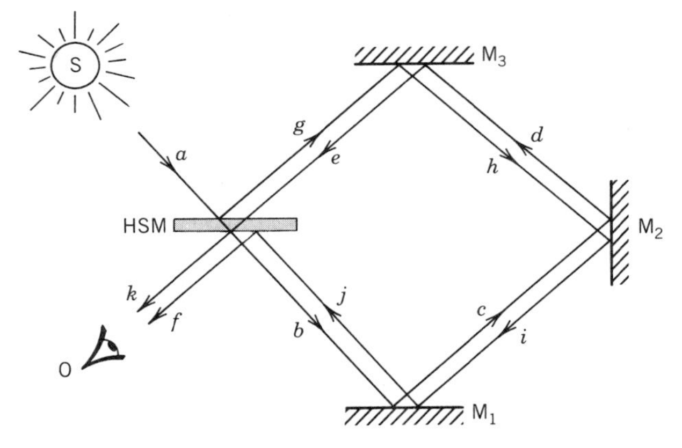

Figure 6.9. Lodge ether drag experiment. Mirrors and light rays on lower disk. S, light source; O, observer; HSM, half-silvered mirror; M_1, fully silvered mirror 1; M_2, fully silvered mirror 2; M_3, fully silvered mirror 3; *a, b, c, d, e, f, g, h, i, j, k,* light rays.

moving object also could not be detected. The contradiction between these two results can be resolved by noting that, in fact, neither of these experiments had succeeded in demonstrating the presence of the ether. Moreover, no other experiments had succeeded in showing ether drift, drag, or the presence of ether. For example, if a pair of parallel plates, one carrying positive charge and the other carrying negative charge, is suspended by a fine thread in an ether wind, then there should be a resulting torque twisting the suspension. Experiment showed no such torque.

D. The Special Theory of Relativity

1. Precursors

In addition to the unresolved experimental problems of detecting the motion of the Earth through the ether or of even detecting the ether, some fundamental theoretical problems existed as well. Galilean–Newtonian relativity requires that forces should be the same (invariant) in all inertial frames of reference. As pointed out in Section B2 above, however, the forces acting on electrical charges in motion would seem to be different in various inertial frames of reference. For example, if two positive charges are both traveling at the same velocity in the same direction, then according to electromagnetic theory, there will be a resulting magnetic field that will act on the moving charges, thereby causing an attractive force between them. This force increases as the velocity increases. Yet velocity is a relative quantity, depending on the frame of reference, so the attractive force is relative and depends on the frame of reference, in contradiction to the requirement of Galilean–Newtonian relativity. Several attempts to solve this and similar problems, as well as the problems arising from the failure of the ether experiments, can be looked upon as precursors to Einstein's solution of the problem.

In 1882, two physicists, G. F. Fitzgerald in Ireland and H. A. Lorentz in the Netherlands, independently suggested a solution to the dilemma posed by the Michelson–Morley experiment. They suggested that the part of the apparatus that carried the upstream–downstream wave-train (the paths *b* and *d* between the half-silvered mirror and the full-silvered mirror 1 in Fig. 6.8d) became shorter by just enough to compensate for the extra time required for the light to travel along the paths *b* and *d*. Thus it would take no longer for light to travel downstream and back as cross-stream, thereby explaining the null result. Such an explanation can be rationalized by saying that moving through the ether generates a resisting pressure, which compresses the apparatus just as a spring is compressed by pushing it against a resisting force. The amount of contraction (shortening) necessary is very small, about half a millionth of 1 percent, so that it would normally not be noticed. The Michelson–Morley experiment, however, was sensitive enough to detect such a small effect.

Lorentz was later (1899) able to justify this postulated contraction by pointing out that it was a consequence of some ideas he had been considering in connection with the problem of the relative nature of the force between moving electric charges. He had been looking for some new transformation equations between different inertial frames of reference (see the end of Section A2 and all of Section A3 above) that would make the formula for the electromagnetic force invariant. One of the consequences of these new transformation equations, known as the **Lorentz transformation,** was that moving objects would contract by an amount depending on the speed. Lorentz also found that time intervals measured at moving objects would become larger than expected. He called such expanded time intervals *local time.*

A French theorist, Henri Poincaré, even suggested in 1904 that it was futile to attempt to measure the motion of the Earth, or anything else, with respect to the ether. Using the idea that it is impossible to measure velocity in an absolute sense, he went beyond Lorentz and showed that the mass of an object (i.e., inertia, that aspect of the mass that plays a role in Newton's second law of motion) increases as the object's speed increases. He also showed that there is a maximum speed that any object can attain—the speed of light.

All of these ideas are included in Einstein's special theory of relativity. The significance of Einstein's work was that he was able to show simply and directly (at least from a physicist's viewpoint) that they were natural consequences of a profound and insightful reexamination of some basic assumptions about the nature of physical measurements, whereas Fitzgerald, Lorentz, and Poincaré introduced them in order to deal with a specific problem. Einstein's thinking was very much influenced by the Austrian physicist-philosopher Ernst Mach, who had carried out a critical and incisive reexamination of Isaac Newton's *Principia.* Mach was one of the founders of the "logical positivist" school of philosophy, and he was particularly critical of Newton's definitions of absolute space and time and of mass.[8]

The circumstances surrounding the development of physics at the beginning of the twentieth century were somewhat similar to those at the time of Isaac Newton. Several very talented scientists were close to making a ''breakthrough'' on a class of problems that were ripe for solution, but only one of them—an Einstein or a Newton—was able to master the situation. As an example, the Lorentz–Fitzgerald contraction ostensibly accounts for the failure of the Michelson–Morley experiment; in fact, there is a modified version of that experiment (called the Kennedy–Thorndyke experiment) that was designed to circumvent the Lorentz–Fitzgerald contraction. This experiment also gives a null result. Moreover, as will be discussed below, Einstein's approach provides an entirely different interpretation of the failure of all the ether experiments.

2. Albert Einstein (1879–1955)

Probably no other scientist, except perhaps Isaac Newton, has been as famous as Albert Einstein. He is as renowned for his white hair, droopy mustache, curved pipe, stooped posture, and gentle informality as he is for his scientific theories. Figure 6.10 shows Einstein as a young man, when he first began to make his mark on the scientific world. As is often true of famous people, there are a number of myths and misconceptions about him and his work. A few brief remarks are apropos. Although he began speaking later than usual as a young child, his considerable abilities were quite evident, even in childhood. He did not do as well in school subjects of no interest to him as in those he enjoyed; but when he found that he had to master some material in order to meet entrance requirements to the school he wanted to attend, he promptly did so. He particularly disliked regimentation, but was capable of considerable self-discipline. He was not an outstanding mathematician, but was competent in the mathematics that he needed.

He made significant contributions in many areas of physics besides relativity theory, publishing over 350 papers in his lifetime. Not all of his work was in theoretical physics; he even filed some patent claims. Unlike Julius Robert Mayer, who had great difficulty communicating his important ideas on energy because he did not frame his ideas in terms his contemporaries could follow, Einstein was quite successful in communicating his most controversial ideas. He readily participated in the scientific debates of his day with good humor. He did not hesitate to speak out on political matters that he considered important. He also deplored the use of the physical theory of relativity to justify moral relativity, if only because such use betrayed a misunderstanding of the true meaning of the theory.

So far as the theory of relativity is concerned, some people have questioned the uniqueness of Einstein's contributions. He himself conceded that, considering

[8]Mach was not the first person to criticize Newton's work. The German philosopher Leibnitz, contemporary and rival of Newton, had been rather critical of these same definitions.

Figure 6.10. Einstein in 1905. (Reproduced by permission of Lotte Jacobi, Hillsboro, N.H.)

the scientific ferment of his time, Poincaré might well have developed the special theory if Einstein had not. He stated, however, that he began thinking about problems related to relativity when he was 16 years old. Moreover, Einstein's approach to the subject was characteristically straightforward and elegant, cutting through to the essential ideas. The general theory of relativity, discussed later in this chapter, is universally recognized as Einstein's unique creation.

3. Special Principles of Relativity

In 1905, while working as an examiner in the Swiss Patent Office, Einstein published three very important papers, one of which was entitled "On the Electrodynamics of Moving Bodies."[9] In this paper he was not so much concerned with relative quantities, whose magnitudes depended on the frame of reference from which they were measured, but rather with invariant quantities, which would be the same in all inertial frames of reference. He enumerated two special principles that should be applicable in all frames of reference:

[9]The other two papers are mentioned in Chapter 4, Section B2, and in Chapter 7, Section C3.

I. The laws of physics are invariant in all inertial reference frames. This means that the formulas expressing the various laws of physics must be calculated in the same way in all inertial reference frames.

II. It is a law of physics that the speed of light in empty space is the same in all inertial reference frames, independent of the speed of the source or detector of light. He pointed out that this meant that the ether could not be detected by any experimental means, and therefore it was a useless concept, which should be discarded.

The first principle states simply that a relativity principle does exist, just as Galileo, Newton, and others had originally indicated. It particularly emphasizes that the principles of physics are the same everywhere. The second principle is the important new physical insight.

The constancy of the speed of light is a significant result of all the experiments aimed at detecting the ether. Equally significant, these experiments confirmed Galileo's idea that it is not possible to determine absolute motion. This also means that measurements in one frame of reference are just as valid as in any other frame of reference. Questions such as "How fast is the Earth really moving?" are therefore meaningless.

4. New Concepts of Space and Time

Under the influence of Mach, Einstein realized that it was necessary to reconsider the meaning of space and time, and how they are measured. He recognized that space and time are not independent concepts, but are necessarily linked with each other. Moreover, these concepts are defined by measurement. Time is measured by watching the movement of the hands of a clock, for example, or the passage of heavenly bodies through the sky. We know time has passed because we see that these objects have changed their positions in space. The speed of light is involved because these observations are made by virtue of the fact that light travels from the moving object to our eyes. (It does not matter if electrical signals are used to tell us that the objects move—such signals also travel with the speed of light, as shown from electromagnetic theory.)

Einstein showed that, in a manner of speaking, time and space are interchangeable, as is illustrated by the following set of statements, which exhibit the symmetry of space and time:

I. A stationary observer of a moving system will observe that events occurring at the SAME PLACE at DIFFERENT TIMES in the moving system occur at DIFFERENT PLACES in the stationary system.

II. A stationary observer of a moving system will observe that events occurring at the SAME TIME at DIFFERENT PLACES in the moving system occur at DIFFERENT TIMES in the stationary system.

III. A stationary observer of a moving system will observe that events occurring at the SAME TIME at the SAME PLACE in the moving system occur at the SAME TIME and SAME PLACE in the stationary system.

Statement II is obtained from statement I by interchanging the words TIMES and PLACE. This interchange changes the meaning of the statement. This will be justified below. On the other hand, interchanging the words TIME and PLACE in statement III does not change the meaning of the statement.

To illustrate these statements, the moving system might be an airplane traveling from New York to Los Angeles, and the stationary system might be the control tower of an airport on the Earth. An airline passenger might be sitting in seat 10C. At 8 A.M. the passenger is served orange juice, while the airplane is above Albany, New York; and at nine o'clock the passenger is drinking a cup of coffee after breakfast, while the airplane is passing over Chicago. In the moving system, the airplane, both events occurred at the same place, seat 10C, but at different times. In the stationary system, the Earth, the two events occurred at different places, over Albany and over Chicago, as would be seen if an observer in the control tower could look inside the airplane.

The foregoing scenario is very plausible, but a scenario based on the second statement may seem implausible: Some time later, when the airplane is over Denver, Colorado, the passenger, who is reading a physics book, looks up and sees a federal marshal at the front of the airplane and a hijacker at the back of the airplane, with guns pointed at each other. Both guns are fired at the same time, as seen by the passenger. As seen by the observer in the control tower on the Earth, however, the shots were not fired simultaneously, but were fired at different times. Implausible as it seems, the scenario based on the second statement is correct, as will be discussed below.

The third scenario is as follows: After both shots miss, the passenger then notices that the flight attendant standing next to him simultaneously gasped and dropped a pot of coffee into his lap, in seat 10C. The president of the airline, watching from the control tower, sees that indeed the flight attendant simultaneously gasped and dropped the pot of coffee into the passenger's lap, over Denver.

The point of all this is to illustrate that because space and time are intertwined with each other, they are relative quantities that are different in different inertial frames of reference. Events that are simultaneous in time in one frame of reference may not be simultaneous in time in another inertial reference frame. Only if the simultaneous events occur at the same place, as in the third scenario, are they simultaneous in all inertial frames of reference.

In particular, statement II asserted that events observed to occur simultaneously in different locations in a moving coordinate system would be observed as not simultaneous in a stationary coordinate system. The essential basis for the statement is that the timing of events depends on the receipt of light signals emitted from the location of the events. (Recall that all electromagnetic radiation travels at the same speed in empty space.)

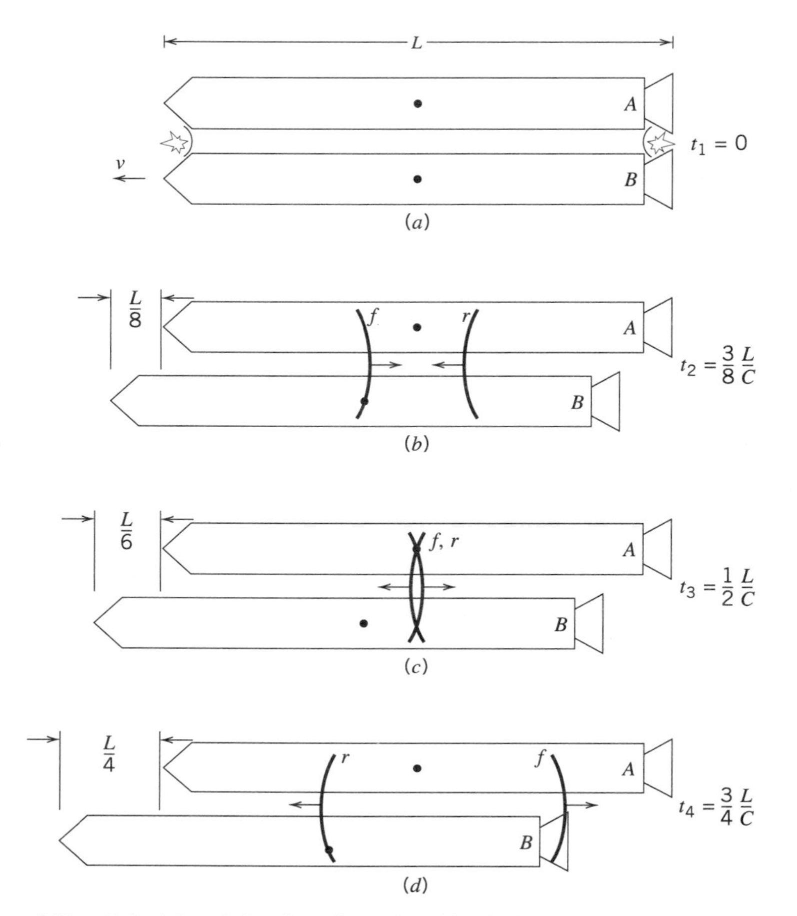

Figure 6.11. Relativity of simultaneity. Two identical space ships, A and B, of length L, are traveling in the same direction. The velocity of B relative to A is 1/3 of the speed of light, i.e., $(1/3)c$. Observers are stationed in the exact center (the midpoint) of each space ship at the position of the heavy dot. The observer in A presses a button that will set off simultaneous explosions at the very front and rear of space ship A, timed to take place just as the midpoints of A and B come abreast of each other. The explosions are therefore equidistant from the two midpoints. In the reference frame of A the following events take place. (a) At time $t_1 = 0$, the explosions take place. (b) At time $t_2 = (3/8)L/c$, the light wave from the front explosion reaches the observer in B, which has moved a distance $L/8$ forward while the light wave moved a distance $3L/8$. (c) At time $t_3 = (1/2)L/c$, the light wave from the front explosion reaches the observer in A, simultaneously with the light wave from the rear explosion, confirming that the explosions occurred simultaneously in the reference frame of A. B has now moved forward a distance $L/6$ while the light waves have each moved a distance $L/2$. (d) At time $t_4 = (3/4)L/c$, the light wave from the rear explosion reaches the observer in B, which has now moved a total distance forward of $L/4$. The observer in A thus observes that the light from the rear explosion reached the midpoint of B after the light from the front explosion by an amount of time equal to $(t_4 - t_3) = (3/8)L/c = 0.375L/c$. (The observer in the space ship B will find that from the reference frame of B the light from the rear explosion arrived $0.354L/c$ after the light from the front explosion, as discussed in connection with Fig. 6.12.) If the space ships are 300 meters in length, then $L/c = 10^{-6}$ seconds = 1 microsecond.

The interdependence of space and time is perhaps most easily seen in what is called the "the problem of simultaneity." This "problem" has to do with whether or not two different observers can agree as to whether two separate events occur at the same time.

The question is "how does one judge that two events occurred simultaneously?" The answer is that the observer "sees" them happening at the same time. But if one event took place very close to an observer and the other occurred far away, then the second event will have occurred before the first event because the light from that event took longer to get to the observer. Only if the observer is equidistant from the two events can it be concluded that seeing them simultaneously implies that they happened simultaneously. A person seated precisely in the middle of a space ship who sees simultaneous flashes of light coming from explosions at the very front and very rear of the space ship concludes that the explosions occurred simultaneously.

However, a second person seated in the precise middle of another identical space ship that is moving forward relative to the first space ship, and who happens to be abreast of the first observer at the time (as judged from the first space ship) of the explosions will not see the light flashes simultaneously. This is illustrated in Fig. 6.11. This second observer will see the light from the front explosion before the light from the rear explosion because the second space ship is moving toward the light coming from the front explosion and away from the light coming from the rear explosion. Since the second observer is seated exactly midway between the location of the two explosions it must be concluded that the front explosion occurred before the second explosion, and not simultaneously with it. The second observer will state that the events seen as simultaneous in the first space ship, which is in relative motion with respect to the second space ship, are not seen to be simultaneous from the second space ship. Einstein realized that if two different observers, in two different inertial frames, cannot even agree as to whether two events are simultaneous, then time is somehow relative.

Einstein showed further that time intervals between two events will be measured differently in different inertial reference frames. Even two otherwise identical clocks will run at different rates in the two reference frames; that is, the time between "tick" and "tock" will be different. This can be illustrated by a rather simple example called a *mirror clock.* Figure 6.12a shows a light bulb inside a box having a coverable hole and a mirror placed 93,000 miles above the hole. If the hole is momentarily uncovered and then recovered, a brief flash of light will arrive at the mirror a half-second later, and then be reflected back to the covered hole, arriving yet another half-second later. Two events have taken place: (1) the emission of the light flash from the hole, and (2) the return of the light flash to the covered hole one second later. This mirror clock is then a one-second timing clock.

If two such clocks are made, one could be placed in inertial reference frame A and the other placed in inertial reference frame B. The two reference frames are moving at constant velocity relative to each other. As seen in its own reference frame, each clock works perfectly (the speed of light is the same in all reference frames). As seen from the other reference frame, however, each clock runs slow.

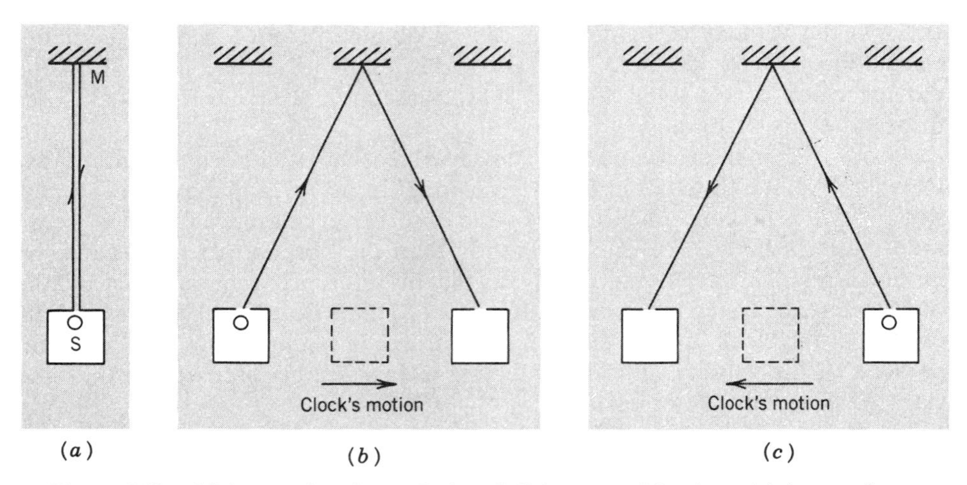

Figure 6.12. Light rays in mirror clock. S, light source; M, mirror. (a) As seen from reference frame in which clock is at rest. (b) Clock in reference frame *A*, as seen from reference frame *B*. (c) Clock in reference frame *B*, as seen from reference frame *A*.

As shown in Fig. 6.12b, the clock in reference frame *A*, as seen from *B*, moves with a speed v. During the half-second it takes the light to travel from the hole in the box to the mirror, the mirror has moved from its original position at the left of the figure to the dotted position in the center of the figure. As seen from *B*, therefore, the ray of light traveled upward along a diagonal path until it was reflected, and then downward along a diagonal path. (Note the similarity between this situation and that of the object dropped from the mast of a moving ship, as discussed in Section A2.) To an observer in reference frame *B*, the flash of light in the clock of reference frame *A* has traveled a greater distance than the flash of light in his own stationary clock. Because the speed of light is constant for all frames of reference (Einstein's second special principle), then the time interval between the two events, the emission of the light flash and its return, is not one second for the moving clock, but some longer time. The moving clock runs slow!

The critical element in this analysis is the assumption that the speed of light is constant in both frames of reference. This is the important new fact introduced in Einstein's special theory of relativity and eventually leads to some particular predictions that can be tested experimentally.

The time interval measured by the moving clock is calculated by dividing one second by $\sqrt{1 - (v/c)^2}$, where v is the relative velocity of the reference frames and c is the velocity of light. (This can be obtained from a simple geometric calculation, given in Optional Section 6.5 below.) If, for example, the ratio v/c is $3/5$, then the time interval on the moving clock as observed from *B* is 1.25 seconds. The moving clock loses $1/4$ second every second.

An observer in reference frame *A*, however, sees the situation quite differently.

Relative to A, it is the clock in B that is moving (Fig. 6.12c), and the clock in B runs slow. There is no contradiction here—the observer in the reference frame A "sees" time intervals between events differently than the observer in B. Because all motion is relative, each observer may say that his reference frame is at rest and the other reference frame is moving with respect to him. Both observers can validly claim that the clock in the other reference frame is running slow. The apparent paradox of the validity of both conflicting claims will be discussed further and clarified in Section D6 below.

The phenomenon exemplified by the slow-running clock is called time dilation. There is another effect observed in moving inertial reference frames, called **length contraction,** which is similar to the Lorentz–Fitzgerald contraction previously mentioned. If a measuring stick is placed in reference frame A and another identical measuring stick is placed in reference frame B, both sticks will be unaffected by the motion as seen by observers in the same reference frame with each stick. However, if the sticks are placed parallel to each other and to the direction of relative motion of the two reference frames, when the observer in A measures the stick in B, in comparison with his own stick in A, the stick in B will appear shorter. Similarly when the observer in B measures the stick in A, the stick in A will appear shorter. Which stick is really shorter? They both are! (An alternative answer is neither one. The question is really meaningless, as will be discussed in the next few paragraphs.)

Both observers will find that the other stick (the one in relative motion with respect to the reference frame) has undergone length contraction. The stick that is at rest with respect to the reference frame maintains its **proper length.** If the sticks are placed perpendicular to the direction of motion of the reference frames, there will be no change in their length (they will be narrower, however, as observed from the other reference frame). Any object in motion in a given reference frame will undergo contraction in that dimension that is parallel to the direction of the motion.

Optional Section 6.5 *Calculation of γ*

The quantity $\gamma = 1/\sqrt{1 - v^2/c^2}$ appears quite frequently in the special theory of relativity, particularly when calculating the values of quantities in one frame of reference from their values in another frame of reference. For example, the time interval t' between two successive events in a moving frame of reference is equal to the product of γ and the proper time interval t (note that t' and t are two different time intervals) between the two events, which is the time interval in a frame of reference in which the two events occur at the same place. It is fairly easy to show how this comes about in the case of the mirror clocks discussed above.

In Fig. 6.12a, let the distance from the light source S to the mirror M be d, then the length of time required for the light to travel from S to M will be d/c, where c is the speed of light. It will take an equal length of time for the light to return to S, so that the total length of time between the two events (the initial emission of the light from S and its return to S) is $t = 2d/c$. This is the proper time, as seen by an observer moving with the mirror clock, that is, an observer for whom the two events occur in the same place. However, an observer who sees that the mirror clock is moving will see that the path length from the initial position of S to the position of M at the time the light is reflected is *greater* than d—it is the diagonal in Fig. 6.12b. Using the Pythagorean theorem, the length of this diagonal is $(d^2 + v^2t''^2)^{1/2}$, where v is the speed of the mirror clock relative to the observer and t'' is the length of time for the light to travel from S to M. Dividing this distance by c gives t''.

$$t'' = \sqrt{d^2 + v^2t''^2}/c$$

This equation can be solved for t'' by squaring both sides and collecting terms involving t''^2 on the left. This gives

$$(1 - v^2/c^2)t''^2 = d^2$$

Solving for t'' and then doubling the result to allow for the length of time for the light to be reflected by the mirror M back to the final position of S gives $t' = (2d/c)/(1 - v^2/c^2)^{1/2}$, which is equal to $\gamma(2d/c)$ as the time interval between the two events seen by this observer.

For length contraction, the relationship between distances L' observed in a moving frame of reference and the same distances L in a frame of reference for which the two points marking the distance are at rest (the proper distance) is $L' = L/\gamma$. Similarly the mass m' of an object at rest in a moving frame of reference is γm, where m is the mass of the object as determined by an observer who is moving at the same velocity as the object.

The interrelationship between time dilation and length contraction is shown most clearly by introducing time as a "fourth dimension": rather than considering space and time as distinct entities, they should be regarded as different aspects of one unified entity called **space–time.** Instead of discussing a particular location in space or a given point in time, it is more useful to consider **events** in space–time. An event is specified by describing both where and when it happened, or is going to happen. Whereas in space alone, the distance between two locations is measured, and in time alone, the elapsed time between two occurrences is measured, in space–time, the **space–time interval** between two events is measured.

One way to understand this is to consider the analogous relationships between a stick and its shadows (see Fig. 6.13). If a stick is held at some angle with the horizontal, above a table and next to a wall, it will cast shadows on the table and on the wall when properly illuminated. If the stick is illuminated from above, a horizontal shadow will be cast on the table; and if it is illuminated from the side, a vertical shadow will be cast on the wall. The length of the shadows on the table and on the wall will depend on the angle at which the stick is held. If the stick is almost vertical, the table shadow will be short and the wall shadow long; whereas if the stick is almost horizontal, the table shadow will be long and the wall shadow short. The length of the stick will be unchanged (invariant); only its shadow will change as its angle is varied.

If it were impossible to measure the length of the stick directly, it would still be possible to calculate its actual length from the lengths of the two shadows. If the length of the table shadow is X and the length of the wall shadow is Y, then according to the Pythagorean theorem, the true length of the stick is $\sqrt{X^2 + Y^2}$. No matter how the stick is angled with resulting changes in X and Y, the length of the stick is invariant.

In terms of space–time, when measurements are made of a distance between two events and an elapsed time between those same two events, the quantities being measured are the ''space-shadow'' and the ''time-shadow'' of the space–time interval between the two events. Einstein showed that if T is the elapsed time, L is the distance between the two events, and c is the speed of light, the space–time interval between the two events is equal to $\sqrt{c^2T^2 - L^2}$. Different observers in different frames of reference may measure T and L to have different values (this corresponds to changing the angles of the stick), but when they calculate the space–time interval using their own measurements, they will all get the same numerical result. The space–time interval is an invariant and does not change from one inertial frame of reference to another inertial frame of reference.

The space–time interval cannot be measured directly, but it can be calculated from measurements. The ''space-shadow'' and the ''time-shadow'' are the quantities that are measured, but the values of the measurements are relative. It is as meaningless to ask what is the correct time interval or the correct distance between two events, as it is to ask what is the ''real length'' of the shadow of an object.

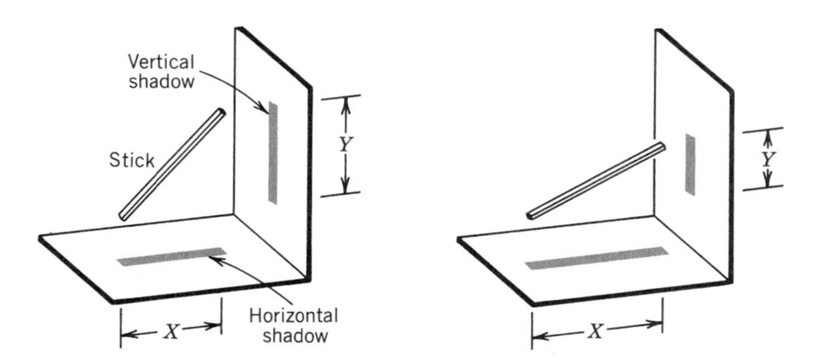

Figure 6.13. Horizontal and vertical shadows, X and Y, of a stick. Regardless of angle of stick, its length can be calculated from $\sqrt{X^2 + Y^2}$.

5. The Maximum Speed Limit

The new concepts of space and time introduced by Einstein have a number of consequences. He was able to use them to show that it is necessary to introduce the new transformation equations that had been suggested by Lorentz (Section D1 above). These equations, the Lorentz transformation, lead to new formulas for calculating relative velocities of objects in different frames of reference. In Section A2 an example was given of a ball thrown forward at 60 mph relative to a ship traveling at 30 mph relative to the ocean. The speed of the ball relative to the ocean is then 90 mph. In algebraic terms, if the velocity of the ball relative to the ship is u and the velocity of the ship relative to the ocean is v, then the speed of the ball relative to the ocean is $u + v$, according to the **Galilean transformation,** which is the name for the transformation equations used by Galileo and Newton and their successors. According to the **Lorentz transformation,** however, the speed of the ball relative to the ocean is given by $(u + v)/(1 + uv/c^2)$, with c being the speed of light. Carrying out the calculation according to the new formula, called the **Lorentz velocity addition formula,** the speed of the ball relative to the ocean will be less than 90 mph, but only by about 0.4 of a millionth of a millionth of a percent. The difference between the two calculations is immeasurably small in this case.

However, if the speed of the ball relative to the ship is 0.6 of the speed of light, and of the ship relative to the ocean is 0.3 of the speed of light, then there is a substantial difference in the speed calculated by the two different formulas. For the Galilean transformation, the speed of the ball relative to the ocean is $0.6 + 0.3 = 0.9$ of the speed of light. In the Lorentz transformation, the speed of the ball relative to the ocean is 0.76 of the speed of light. If the component speeds are 0.9 and 0.6 of the speed of light, the Galilean formula gives $1.5c$, whereas the Lorentz formula gives $0.97c$. In fact, as Table 6.2 illustrates, it is impossible to add two speeds that are even only slightly less than the speed of light and obtain a result that will be equal to the speed of light. If one of the speeds is exactly equal to the speed of light, then adding it to the other speed according to the Lorentz velocity addition formula will yield the speed of light. This means that the speed of light is constant for all observers, as demanded by Einstein's relativity principles.

Thus the speed of light turns out to be a **natural speed limit** because it is impossible, according to the Lorentz velocity addition formula, to add some additional increment of speed to an object to get it over the speed of light. This means that Newton's laws of motion as usually discussed, in particular Newton's second law, are incorrect. As discussed in Section D2 of Chapter 3, if a constant force is applied to an object, the acceleration of the object is constant; that is, a graph of velocity versus time gives a straight line increasing continually with time. However, the imposition of a maximum speed limit means that the velocity cannot increase indefinitely. The velocity may become very close to the speed of light, but may not exceed it, as shown in Fig. 6.14. As the velocity gets closer to the speed of light, it increases at a smaller rate; that is, its acceleration becomes

Table 6.2. Comparison of Results for Velocity Addition According to Galilean and Lorentz Transformations

u	v	Galilean u + v	Lorentz $\left(\frac{u + v}{1 + uv/c^2}\right)$
60 mph	30 mph	90 mph	90 mph
186 mps (0.001*c*)	18.6 mps (0.0001*c*)	204.6 mps	204.59998 mps
0.6*c*	0.3*c*	0.9*c*	0.763*c*
0.5*c*	0.5*c*	*c*	0.800*c*
0.75*c*	0.75*c*	1.5*c*	0.960*c*
0.9*c*	0.6*c*	1.6*c*	0.974*c*
c	0.1*c*	1.1*c*	1.000*c*
c	*c*	2*c*	*c*

smaller. The object behaves as if its inertia is increasing because the same force is giving less acceleration. But because mass is the measure of inertia, it can be said that the mass of the object becomes greater as its speed increases. The mass increase becomes very noticeable as the speed comes very close to the speed of light. Velocity is a relative quantity, however, so that if the object were observed from an inertial frame of reference that is traveling at the same speed as the object, then its velocity would be zero and in that frame of reference its mass will be unchanged. Thus mass is also a relative quantity.

Actually, Newton's laws of motion are not irreparably invalidated by Einstein's relativity theory. Rather, the definitions of some of the quantities that are calculated in mechanics must be changed. It is necessary to restate Newton's second law in its original form, making force proportional to the time rate of change of momentum, and specifically recognize that in calculating the momentum as the product of mass and velocity the mass is not independent of velocity.

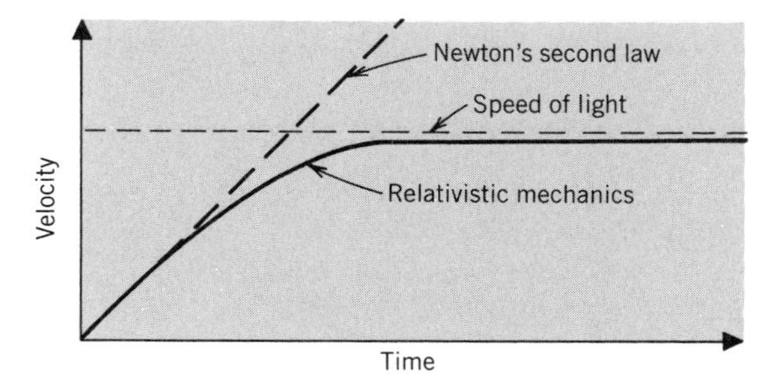

Figure 6.14. Graph of velocity versus time for constant force. In relativistic mechanics, the velocity cannot increase indefinitely, but rather is limited by the speed of light.

Similarly, it is necessary to reconsider and modify the definition of kinetic energy to include the fact that increased kinetic energy does not come from increased velocity alone, but comes from increased mass as well. The work expended to give an object kinetic energy not only increases the speed of the object, but also its mass.[10] In other words, *mass is a form of energy,* just as heat is a form of energy. Where Poincaré and others had shown that electromagnetic energy can be regarded as having mass, Einstein boldly asserted that all mass is equivalent to energy, by his formula $E = mc^2$. He pointed out that not only did the relativistic increase in mass of an object as its speed became closer to the speed of light represent an increase of energy, but even the mass of an object at zero velocity, called its **proper mass,** or **rest mass,** is a form of energy.

In principle, according to Chapters 4 and 5, energy can be converted from one form to another. Thus mass energy can be converted into light energy or gamma ray energy or vice versa. Such conversions are involved in nuclear energy and in the creation and annihilation of elementary particles of matter, as will be discussed in Chapter 8.

One important point needs to be emphasized regarding the implications of the theory of relativity for other theories of physics. Relativistic effects usually depend on extremely high velocities to be observable. Typically a speed equal to one-tenth of the speed of light will result in a one-half percent effect. The effect of a speed which is 1 percent of the speed of light will be 100 times smaller, and not measurable in most cases. Even such a speed is extremely high, 1860 miles per second, compared with the speed of supersonic rockets, which conceivably could be 1860 miles per hour—that is, slower by a factor of 3600. As a result, in most common cases, relativistic effects are not noticeable. On the other hand, relativity plays a large role in astrophysics and in atomic and nuclear physics.

It is interesting to speculate how things might appear to us if the speed of light were much less than it actually is—say 20 mph. The consequences would be interesting: For example, traffic citations for speeding would be almost nonexistent because it would be impossible to travel faster than 20 mph. Such a situation is the subject of a series of short stories written mostly in the 1930s and 1940s by G. W. Gamow.[11] In these stories, bicycles and their riders become extremely short and thin, respectively, as they pedal along, as seen by pedestrians, whereas city blocks become extremely short as seen by the bicycle riders.

6. Twin Paradox and Space Travel; Minkowski Diagrams

Sometime after he proposed the theory of relativity, Einstein was asked to consider the following problem: Suppose two identical twins are born, and at a

[10]The new formula is not just $\frac{1}{2}mv^2$ with m also increasing, but is somewhat more complicated.

[11]Some aspects of these stories distort relativity theory. In particular, three-dimensional objects moving at speeds close to the speed of light appear rotated, rather than foreshortened in the direction of motion. Nevertheless, the general thrust of the stories is useful in visualizing the effects of relativity, as well as those of quantum theory (the subject of the next chapter).

certain age twin A is sent off into space at a very high speed, so that relativistic effects become apparent. Because aging is a process that depends on time, twin B, who stays on Earth, would soon be able to tell that time is passing more slowly for twin A (time dilation, Section D4), and thus twin A becomes younger than twin B. Twin A, however, knowing that all motion is relative considers that it is B who is moving, and therefore B is younger. Of course, the way to determine with certainty which twin is younger is to bring them back together again and compare them with one another. A's spaceship reverses course and returns to Earth, so that the comparison can be made. Einstein was asked which twin was younger and why. After some thought, Einstein replied that the traveling twin, A, would indeed be younger.[12] This problem is called the "twin paradox."

The crux of the distinction between the ages of the twins is that twin A was subject to accelerations (1) when first sent off into space, and (2) when the direction was reversed for the return to Earth. In Einstein's general theory of relativity (to be discussed in Section E below, which deals with all reference frames and not just inertial reference frames) it is shown that accelerations result in a slowing of time. Without going into great detail, we can see that there is a difference in the history of twin A as compared with twin B. This is nicely illustrated with the aid of a *Minkowski diagram.* The Minkowski diagram is simply a graph of time versus distance, as shown in Fig. 6.15a. Note that in this graph, unlike those shown in Section B1 of Chapter 3, the X-axis is plotted horizontally, and the vertical axis is the cT axis rather than the T axis. (The cT axis is used to emphasize the unity of space–time, and to give the so-called fourth dimension the same kind of measurement units as the other three dimensions.) The data plotted on such a diagram are the positions of an object, and the particular times it has those positions. An object at rest in a frame of reference will keep the same value of X as time goes on, and the resulting graph will be a vertical straight line, as indicated by the line B in the diagram. An object traveling to the right is represented by the lower part of line A in the diagram, and one traveling to the left by the upper part of line A.

Such a graph is called a **world line,** and in effect tells the history of the object because it records where the object has been and when it has been there. Note that an object traveling at the speed of light would be shown by the dashed line which is at a 45 degree angle. Thus it is impossible, according to Einstein's relativity principles, for a world line to have a slope less than 45 degrees. If A and B are the world lines of the two identical twins, then Fig. 6.15a shows that their histories are different, so we can predict that their ages will be different when they are brought together again.

Figure 6.15b portrays a somewhat different situation. Instead of identical

[12]An experiment to test this has actually been carried out. Carefully stabilized *atomic clocks* have been sent on long airplane trips and returned for comparison with identical clocks that were kept in a laboratory. The traveling clocks turned out to be younger, in that they had counted off fewer time intervals (because of time dilation) than the fixed clocks.

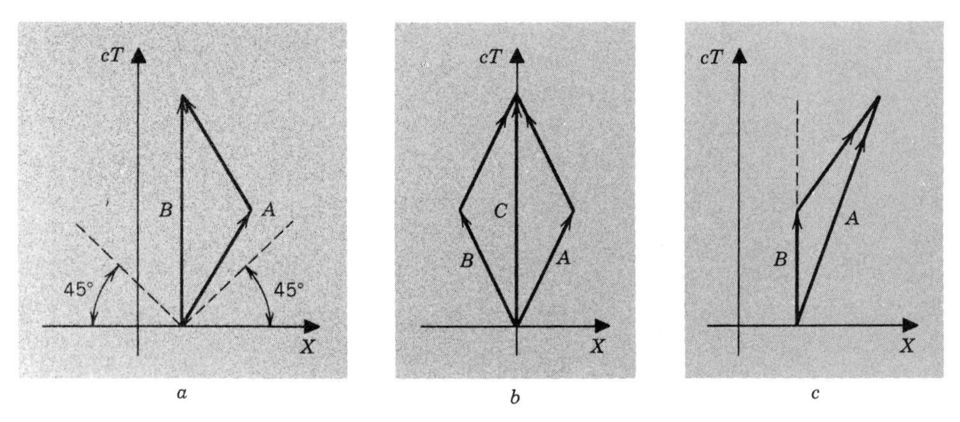

Figure 6.15. Minkowski diagrams. *cT,* time axis; *X,* space axis. (a) *A* leaves Earth and returns later. (b) Both *A* and *B* leave Earth and return later. (c) *A* leaves Earth, followed by *B* sometime later, who "catches up" with *A*.

twins, it shows the case of identical triplets, *A, B,* and *C.* While *C* stays on Earth, *A* travels to the right as before and *B* travels to the left, with respect to the Earth. When both *A* and *B* return to Earth, their ages will be identical, but less than that of *C*. Yet another situation is shown in Fig. 6.15c, which deals only with twins. In this case, instead of returning *A* to Earth, *B* is sent off on a spaceship to catch up with *A*, and this time, *B* turns out to be younger when they meet.

Relativity and space travel have greatly influenced the popular imagination. Novels, motion pictures, and television shows often depict space travel, and have done so for well over a century, starting with Jules Verne. Similarly, the idea of time as a fourth dimension, and of travel through time, goes back at least as far as Samuel Langhorne Clemens (Mark Twain), even before Einstein appeared on the scene. Two questions naturally arise: Is it possible to travel to distant galaxies? Can one travel in time?

Astronomical distances are measured in light-years, that is, the distance that light would travel in one year. In order to travel to a galaxy that is 100 light-years away, a trip of at least 100 years is required, even for a spaceship traveling at the maximum speed, the speed of light. If the crew of the ship sent back a radio message immediately on arriving, it would take another 100 years for the mission controllers on Earth to receive the message, so both the crew and the mission controllers would have to be extremely long-lived. Thus it would seem that relativity theory, through the imposition of a maximum speed limit, makes such extended space travel impossible.

The problem, however, is worth a second look. If the space travelers were traveling at 99.9 percent of the speed of light, their biological clocks would run slow, as seen from the Earth. During the entire round trip, they would have aged only 9.94 years (time dilation) even though they had been gone slightly more than 200 years. Of course, as seen from their frame of reference, the space travelers find that time passes in the normal way. They would observe, however, that the distance to the galaxy was not 100 light-years, but rather only 4.47 light-years, so that at their speed of travel the round trip should take only 9.94 years. They would find, of course, that things might well have changed considerably when they returned to Earth. For example, the Earth calendar would have advanced by 200 years, and there might be no record of their ever having departed, other than an obscure footnote in a history book.

While space travel is in some sense feasible, time travel, however, is another story. There is no question that people travel forward in time, but traveling backward in time does not seem to make sense, particularly if time travelers could participate in events of an earlier time. Cause-and-effect relationships would be destroyed. For example, children might be living before their grandparents ever met. Of course, when observations are made of occurrences in deep space, say in a galaxy 100 light-years away, what is observed is something that happened 100 years ago. The crew of the spaceship discussed above cannot affect galactic events that are observed on Earth, because they have already happened. Moreover, they cannot affect events happening now or even 5 or 10 years from now, because in terms of observations from Earth, it will take them a little more than 100 years to arrive at the galaxy.

Referring back to the Minkowski diagram of Fig. 6.15a, only those events in the part of the diagram above the two 45 degree dashed lines are accessible to travelers starting out from a given event in space–time. All other events, especially those in the negative cT direction that involve going backward in time, are not accessible.

E. General Theory of Relativity

The theory of relativity outlined and discussed above is known as the ''special'' theory of relativity, because it applies to only one special case. It indicates the correct transformation equations between nonaccelerated (inertial) reference frames. With these transformation equations, the laws of physics appear in the same form in all such inertial reference frames and are seen to satisfy a relativity principle as originally expected by Galileo, Newton, and others. Using the results of these new transformation equations, we have seen that certain unexpected consequences follow in comparing times, lengths, and masses in different inertial reference frames. These transformation equations apply only between inertial reference frames, however, and Einstein believed that one should be able to use any reference frame, including accelerated reference frames. A theory that could specify the transformation equations using accelerated reference frames would not be limited to the special case of inertial reference frames; such a theory would be a ''general'' theory of relativity.

Einstein spent the ten years or so after publishing his special theory of relativity working primarily on a general theory of relativity. In 1916 he published his complete description of such a theory, with considerable reaction from the scientific community. Although it is said that any one of several other physicists were about to arrive at the special theory of relativity when Einstein published his work (in 1905), it is generally accepted that Einstein's work on general relativity was far ahead of any of his contemporaries. Einstein's general theory of relativity stands clearly as one of the greatest achievements of the human intellect in the area of scientific reasoning.

1. The Equivalence Postulate

Einstein began his work on the general theory of relativity with an observation which, at first, seems like an aside. He noted that the same quantity—mass—is involved in both Newton's second law of motion *and* in the universal law of gravitation. He observed that if these two laws actually were independent, they would *define* two different kinds of mass. Newton's second law ($\vec{F} = m\vec{a}$) would define an **inertial mass,** and the law of gravity ($\vec{F} = Gm_1m_2/r^2$) would define a **gravitational mass.** In fact, a number of precision experiments were carried out at about that time to see whether there was any difference between the inertial and gravitational masses of an object. No difference between the two masses could be measured, and it seemed a remarkable coincidence that they should be the same. Einstein questioned whether it is coincidental that the same quantity is involved in both these laws. He decided that acceleration (as involved in Newton's second law) and gravity must be related. He then proceeded to indicate that it is impossible, in fact, to tell the difference between a gravitational force and an "equivalent" acceleration. This relationship has become known as Einstein's **equivalence postulate.**

As a simple example of the equivalence of gravity and acceleration, consider the hypothetical situation of a person standing in a spaceship in outer space, far away from any star or planet. If the spaceship is moving at a constant velocity, the person would experience "weightlessness," just as the astronauts did on their trips to the Moon. If the spaceship were resting on a planet, the person would experience a force of gravity, giving the person weight. Einstein's postulate indicates that the same effect could be produced by simply accelerating the spaceship. We are all aware of the basic idea involved. When an elevator car starts moving upward, one feels the acceleration and is pushed down toward the floor. (This is Newton's third law of motion. The bottom of the elevator pushes up on the passenger and the passenger pushes down on the elevator.) If one were to make the acceleration of a spaceship in outer space continuous and exactly equal to the acceleration of gravity on the Earth's surface, it would create a "downward" force that would feel like gravity. If the spaceship had no windows, it would be impossible for the person inside, either by "feel" or even by careful experiment, to tell whether the force was due to an acceleration or to a gravitational force from a large mass.

Thus we have a situation where a force can be simulated by an acceleration.

Many other examples exist. The acceleration does not have to be in a straight line. When a fast-moving car goes around a corner, the riders all feel a "force" pushing them toward the side of the car. This force is referred to as a "centrifugal" force; it is not a force at all, but rather is just the effect we feel as our inertial mass tries to continue in straight-line, uniform motion while the car goes around the corner. If there is no restraining force (a so-called *centripetal* force, as discussed in Chapter 3), the riders will accelerate relative to the car. Relative to the ground, however, they continue in straight-line uniform motion.

The centrifugal force is a "fictitious" force in that it is not caused by some external force exerted on a body, but rather is due to the acceleration of the car and the inertia of the object. Such fictitious forces are often called **inertial forces.** Note that the centrifugal force can also be used to simulate gravity. One suggested way to provide a "gravitylike" environment in space is to make a space station in the shape of a spinning torus (doughnut-shaped). Such a space station was portrayed in the popular science-fiction movie *2001,* for example. The centrifugal force resulting from the spinning torus would push everything (and everyone) toward the outer edge of the torus. If the torus were to spin at just the right rate, it could provide a centrifugal force just equal to the force of gravity on the Earth's surface.

The important thing about a centrifugal force, for the present discussion, is that it is actually a fictitious force, which arises only when one tries to use an accelerated reference frame. In an inertial (nonaccelerated) reference frame, this force does not appear. Because it is mass that determines the magnitude of this fictitious force (i.e., inertial mass), Einstein wondered if it could be true that gravity, which is also due to mass (gravitational mass), could also be regarded as a "fictitious" force, whose existence depended on the choice of reference frame. Basically, this idea arises from his equivalence postulate—that inertial and gravitational masses are the same. Thus Einstein felt it was not coincidental that mass is the primary thing responsible for both inertia and gravity, but that they must be related.

In fact, it is possible, by suitable choice of reference frame, to "get rid of gravity" even on the surface of the Earth. Suppose a person stands on a scale in an elevator that is moving with constant velocity. The scale will indicate his weight. If the cable attached to the elevator breaks, the elevator falls with the acceleration of gravity. While the car is falling, the scale will actually read *zero,* indicating that the person is weightless. *In the frame of the falling elevator, the force of gravity has vanished.*

The equivalence principle does much more than assert the identity of gravitational and inertial mass. It states that any effect that can be ascribed to an accelerated reference frame could equally well be called a gravitational effect. Figure 6.16 shows an accelerating spaceship in outer space. (The dotted line running through parts *a, b, c* of the figure shows the location of the midpoint of the floor of the spaceship as time goes on. See Fig. 3.8b in Chapter 3.) A ball is thrown across the ship and takes two seconds to hit the other wall. Figure 6.16b shows the location of the ball and the spaceship after 1 second and Fig. 6.16c

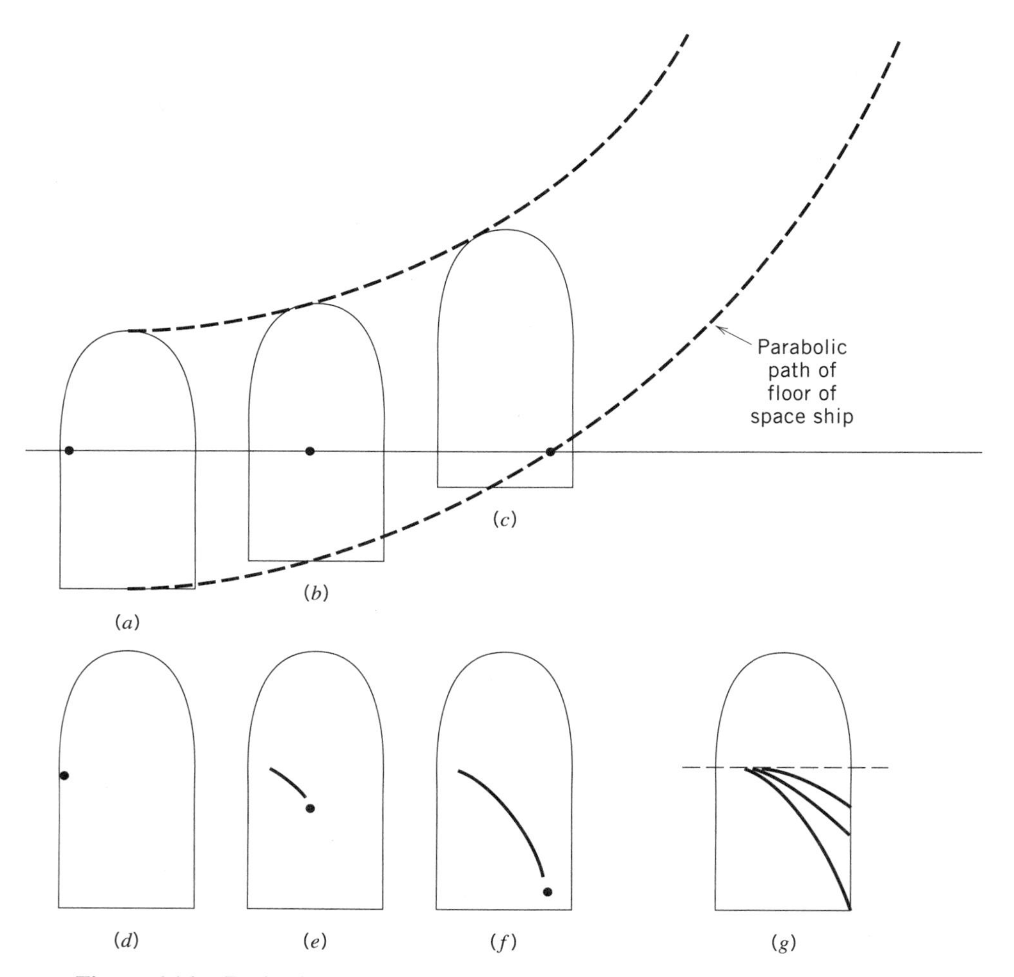

Figure 6.16. Projectile trajectory in accelerating spaceship. (a,b,c) As seen from outside the spaceship. (d,e,f) As seen from inside the spaceship. (g) Internally viewed trajectories for different horizontal velocities.

after two seconds, as seen from outside the spaceship. The ball has traveled in a straight line. Figures 6.16d, 6.16e, and 6.16f show the location of the ball as seen from *inside* the spaceship. As seen from inside the spaceship, the ball has traversed a parabolic path. Observers inside the spaceship can describe the trajectory of the ball as resulting from combination of horizontal motion and falling motion due to gravity, as did Galileo in his analysis of projectile motion (Chapter 3, Section B2 and Fig. 3.11).

If the horizontal speed of the ball is increased, it will travel a straight line as seen from outside the spaceship and a parabola as seen from inside the spaceship, although it will not ''fall'' as far. Even if the speed of the ball were equal to the speed of light, as seen from inside the spaceship it would *still* travel a parabolic trajectory and seem to *fall* a little. Figure 6.16g shows the effect of increasing the horizontal speed of the ball, as seen from inside the spaceship.

According to the equivalence principle, all of the effects shown in Fig. 6.16, as seen from inside the spaceship, can be ascribed *either to the acceleration of the spaceship or the effect of a gravitational field.* If a ray of light were to travel across the spaceship, it too would follow the same path as the ball traveling at the speed of light. Thus according to the equivalence principle, a ray of light can be ''bent'' by a gravitational field.

Einstein interpreted the bending of a ray of light as representing a **curvature of space** itself. He reasoned that we only know about space through measurements with light. A carpenter or machinist determines whether a surface is flat or curved by measuring it with a straightedge or ruler. But how does one know whether the ruler itself is straight? One ''sights'' along the ruler—if it deviates from the line of sight, then it is not straight. It is the line of sight (i.e., the path of a ray of light) that determines whether the ruler is straight. If the path of a light ray itself is not ''straight'' but ''curved,'' then space itself is curved.

Einstein concluded that gravity must be a fictitious force due simply to accelerated motion of a reference frame. This accelerated motion, he reasoned, was movement through ''curved'' space. His general theory of relativity states that large concentrations of mass cause space near them to be curved. The motion of any object through this curved space is necessarily accelerated, and it will feel a force simply because of the curvature of the space. Thus Einstein's general theory of relativity indicates that space can be curved, and we must pause and consider this idea briefly.

2. Curved Space and Non-Euclidean Geometry

It seems peculiar that space can be thought to be curved. It is hard to even imagine what this means. Perhaps it is best to start by considering curved two-dimensional spaces and then to return to three-dimensional (i.e., normal) space. A two-dimensional space is a surface. The surface may be flat, such as a smooth table top, or curved like the surface of a sphere. A flat, two-dimensional surface is referred to as a Euclidean surface (after the famous mathematician who developed many of the ideas of geometry). A curved surface is said to be non-Euclidean, because Euclid's theorems on geometry do not apply. On a flat surface, as Euclid noted, parallel lines never meet and the sum of the three angles of a triangle is 180 degrees. On a non-Euclidean surface, these simple theorems do not apply. For example, the sum of the angles of a triangle drawn on the surface of a sphere is *greater* than 180 degrees. Consider a triangle drawn on the Earth's surface, with the base along the equator and the two sides due-north directed lines meeting at the North Pole. The angles between the equator and the sides are both 90 degree angles. Thus these two angles total 180 degrees by themselves. The angle at the North Pole can be of any size from 0 degrees to 180 degrees; thus the total sum of the angles can be from 180 degrees to 360 degrees.

In general, a curved surface where the sum of the three angles of a triangle is greater than 180 degrees is said to have positive curvature. Curved surfaces on which the sum of the three angles of a triangle is less than 180 degrees are said to have negative curvature. (An example of a surface with negative curvature is a riding saddle.)

Because we wish to know whether or not our three-dimensional space is curved, let us first consider whether a person confined to a two-dimensional space can tell if it is curved. We do not want to determine if the two-dimensional space is curved by climbing a tower and looking out toward the horizon, or by traveling upward in a spacecraft and looking back. These methods involve moving into the third dimension to look back at the two-dimensional surface, and we cannot do the equivalent of this to look at three-dimensional space. We must confine ourselves to the two-dimensional surface and try to determine if it is flat or curved.

As already indicated, it is possible to determine if the two-dimensional surface is curved by doing geometry. We can, while confined to the surface, study parallel lines, triangles, and other constructions formed of (what we believe to be) straight lines. If the parallel lines eventually meet or intersect, or if triangles have angles that total other than 180 degrees, we know that we are on a curved surface. Similarly in order to study three-dimensional space, we must study "straight lines" and their geometry.

Actually, it is not "straight lines" that we must study. As every student in beginning geometry learns, a straight line is the shortest distance between two points. However, this is true only in a "flat" surface or three-dimensional space without curvature. On a sphere, the shortest distance between two points is a "great circle" (the line formed by intersecting the sphere with a plane passing through both points and the center of the sphere). In general, the shortest distance between two points in a given space is called a **geodesic.** It is generally considered that light will travel along the shortest path possible between two points; that is, a light ray will define the geodesic. Thus one possible way to determine if three-dimensional space is curved is to determine if light always travels in straight lines.

3. Experimental Tests of General Relativity

Einstein's consideration of the effects of transforming to accelerated reference systems led him to the equivalence postulate, and hence to the prediction that gravity was due to motion through curved space. We wish now to review some of the experimental evidence that proves that this reasoning is basically correct.

Perhaps the most convincing experimental evidence for the curvature of space is the deflection of light rays passing by the Sun. Einstein's prediction that mass causes space to be curved requires large masses to produce greater curvature. Thus the largest curvature of space close to the Earth would be expected near the Sun. If we accept that light rays will define the geodesics in space, then we can test Einstein's prediction if we can carefully observe light rays passing by the Sun—for example, from a distant star. The only way to do this is during a total eclipse of the Sun, when the Moon passes between the Earth and the Sun and momentarily blocks out the light from the Sun.

If the light rays from different stars can be observed and compared during an eclipse, one could test whether the light rays passing near the Sun are deflected. The situation is depicted in Fig. 6.17. If the Sun does cause space to be curved, then the starlight passing near the Sun will be deflected, and the apparent position of the star, as determined by looking back from the Earth, will be shifted from its true position. The shift could be detected by carefully comparing a photograph taken on another date of the same star field when visible at night (and hence when the Sun is not near the line of sight).

The prediction for the deflection of light rays near the Sun was confirmed in a famous experiment performed by a team led by the English physicist Arthur Eddington during a total eclipse of the Sun in 1919. Eddington traveled to an island off the west coast of Africa where the eclipse was visible long enough to obtain good photographs of the star field near the Sun. When he studied the photographs of the same star field taken months earlier, he discovered that the stars observed near the Sun appeared to have moved as compared with stars not near the Sun during the eclipse—and in about the amount predicted by Einstein. This experiment was quite convincing.

Newton's universal law of gravitation predicts that light, which is believed to be massless, should be completely unaffected by the mass of the Sun. Einstein predicted, in contrast, that mass would curve space itself, and light rays traveling along geodesics through space would be deflected by a large mass. The verification of Einstein's prediction by Eddington's experiment received considerable attention in the media at that time and contributed significantly to Einstein's reputation as the new Newton of science.

Besides the deflection of starlight by the Sun, there is other experimental evidence for general relativity. In Chapter 3 (Newtonian Mechanics) we discussed the fact that the mutual interaction of all the other planets with the planet Mercury causes the elliptical orbit of Mercury to rotate slowly, or precess, around the Sun.

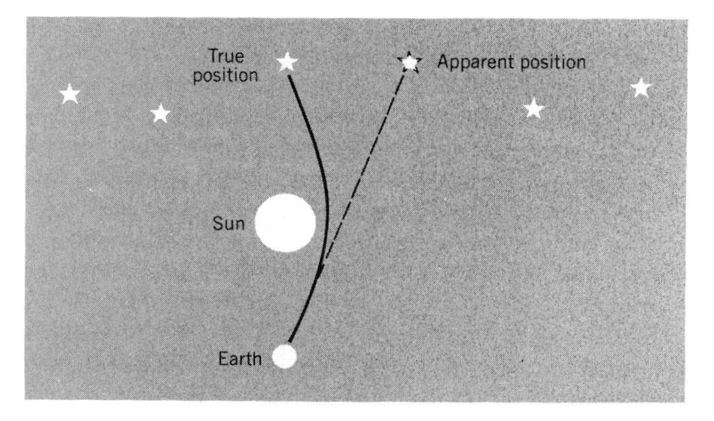

Figure 6.17. Gravitational deflection of starlight by the Sun.

This slow precession of the orbit of Mercury has been known for some time and amounts to 574 seconds of arc per century. By about 1900, the gravitational attractions of the other planets had been calculated carefully and were known to be responsible for about 531 seconds of arc per century. The discrepancy of about 43 seconds of arc was not understood until Einstein's general theory of relativity was considered. Because the orbit of Mercury is very close to the Sun, the planet is actually traveling through a portion of space with a small but significant amount of curvature. The effects of this curved space can be shown to account for essentially all of the remaining 43 seconds of arc. This discrepancy for the precession of the orbit of Mercury is sometimes referred to as one of the "minor" problems of nineteenth-century physics. Most physicists believed the problem could be solved by some small, but unimportant, correction to the analysis of the situation. Instead, we see that a revolutionary new view of the nature of space is required to solve the problem.

There exist other, somewhat more complicated, experimental verifications of general relativity as well. These include the slowing of clocks in accelerated reference frames or large gravitational fields, and the "red shift" of light falling in a gravitational field. It is not necessary for us to consider these other tests here, but only to note that the basic principles of general relativity are now well established. At the same time, we must note that there exist several *forms* of the general theory of relativity, including the form originally proposed by Einstein. The results of these theories do not differ from Einstein's version insofar as the matters discussed in this text are concerned.

4. Cosmological Consequences

The acceptance of the need for the general theory of relativity and its prediction of curved space leads to some important specific consequences regarding the nature of our physical universe. These consequences include an overall curvature of space and the possibility of so-called *black holes.* If a large concentration of mass causes space near it to be curved, then there is a possibility that a sufficiently large mass may cause space to curve enough to actually fold back onto itself. This may occur either for the universe as a whole, or locally near a very concentrated mass. If there exists enough mass in the universe, and if the universe is not too large, then the space around the universe will eventually fold back onto itself. In such a situation, a light ray sent out in any direction would not continue outward in a straight line forever, but would eventually "fall" back into the mass of the universe. The mass density of the universe would determine the actual size of the space associated with the universe. If the total mass density of the universe is not great enough, however, the curvature of space would not be sufficient to cause space to eventually fold back on itself. Such a situation would produce an infinite universe—that is, without spatial limit.

There is an ongoing controversy whether the actual physical universe is "closed" (i.e., does fold back on itself) or "open" (i.e., never folds back on itself). Although general relativity predicts that space must be curved (because mass exists), it cannot say whether it is closed or open. The answer depends on the total mass and size of the universe, but astrophysicists find it difficult to make an accurate determination of the total mass or size of the universe.

It *is* known that the universe is expanding in size. This is presumably due to a great explosion, known as the *Big Bang,* which apparently began the universe as we now know it. Astronomers can observe that light received on Earth from distant galaxies is always shifted in color toward the red part of the light spectrum, indicating large velocities directed away from the Earth.[13] The farthest known galaxies are believed to be about 15 billion light-years away (the distance light would travel in 15 billion years) and are receding at a tremendous rate. Although there are ambiguities in the determination of the actual size of the universe and the rate of expansion, there are several pieces of evidence that convince astrophysicists that the Big Bang occurred. These include the distribution and velocities of galaxies in the universe and the observation of a low-energy microwave background, which is believed to be the direct result of the Big Bang explosion. Because it is impossible to know anything about the state of the universe before such a Big Bang explosion, this event could be considered to be the ''beginning of time.''

Since the universe is expanding, the question arises as to whether the universe will continue to expand forever, or will eventually fall back upon itself. The answer depends on whether the universe is open or closed. If the mass density is great enough, the space of the universe is closed and the expansion of the universe eventually will stop and the universe will fall back upon itself. If the density is not great enough, the universe is open and the expansion will continue forever. If one considers all the ''bright'' mass of the universe (i.e., mass in stars and any other objects emitting light), then there is estimated to be less than one percent of the so-called ''critical'' mass that would be required to close the universe; however, recent studies of the motions of galaxies and clusters (and ''superclusters'') of galaxies seem to indicate the presence of just enough matter to make the universe eventually stop expanding. Since most of this mass is apparently not emitting light, it is referred to as ''dark'' matter. This dark matter may include giant ''halos'' of mostly hydrogen gas surrounding galaxies and invisible black holes. A number of astrophysicists note that the initial conditions of the Big Bang (i.e., the total energy of the ''explosion'') needed to be ''adjusted by God'' very carefully in order for the universe to be expanding as we now see it. If the energy had been a little less, the universe would have quickly collapsed back on itself. If the energy had been a little more, it would have blown completely apart quickly. Astrophysicists increasingly are struck by just how close to the ''critical'' mass the total mass of the universe appears to be and wonder if the mass isn't *exactly* equal to this critical amount. If this is the case, then the universe will eventually just stop expanding, but never fall back on itself (i.e., recollapse).

On a local scale, general relativity indicates that if there exists a sufficiently concentrated mass, it could curve space back upon itself in the immediate vicinity.

[13]This is known from the Doppler effect, which is a well-known phenomenon associated with wave phenomena, involving relative motion of source and receiver.

Such a situation has become popularly known as a *black hole.* The mass density required is tremendous. It is greater than that in a normal star, or even in a "neutron" star. Black holes are expected when a massive star (greater than several times the mass of the Sun) eventually burns up its nuclear fuel and collapses. When the collapsing star reaches a size of only about 30 kilometers (18 miles), it curves space around itself enough to become a black hole. Because nothing can escape from such a folded-space object, including light rays that will follow the curved space, the object is called a black hole.[14] Because nothing can escape from the black hole, such an object cannot be seen, but has to be detected by its effects on other nearby objects. Several strong candidates for black holes are known by astronomers. They exist as partners with normal visible stars (binary systems). The normal stars can be observed to be rotating around unseen companions, and radiation (with particular characteristics) is emitted, indicating mass is being pulled from the normal stars into the black hole. It is now believed that the ultimate fate of most massive stars (with masses greater than about seven times the mass of the Sun) is a black hole; furthermore, it is also believed that there may be a black hole at the center of most galaxies, including our own galaxy, the Milky Way.

OPTIONAL SECTION 6.6 *The Size of Black Holes*

Using the ideas developed in Chapter 3, we can make a simple estimate of the "size" of a black hole. By its size, we mean that radius within which nothing can escape, including light. This radius is usually referred to as the *Schwarzchild radius* of the black hole.

In order for an object to escape from a body with mass, the object must have a large enough velocity to overcome the gravitational pull by the body. More specifically, the object must have a kinetic energy great enough to roll "up the hill" formed by the gravitational energy. Hence, from Chapter 4 (see Optional Section 4.1),

$$\frac{1}{2}mv^2 = mgh$$

Now mgh is the gravitational potential energy for an object on the surface of the Earth, where g is the acceleration of gravity and h is height of the hill. More generally, from Optional Section 3.7, we saw that

$$g = \frac{GM}{R^2}$$

[14]Actually, the application of the ideas of quantum theory (the subject of the next chapter) may allow a black hole to slowly emit electromagnetic radiation. See the article and book by Stephen Hawking listed in the references at the end of this chapter.

where M is the mass of the body and R is its radius. Furthermore, h is the height of the hill, which in this case is really just R, the radius of the body. Hence,

$$mgh = m\left(\frac{GM}{R^2}\right)R = \frac{GmM}{R}$$

so we have

$$\frac{1}{2}mv^2 = \frac{GmM}{R}$$

or

$$v^2 = 2\frac{GM}{R}$$

The velocity, v, is known as the "escape velocity." If we want the escape velocity to be equal to the velocity of light ($c = 3 \times 10^8$ m/sec), we have

$$c^2 = 2\frac{GM}{R}$$

or, solving for R,

$$R_s = 2\frac{GM}{c^2}$$

This is called the Schwarzchild radius for an object with mass M.

As an example, let us calculate this radius for a mass equal to the mass of the Sun ($M = 2 \times 10^{30}$ kg). We then have

$$\begin{aligned} R_s &= (2)\left(6.7 \times 10^{-11} Nt \cdot \frac{\text{m}^2}{\text{kg}^2}\right)(2 \times 10^{30}\text{kg}) / (3 \times 10^8 \text{m/sec})^2 \\ &= 2950 \text{ m} \\ &\simeq 3 \text{ km} \end{aligned}$$

This would be the radius from within which nothing could escape from a black hole having the mass of our Sun. (Our Sun is not a black hole because its mass cannot collapse to such a small size because of the pressure from the energy released in the nuclear reactions.) Actually, it is believed that the fate of a star depends on its mass: if the star has a mass less than about 1.4 times the mass of our Sun, the star, when it finally exhausts its nuclear fuel, will contract to form a very dense star known as a "white dwarf." If the star is more massive, it will end up as a neutron star, also known as a pulsar. Finally, if the mass of the star is at least several times the mass of our Sun, when the nuclear fuel is gone, it is believed that the star will collapse to form a black hole. Note, as seen from the formula for the Schwarzchild radius obtained above, that the radius of a black hole is directly proportional to the mass (M); hence the radius of a black hole with mass 10× that of the Sun would be 30 km. This is still very small, for such a large mass!

5. Unified Field Theories

General relativity indicates that the properties of space (and time) are dependent on gravitational forces and the presence of matter. The properties of space–time are determined by light rays, which we have seen depend on electromagnetic fields. Einstein felt that these electromagnetic fields should also modify the curvature of space, and he proceeded to try to incorporate electromagnetic forces into general relativity to obtain what became known as a *unified-field theory.* Although Einstein and others made considerable progress in this direction, it is now clear that a final understanding of the basic forces in nature requires one to incorporate the basic ideas of quantum mechanics. These ideas will be presented in the next two chapters, and we will return to the subject of unified field theories later.

F. Influence of Relativity Theory on Philosophy, Literature, and Art

The theory of relativity has had a significant impact on philosophy and literature and the visual arts. In some respects, it has been used to validate or support ideas that were developed anyway; in other respects it has inspired new modes of thought, although not always from a correct understanding of the theory. The idea of frame of reference in physics has been transferred to philosophy and morality as defining basic viewpoints and outlooks. Thus the physical concept of no absolute frame of reference has suggested to some individuals moral relativism, without consideration of the idea of invariant quantities, which is really an essential component of the theory of relativity.

At the same time that the theory of relativity was being developed, new modes of expression were being developed in the arts and literature. Some of the Cubist painters were familiar in some fashion with the relativity theory and incorporated and interpreted its themes in their work. Poets such as William Carlos Williams and Archibald MacLeish celebrated both relativity and Einstein, and limericks were composed about the Stein family (Gertrude, Ep-, and Ein-) and relativistic time travel. Vladimir Nabokov and William Faulkner used relativistic concepts metaphorically. The role of Einstein as a Muse for the arts and literature is discussed in a book by Friedman and Donley, listed in the references for this chapter.

STUDY QUESTIONS

1. How is the theory of relativity related to the question of the validity of the geocentric versus the heliocentric theory of the universe?
2. Which quantities are considered to be relative in Galilean–Newtonian relativity?
3. What is a frame of reference? Why is it necessary to have a frame of reference? What are coordinates? Give an example of coordinates.
4. What is an inertial reference frame? A noninertial reference frame?
5. How can a noninertial reference frame be recognized?
6. What are transformation equations?
7. What are invariants? Name some invariant quantities in Galilean–Newtonian relativity.
8. What is the triboelectric effect?
9. What is Coulomb's law?
10. What is an electric field? How can one tell whether an electric field is present?
11. How are electric fields represented graphically?
12. What is the name given to magnetic charges?
13. What is an electric current?
14. What is a magnetic field?
15. How are magnetic fields generated?
16. What is the effect of a magnetic field on a stationary (at rest) charge?
17. What is the effect of a magnetic field on an electric current?
18. In what way is a magnetic field similar to an electric field? In what way is it different?
19. How did the study of magnetic fields and their interactions with electric charge lead to some apparent paradoxes?
20. What are electromagnetic fields and how are they generated?
21. What do electromagnetic fields have to do with light? What is the electromagnetic spectrum?
22. What is the electromagnetic ether?
23. Define for a wave: frequency, wavelength, velocity. How are these related to each other?
24. What is a transverse wave? A polarized wave?
25. What were thought to be the properties of the electromagnetic ether?
26. What is star aberration?

27. What is ether drift? Why was it important to measure ether drift?
28. Describe some experiments to measure ether drift.
29. Describe an experiment to measure ether drag.
30. Why did all the ether experiments fail?
31. What is so "special" about Einstein's special theory of relativity?
32. What are the two principles governing Einstein's special theory and why did they lead to a reexamination of concepts of space and time?
33. What were Newton's concepts of space and time? What were the criticisms leveled against them?
34. How are time and space related to each other?
35. What does it mean to say that "concepts are defined by measurements"?
36. What does it mean to say that time represents a "fourth dimension"?
37. How does the introduction of the concept of space–time affect our concepts of space and time?
38. What is time dilation? Length contraction?
39. What are the Galilean transformation and Lorentz transformation and how are they related to space–time concepts?
40. Why is the speed of light (in empty space) a "natural" speed limit?
41. How does a "natural" speed limit lead to the idea that mass is a form of energy?
42. If the speed of light in empty space were smaller by a factor of a million, how would our physical observations be different?
43. If the speed of light in empty space were larger by a factor of a million, how would our physical observations be different?
44. What is the "twin paradox," and how is it resolved?
45. How does the theory of relativity affect the possibilities of space travel?
46. How does the theory of relativity affect the possibilities of time travel?
47. What is Einstein's equivalence postulate?
48. Give a few examples of inertial "forces."
49. Give examples of both positive and negative curved two-dimensional spaces. How are they defined?
50. What is a geodesic?
51. State some of the experimental evidence for space being curved.
52. Is Einstein's general theory of relativity accepted today as perfectly correct?
53. What does it mean if the universe is "open"? "closed"?
54. What is the physical origin of a black hole? Has one been observed?

PROBLEMS

1. Automobile batteries are sometimes rated in *ampere-hours.* For example a battery rated at 200 ampere-hours will deliver 200 amperes for 1 hour or 50 amperes for four hours. This really represents the number of coulombs of electric charge that can be moved through an external circuit. How many coulombs will a battery rated at 200 ampere-hours deliver to a circuit?

2. When a 300-watt light bulb is used in an electric circuit for which the voltage or potential difference is 115 volts, (a) how much current flows through the light bulb? (b) How much is the resistance of the bulb under these conditions?

3. People sometimes connect too many appliances to one wall outlet, and as a result a fuse is ''blown.'' If a particular outlet is on a line that is fused at 15 amperes, how much is the maximum total power of all the appliances that may be connected without blowing the fuse? Assume the voltage at the outlet is 110 volts.

4. Many travelers use an immersion heater to prepare cups of hot beverages in their hotel room. In a particular instance, a heater rated at 240 watts when plugged into a 120-volt outlet heats a cup of coffee in 5 minutes. (a) How much energy, in kWh, is used to heat the coffee? (b) At 20 cents per kWh, how much does it cost to heat the coffee? (c) How much is the current in the heater when it is in operation? (d) How much is the electrical resistance of the heater?

5. The owner of the heater in problem 4 attempts to connect the heater to an automobile electric system, which operates at 12 volts. (a) Assuming the electrical resistance of the heater is unchanged, how much will be the current through the heater? (b) How many watts of heat will be generated? (c) How long will it take to heat a cup of coffee under these circumstances?

6. Calvin and Hobbes are traveling in a space ship at 90 percent of the speed of light. Normally, Calvin's heart runs at 75 beats per minute. (a) After getting over his initial excitement, Calvin relaxes and checks his heartbeat rate. How much should it be? (b) His parents at home are also able to check his heartbeat rate. How much do they find it to be?

7. In Section D6 of this chapter it was stated that space travelers traveling at 99.9 percent of the speed of light would find that a galaxy seen from the Earth to be at a distance of 100 light-years to them would seem to be only 4.47 light-years distant. Make the calculation to verify this.

8. There are 28.35 grams in an ounce. (a) If an amount of gold weighing one ounce were completely converted to heat energy, how many joules of energy would be ''produced?'' Remember that the $E = mc^2$ formula requires that m be expressed in kilograms (1000 grams = 1 kilogram) and c be expressed in meters/sec). (b) If this amount of heat were converted into electricity at an efficiency of 30 percent, how many kilowatt hours of electricity would be

generated? (c) If the power company sold this electricity at 20 cents per kilowatt hour, what would be the cost to the consumer?

9. Space travel to planets in a distant galaxy at a speed of 99 percent of the speed of light is at present in the realm of science fiction. However, subatomic particles do travel at such speeds. Such a particle is the muon, which has a proper lifetime of 2.2 millionths of a second, and which travels at 99 percent of the speed of light. (a) As observed in a laboratory fixed to the Earth, how much is its lifetime? (b) During that lifetime how far will the muon travel? (c) As seen by the muon, how far should it travel? (d) Do parts (b) and (c) give the same answer? Why?

REFERENCES

Hermann Bondi, *Relativity and Common Sense,* Dover Publications, New York, 1980.
An introductory, essentially nonmathematical discussion of relativity, written for the lay public.

Max Born, *Einstein's Theory of Relativity,* Dover Publications, New York, 1962.
A thorough discussion, using only algebra and simple geometry, of mechanics, optics, and electrodynamics as involved in relativity theory.

Barry M. Casper and Richard J. Noer, *Revolutions in Physics,* W. W. Norton, New York, 1972.
Chapters 12–15 provide a good low-level discussion of relativity using very simple mathematics.

A. J. Friedman and Carol Donley, *Einstein as Myth and Muse,* Cambridge: Cambridge University Press, 1985.

Kenneth W. Ford, *Basic Physics,* John Wiley & Sons, New York, 1968.
A physics textbook using only algebra and trigonometry, with good discussion. See Chapters 1, 4, 15, 16, and 19–22.

G. W. Gamow, *Mr. Tompkins in Paperback,* Cambridge University Press, New York, 1971.
See footnote 11 in Section D5 of this chapter. Chapters 1–6 deal with relativity.

S. W. Hawking, "The Edge of Spacetime," *The American Scientist,* Vol. 72, no. 4, July–August, 1984, pp. 355–359.
An interesting discussion of some recent ideas in general relativity, especially concerning the application of the ideas of quantum theory (the subject of Chapter 7 of this text).

S. W. Hawking, *A Brief History of Time,* Bantam Books, New York, 1988.
A lively book written for the lay public containing various thoughts and explanations of modern physics and cosmology. Contains some speculative ideas.

Gerald Holton and Stephen G. Brush, *Introduction to Concepts and Theories in Physical Science,* 2nd ed., Addison-Wesley, Reading, Mass., 1973, Chapter 31.
See comment in references for Chapter 1.

Robert Bruce Lindsay and Henry Margenau, *Foundations of Physics,* John Wiley & Sons, New York, 1936.

An advanced textbook with discussions of basic metaphysical assumptions in physics. See Chapters I, II, VII, and VIII.

Robert Resnick, ''Misconceptions About Einstein,'' *Journal of Chemical Education,* Vol. 57, December 1980, pp. 854–862.
An article containing interesting biographical information.

D. W. Sciama, *The Physical Foundations of General Relativity,* Anchor Books, Garden City, New York, 1969.
Written for the lay public, but contains rather sophisticated, subtle discussions.

Edmund Whittaker, *A History of the Theories of Aether and Electricity: Vol. I, The Classical Theories; Vol. II, The Modern Theories,* Philosophical Library, New York, 1941; Tomash Publishers, Los Angeles and American Institute of Physics, New York, 1987.
Volume I of the 1987 printing of these books has a special introductory note commenting on limitations of the author's style and viewpoint, in particular, the author's suggestion that Einstein's contribution to relativity theory was minimal!

Max Planck. (American Institute of Physics Niels Bohr Library
W.F. Meggers Collection.)

Chapter 7

QUANTUM THEORY AND THE END OF CAUSALITY

You can't predict or know everything

A. INTRODUCTION

In this chapter we discuss a physical theory that is, in many ways, far more revolutionary and extensive in its implications than Einstein's relativity theories. Initially, a number of shortcomings of theories discussed so far (Newtonian mechanics, Maxwellian electromagnetism, thermodynamics) will be pointed out, and then the quantum theory will be discussed. It is fair to ask why old theories should be discussed, if they are to be immediately discarded. Aside from the fact that one can best appreciate a new concept if one knows just how it is better than the concept it replaces, it is also true that old theories have a certain degree of validity and are often extremely useful, and many applications are based on them. As will be pointed out below, in their fullest application the new theories are quite complex, and it is simply easier to use the old theories, provided their limitations are recognized.

Moreover, we have now become sufficiently sophisticated that we can recognize that at some future time even the latest theories or concepts may turn out to be flawed. It is therefore useful to know something about the metaphysical soul-searching that scientists and philosophers have carried out over the past one hundred years or so in their efforts to achieve a better understanding of how physical knowledge is acquired and theories validated.

1. Physical Theories, Common Sense, and Extreme Conditions

Scientific theories are usually expected to be logical and to make sense. Assuming that one can keep up with and assimilate all the numerical analyses and intricate mathematical relationships involved in a theory that describes scientific phenomena, one also expects a theory to be reasonable and not violate "common sense." Of course, common sense is a subjective concept, and depends very much on the range of experience of an individual or groups of individuals. Nevertheless, a new theory should not contradict major theories or ideas that are already accepted and proven, unless it can be shown that the accepted ideas contain flaws, which the new theory does not.

In considering a theory, one prefers to take very simple cases and verify that the predictions of the theory for them are reasonable and not contradictory. If the theory is to have any validity for a complicated case, surely it should be valid for a simple case. Similarly, the *range* of validity of a theory has to be tested; it must satisfy extreme cases. Indeed, it is in the attempt to test the applicability of theories to extreme cases that their limits of validity are often determined, and the need for new theories is often recognized. Sometimes it is possible to introduce modifications into the existing theories, and thereby handle the extreme cases.

Quantum theory, the subject of this chapter, appears to violate common sense. But quantum theory was developed to deal with very small objects (i.e., objects that are the size of atoms and molecules or even smaller) for which previous experience does not apply. Thus it is not surprising that Newtonian theory failed for such objects. Newtonian theory was developed to account for the motion of very large objects, relatively speaking—bits of dust, bullets, cannonballs, and planets, all visible to the naked eye or with ordinary microscopes—and there was no sufficient proof that its domain of validity extended to very small objects. It is also true that Newtonian theory does not work very well at very low temperatures (approaching absolute zero), because here again the small scale motion of matter becomes significant.

2. Relationship of Quantum Theory to Other Theories

It is customary to refer to theories derived from Newtonian mechanics and from Maxwell's electromagnetism as "classical" physics, whereas Einstein's relativity theory and quantum theory are referred to as "modern" physics. (It is almost a hundred years since the inception of modern physics, but we still refer to it as "modern.") As already mentioned, classical physics fails when it is used to describe phenomena taking place under extreme conditions—very high speeds or very small (atomic) dimensions, or very low or very high temperatures.

Figure 7.1 shows the domains of applicability of classical physics and the two great theories of modern physics, relativistic mechanics and quantum mechanics. Note that the horizontal scale of this figure is distorted in such a way as to give more emphasis to smaller scale phenomena. The letters N, A, H, E on the scale denote, respectively, the size of the nucleus of an atom, an atom, human beings, and the Earth. Similarly, the vertical scale is distorted to exaggerate lower speed phenomena. The letters *So* and *c* denote the speed of sound and the speed of light in empty space. Although the figure seems to be divided into four parts, it should

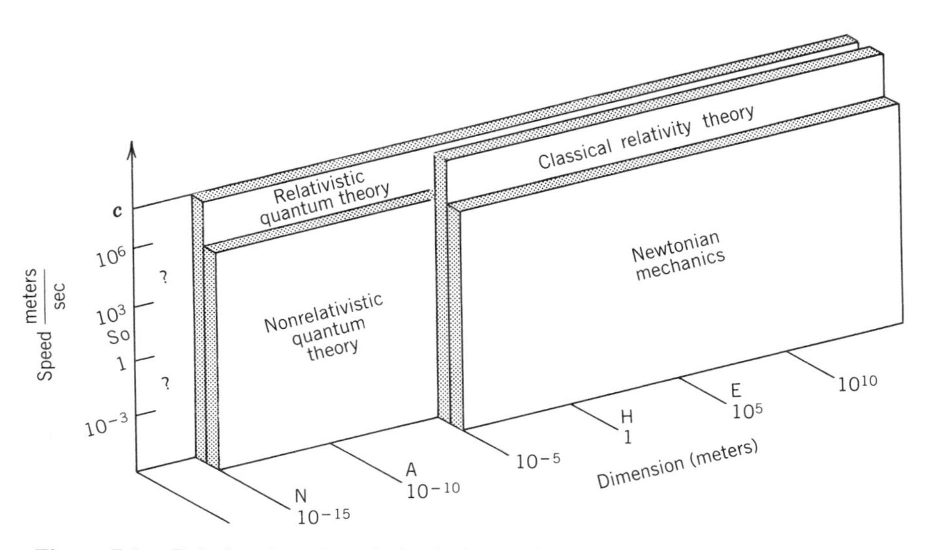

Figure 7.1. Relative domains of physical theories. N, typical nuclear diameter; A, typical atomic diameter; H, size of humans; E, Earth diameter; So, speed of sound; *c*, speed of light.

be looked upon as representing an overlay of four regions, each of which is smaller than the one underlying it.

Relativistic quantum physics, which encompasses the material in Chapters 6, 7, and 8, is considered to be the most general theory currently available. It covers almost the entire field of the figure, and it is valid for all dimensions and all speeds presently accessible to experiment. It is not actually known whether it is fully applicable to dimensions very much smaller than the nucleus of an atom; it is possible that it may fail for such dimensions and have to be replaced by a more general theory. Relativistic quantum physics is rather difficult to use, so for many purposes it is sufficiently accurate to use nonrelativistic quantum physics, provided we are aware of its limitations. Nonrelativistic quantum physics covers a goodly portion of the region covered by relativistic quantum physics, particularly at speeds below about 10,000 or 100,000 meters/second. Classical relativity theory covers the right-hand portion of Fig. 7.1, that is, objects larger than molecules at all speeds. Classical physics, designated in the figure as Newtonian mechanics, is generally more convenient and easier to use and understand than quantum physics, and so it is used within its domain of applicability (i.e., objects that are larger than molecules and slower than a few tenths of a percent of the speed of light), even though we know it is not quite right.[1]

Figure 7.1 reflects the idea that all of our physical theories are but approximations of a true understanding of physics. In the domains where new and old theories can both lay claim to validity, the new theory can only claim greater accuracy (the degree to which the accuracy is greater may be extremely small, even immeasurable). Thus we continue to use classical (Newtonian) physics in the design of bridges and automobiles because it is more convenient and is sufficiently accurate, but for studies of the nuclei of atoms or the electronic properties of solids we must use quantum theory, because classical theory gives wrong answers in these cases. We could use quantum theory to design a bridge or an automobile, but the answers would be the same as if we used classical theory.

3. Historical Origin and Content of Quantum Theory

By the end of the nineteenth century the general intellectual consensus was that basic scientific knowledge was fairly complete. Many individuals believed that the major theories of physics were established and that possibly all that was unknown in the universe eventually could be explained on the basis of these theories. It was recognized, however, that some ''residual'' problems had to be solved.

The quantum theory arose out of the effort to apply the electromagnetic theory of light in combination with theories of matter, energy, and thermodynamics to the problems of the interaction of electromagnetic radiation with matter. It was

[1]Table 6.2 gives some examples of how classical physics is in error. Some other examples are discussed in this chapter. Actually Fig. 7.1 is misleading in that the domains of applicability are not as solid and uniform as represented. There are ''holes'' in the overlays. Figure 7.7 represents such a ''hole.''

the attempt to meld these major nineteenth-century theories in a coherent manner to understand these "residual" problems that ultimately led to the revolutionary developments of quantum physics. It is perhaps ironic that one of the greatest achievements of nineteen-century physics—the electromagnetic theory of light—was so flawed as to necessitate two major reformulations, namely, Einstein's relativity theory and the quantum theory. Although Einstein played a pivotal role in the early developments of quantum theory, no one person can be singled out as the outstanding genius who led all others in the development of the new theory. As the twentieth century developed, science became a very large-scale endeavor and scientists were more aware of each other's efforts. As a result many more individuals were in a position to make significant contributions.

We will approach quantum theory through an examination of some of these problems of nineteenth-century science and their context: (1) the blackbody radiation problem, (2) the photoelectric effect, and (3) atomic spectra and structure. These problems are listed in order of increasing importance as perceived by most physicists at the end of the nineteenth century. Indeed the blackbody problem was considered to be rather minor; yet in the solution of this problem lay the seed that would lead to the full quantum theory. In fact, the three problems are intimately related to each other, and their solution became important not only for pure science but for large areas of applied science as well.

B. Cavity or Blackbody Radiation

1. Temperature and Radiation from Hot Bodies

It is common knowledge that metal objects change color when heated. As an iron rod is heated, for example, it eventually becomes hot enough to glow a dull red, then a cherry red, and then a bright orange or yellow. Ultimately the iron rod melts, but if we use instead a piece of tungsten wire, enclosed in a vacuum or an inert atmosphere to prevent chemical reactions with the air, then it can be raised to quite a high temperature. The hotter it is, the more the color of the emitted light changes, so that it goes on from bright yellow to "white hot." If the wire is enclosed in a glass bulb and the heating done electrically, then we have an incandescent light bulb. Actually, there is not just one color of light emitted, but a range of colors of various intensities. The white-hot incandescent wire in the light bulb emits violet, blue, green, yellow, orange, and red light (and all colors in between) that we can see, as well as "colors" of the electromagnetic spectrum that we cannot see, such as infrared and ultraviolet, as can be verified by using appropriate instruments.

Certainly, in view of the various thermodynamic concepts discussed in earlier chapters, as well as the concepts of the kinetic-molecular theory of matter and the electromagnetic theory of light, it should be possible to understand the relationship between the heat put into the iron rod or tungsten wire, the temperature reached, and the range and intensity of the spectrum of electromagnetic radiation emitted. As heat energy is put into the solid, the kinetic energy of motion of the

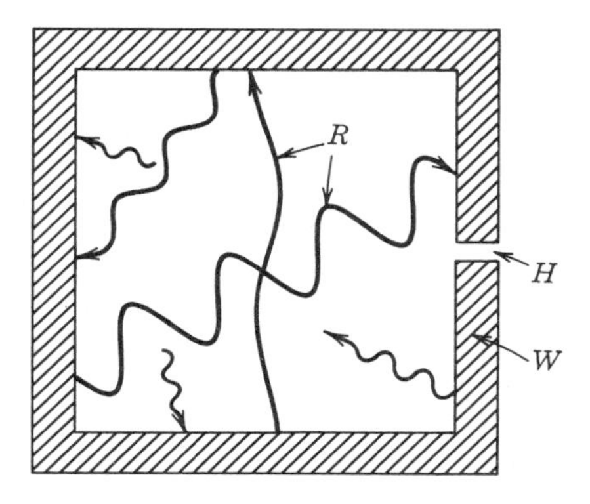

Figure 7.2. The ideal blackbody or cavity radiator. R, radiation of various wavelengths being exchanged by oscillators; W, cavity walls at high temperature; H, hole through which radiation is observed.

molecules and their constituent parts increases. In particular, the electrically charged parts of the atoms have increased energies of to-and-fro motion. But we know from the electromagnetic theory of light that to-and-fro (oscillatory) motion of electric charge results in the radiation of electromagnetic energy. If the oscillations are of high enough frequency, visible light will be radiated. As the temperature of the solid increases, the range of amplitudes and frequencies of the atomic or molecular oscillators increases, and the range and intensity of the emitted electromagnetic radiation increases. Thus it can be understood how a heated body can emit light.

The next step in understanding the radiation from hot bodies consists in taking the ideas mentioned above and putting them on a more quantitative basis and making detailed experimental measurements to test the accuracy of the calculations. It was quickly realized that the amount of radiation emitted from heated bodies depends on the condition and the nature of the *surface* of the bodies, as well as on their bulk nature and temperature. Thus it is necessary to consider an "ideal" case, just as the freely falling body of Galileo and Newton is an "ideal" case, and just as the Carnot engine represents an "ideal" engine.

Analysis of the situation shows that the best or ideal emitter of electromagnetic radiation at elevated temperatures is also the best *absorber* of radiation. A surface that can easily absorb all frequencies (colors) of light can also easily emit all frequencies. Consequently, the best emitter should have a black surface. Moreover, a white (or even more so a reflecting) surface will be both a poor absorber and a poor emitter of radiation.[2]

How does one make a truly black surface? One must first ask what is meant by a black surface. A truly black surface is one from which absolutely no incident light can escape. Any light that is used to illuminate such a surface cannot be seen. We can visualize a portion of such a surface (as shown in Fig. 7.2) by considering a hollow body that has a tiny hole connecting its interior to the exterior surface. Any light that is incident on the hole will penetrate into the

[2]These facts may be used by travelers in a desert. White robes reflect rather than absorb the sun's heat during the day and will retain their wearer's body heat during the night. Similarly, the use of reflecting aluminum foil on building insulation decreases heat loss during the winter and heat gain during the summer.

interior, and even though it may be reflected many times from the interior walls, it will not find its way back to the hole to escape. The hole represents a ''surface'' that is an almost ideal absorber of radiation. On the other hand, if the interior walls of the cavity are heated, then the radiation coming from the hole will correspond to the radiation emitted by the ideal emitter of radiation. Thus the ideal radiator or blackbody radiator is also called a *cavity radiator.* One could ''approximate'' a cavity radiator by placing a tightly fitting opaque screen, in which a small hole is drilled, over the opening of a fireplace. With a hot fire burning in the fireplace, the hole will be the ''surface'' of the blackbody.

2. Analysis of the Cavity Radiator

In the analysis of the cavity radiator, it is assumed that the atoms in the walls of the cavity are in thermal equilibrium with each other, and therefore they have on the average the same motional energy per atom. At any specific instant, some have more and some less energy, and the individual atoms that have more or less energy may change as time goes on; but the average energy is constant. The changes in energy of individual atoms result from interactions of the atoms with each other, either by coupling through their chemical bonds or by emitting or absorbing radiation from other atoms across the space of the cavity. In fact, the spectrum of the electromagnetic radiation emitted and absorbed—the range of frequencies and the intensity of the radiation at the various frequencies—is representative of the distribution or relative share of the total energy among the different frequencies or modes of motion. Because radiation of electromagnetic energy depends on *oscillating* electric charges, it is useful to discuss the emission of electromagnetic energy in terms of atomic or molecular ''oscillators,'' which are located within the individual atoms or molecules.

The possible motions of the individual atoms or molecules are quite complex, but they can be analyzed in terms of simpler springlike motions of a number of different oscillators, each of which oscillates (vibrates) at its own characteristic frequency. The total motion of the atom or the molecule is then the sum of the different motions of the individual oscillators, just as the bouncing motion of an automobile passenger on a bumpy road is the result of the up-and-down and sideways motions due to the various springs on the wheels, the bouncing of the tires, and the cushions in the seat.

Thus it is the number of active oscillators at each given frequency that determines the electromagnetic spectrum emitted by the cavity. Conversely, if the emitted electromagnetic spectrum were studied and measured, and the proper mathematical analysis carried out, it is possible to deduce the manner in which the total thermal energy of the system is shared among the various oscillators at any given temperature.

This distribution of energy, or **partition function** as it is called, can also be calculated according to the principles of thermodynamics, using the various concepts of energy and entropy discussed in the previous chapters. The results of the calculation as compared with the actual experimental measurements of the spectrum are shown in Fig. 7.3, where the calculated spectrum for 7000 kelvins is shown by the dashed lines, and the actual spectra for 7000, 6000, and 4000 kelvins

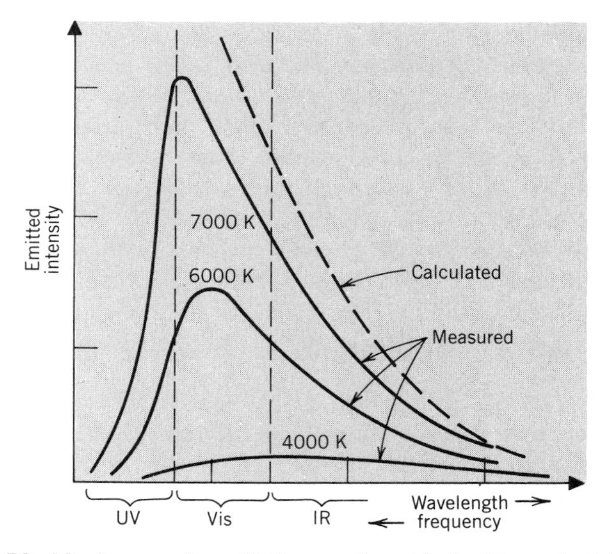

Figure 7.3. Blackbody or cavity radiation spectra. Dashed line, calculated according to classical theory for 7000 kelvins. Solid lines, measurements at indicated temperatures. Note the spectral regions indicated: UV, ultraviolet; Vis, visible light; IR infrared.

are shown by the solid lines. Note that in this graph, the horizontal axis is the wavelength, which is inversely proportional to frequency. Thus wavelength increases to the right and frequency to the left. The size of the area underneath a particular temperature graph is proportional to the total energy emitted by the cavity. The graphs shown in Fig. 7.3 depend only on temperature and are independent of specific details of atomic or molecular structure, just as the behavior of all Carnot engines is independent of all detailed design and depends on temperature only.

It is clear that the calculations agree with the experiment only at long wavelengths, and the disagreement becomes more pronounced as the wavelength decreases (frequency increases). In fact, at short wavelengths (high frequencies), which would correspond to the ultraviolet portion of the spectrum, the disagreement between theory and experiment is so violent as to be rather catastrophic. This disagreement became known as the **ultraviolet catastrophe.** (Of course, it was catastrophic only for the theory and those who wanted to believe the theory.) To illustrate how great the disagreement was, it became apparent that the theoretical results contradicted the principle of conservation of energy. Yet conservation of energy is one of the cornerstones of the theory of thermodynamics. So the theory even contradicted itself!

3. Planck's Analysis of the Problem.

In 1899 a German theoretical physicist, Max Planck, published an analysis of the problem, in which he modified the theory to avoid the ultraviolet catastrophe. It was this modification that started the development of the quantum theory.

Planck realized that the existing theory, with its requirement that all oscillators should on the average have the same energy per oscillator, forced the ultraviolet catastrophe, because there were many more oscillators at high frequency than at low frequency. But the experimental results (Fig. 7.3) clearly show that the average energy per oscillator must *decrease* as the frequency of the oscillator increases. This means that only a few of the available high-frequency oscillators will be active. There must be some way of discriminating against the high-frequency oscillators to keep them from having the same average energy as the low-frequency oscillators. Yet the oscillators, because they reside in the atoms of the hot body, must interact and share energy with each other. So the question remains as to how the oscillators can exchange energy and still have the higher frequency oscillators on the average with less energy per oscillator than the lower frequency oscillators.

Planck's ingenious solution to this problem was to propose that in any situation where an oscillator gains or loses energy, its change of energy could only be in units of a certain minimum amount of energy, which he called a quantum of energy. Thus an oscillator could gain (or lose) one quantum of energy, or two quanta (the plural of quantum) of energy, or three quanta, or four quanta, and so on, but never a half quantum or a quarter quantum or any fraction of a quantum. Moreover, each oscillator has its own characteristic size quantum, proportional to its frequency. A high-frequency oscillator would have a larger quantum than a low-frequency oscillator. He showed that this would result in more energy being received on the average by the low-frequency oscillators than the high-frequency oscillators.

For example, if two different oscillators interact with each other, and one of the oscillators happens to have a frequency that is exactly twice as high as the frequency of the other oscillator, then its quantum will be twice as large as the quantum of the second oscillator. If the high-frequency oscillator "wants" to lose some energy to the low-frequency oscillator as a result of the interaction, its quantum will be equivalent to exactly two of the quanta of the low-frequency oscillator, and the low-frequency oscillator will accept the energy, thereby increasing its energy by two of its own quanta. But suppose now the low-frequency oscillator "wants" to lose some energy. One of its quanta is only *half* the size of the required quantum for the high-frequency oscillator, and the high-frequency oscillator cannot accept the energy. The low-frequency oscillator cannot lose just one quantum to the high-frequency oscillator, and so it must lose the quantum to some more compatible oscillator. Only if the low-frequency oscillator gives up two (or some multiple of two) of its quanta can the high-frequency oscillator accept the energy. As a result, the probability of the high-frequency oscillator gaining energy from a low-frequency oscillator decreases, and therefore its average energy must decrease.

We can make a fanciful analogy to this situation, by imagining a community in which people interact by buying and selling various goods and services to each other. Some members of this community are willing to participate in transactions involving one or more $1 bills, others will deal only with one or more $2 bills, others will deal only with $5 bills, some only $1000 bills, and so on. If the "big spenders" want to buy a glass of milk or a pair of shoes, they must pay $1000, for example, and they get no change. On the other hand, the items they have for sale cost $1000. Unfortunately, $1000 transactions do not take place very often, and as a result the "big spenders" will not have much money left after a few transactions. The "small spenders," on the other hand, can participate in a large number of transactions and accumulate a fair amount of wealth, relatively speaking.

By incorporating this idea into the theory, Planck was able to calculate a spectral distribution function that was in exact agreement with the experimental values.

Planck was not entirely pleased with his quantum idea because it violated some of his commonsense feelings about energy. A simple example will illustrate what bothered him.

Imagine an oscillator consisting of a weight hung by a spring, as in Fig. 7.4. If the weight is pulled down from its equilibrium position (say by an inch) and then released, it will oscillate up and down with a frequency that is based on the weight and on the stiffness of the spring, and with an amplitude of one inch (i.e., it will oscillate between two extreme positions, each of which is one inch from the equilibrium position). If it is pulled down two inches, the frequency will be the same, but the extreme positions will be two inches from the equilibrium position. The total energy (kinetic plus potential) associated with the oscillations in the second case will be four times as much as in the first case (the energy is proportional to the square of the amplitude of the oscillations). If the weight is pulled down 1.2 inches, then the energy should be 1.44 times as much as in the first case.

According to Planck's hypothesis, however, this is not possible, because the energy of oscillation must be exactly twice, or three times, or four times, and so on, the original amount. If the weight were pulled down a distance equal to 1.4142135 . . . inches (the square root of 2), then the energy of oscillation will be twice the original energy and will be allowed. No amplitudes between 1 and 1.4142135 . . . inches are permitted. Similarly no amplitudes between 1.4142135

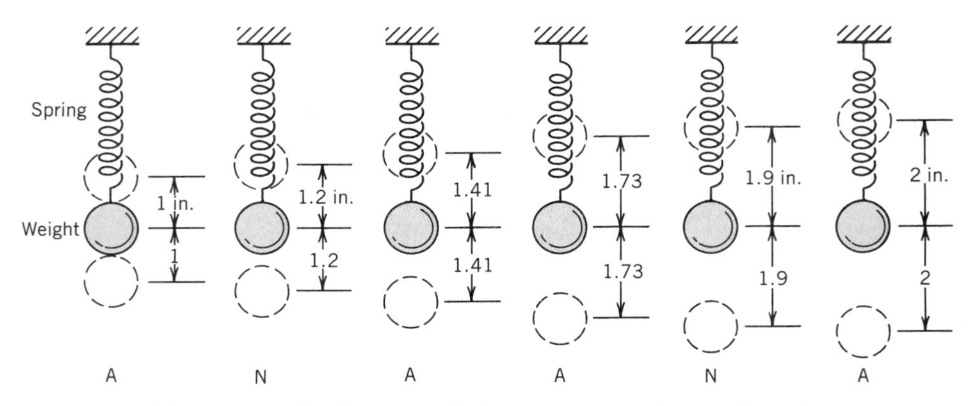

Figure 7.4. Oscillator amplitudes. A, allowed; N, not allowed.

. . . inches and 1.7320508 . . . (square root of 3) inches are allowed, but an amplitude of exactly 1.7320508 . . . is allowed. No amplitudes between 1.7320508 . . . inches and 2 (square root of 4) are allowed, but an amplitude of 2 inches is allowed. No amplitudes between 2 inches . . . , and so on.

Planck knew this was wrong on a large scale, and he could see no reason why it should be right on an atomic scale. In fact, he spent considerable time trying to find another way of eliminating the ultraviolet catastrophe without introducing the quantum concept, but to no avail.

Actually, it can be shown that the quantum concepts do not violate common sense on the large scale. To see this, it is necessary to reexamine the relationship between the size of a quantum and the frequency of the oscillator. The size of a quantum is proportional to the frequency of the oscillator, but the proportionality constant (which is called *Planck's constant*) is such a tiny number (a decimal point followed by 33 zeros and then a six in the thirty-fourth place in the proper system of units), that the quantum of energy for the spring is a very small fraction of the total energy of the spring. As a result, the addition of one more quantum to the large number required for a one-inch amplitude results in such a tiny change in the allowed amplitude that it is impossible to recognize that there are even smaller changes in amplitude, which are not allowed. Thus for all ''practical'' purposes the allowed amplitudes vary smoothly. This is an example of the idea expressed in Section A2 and Fig. 7.1, that the quantum theory gives results for large objects that are indistinguishable from the classical theory. However, for the atomic oscillators, quantum theory gives quite different and correct results. It is the attempt to apply large-scale ''common sense'' to submicroscopic phenomena that causes classical theory to fail.

The value of Planck's constant, which is almost universally designated by the symbol h, was determined by Planck by a direct comparison of his theory with the experimental measurements of the cavity radiation spectrum. Because of the very seminal nature of his work, even though he was uneasy with his results, Planck received the Nobel Prize for his solution of the cavity radiation problem.

C. The Photoelectric Effect

1. The Nature of the Photoelectric Effect

The photoelectric effect was first recognized as such by Heinrich Hertz in 1887 in the course of his experimentation to verify Maxwell's electromagnetic theory of radiation. Essentially it is based on the fact that an electric current can be caused to traverse the empty space between two objects in a vacuum without a connecting wire, when one of the objects is illuminated by light. This is shown schematically in Fig. 7.5, which shows an electric battery connected to two plates sealed in a vacuum envelope. One of the plates, called the *photocathode,* is connected to the negative terminal of the battery, and the other plate, called the *anode,* is connected to the positive terminal. If, and only if, the photocathode is illumi-

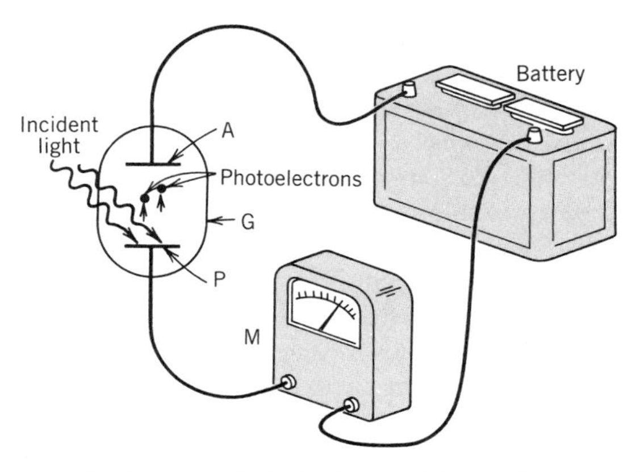

Figure 7.5. Schematic diagram of photoelectric effect. A, anode; P, photocathode; M, meter to indicate current; G, glass bulb with vacuum inside.

nated by light, the meter in the circuit will show that an electric current is flowing. Various experiments prove that the electric current between the two plates consists of tiny bits of matter, carrying a negative electric charge, originating in the photocathode and accelerating toward the anode. These tiny bits of matter are **electrons** (discovered by an English physicist, J. J. Thomson, in 1897). In the photoelectric effect, they are called **photoelectrons.** The photoelectric effect is used in various devices such as automatic door openers, sound tracks on motion picture films, shutter controls on cameras, burglar alarms, and numerous other devices for detecting and measuring light levels and changes in light levels.

2. Qualitative Explanation by Classical Electromagnetic Theory

Qualitatively, a plausible explanation of the photoelectric effect can be given in terms of the electromagnetic theory of light. As discussed in Chapter 6, an electromagnetic light wave consists of oscillating and propagating electric and magnetic fields. The electric field forces the electron into oscillation and gives it sufficient energy, if the amplitude of the wave is large enough, to break its chemical bond to the surface, thereby permitting the electron to be pulled out of the surface when a voltage is applied.

Unfortunately, however, it was not possible to explain some seemingly paradoxical results. Some colors of light, no matter how bright, did not cause the photoelectric effect when certain materials were used as the photocathode, whereas other colors did cause the effect even though their intensity was much less than the first colors. For example, the most intense yellow light causes no effect from a metal such as copper, whereas ultraviolet light causes the photoelectric effect in copper no matter how weak the light. It also seemed that some sort of "triggering" mechanism was needed to initiate the effect. (There were also effects caused by the conditions of the surface of the photocathode, and so some experiments gave inconsistent results.)

The paradox may be visualized by using waves pounding on a beach as an analogy. As the waves come in, they pick up pebbles, wood, and other debris and cast them far up on the beach. But at certain beaches, on certain days, if the distance between crests of the waves is too large, they will not move a single pebble. On the other hand, if the distance between the crests of the wave is small enough, even a tiny ripple will cast pebbles on the beach!

3. Einstein's "Heuristic" Explanation

Finally, Einstein, in 1905 (the same year in which he published his first paper on special relativity theory, together with several other significant contributions), proposed a theory of the photoelectric effect based on a thermodynamic analysis of certain aspects of the cavity radiation problem. He called this theory *heuristic,* because although he could not justify it from accepted fundamental principles, it seemed to work.

Einstein's analysis showed that the radiation within a cavity behaved in the same manner as a gas of particles having energies equal to hf, the quanta of the oscillators in the cavity radiator. He then suggested that if radiation in the cavity behaved in that manner, then all electromagnetic radiation could be considered to be such particles. This means that light energy is transported in the form of bundles or quanta of energy. (At the same time, he kept the wave as the means of transport.) Much later it was suggested that a quantum of light should have a special name, a **photon.** The energy in each photon of light depended on the color of the light, in particular, on its frequency f. (Recall that according to the wave theory of light, a light wave has a wavelength λ, the distance between successive crests of the wave, and a frequency f, the number of times per second the wave oscillates to and fro. Recall also that wavelength multiplied by frequency gives the speed of light, 186,000 miles per second.) Using Planck's constant h, the energy of the photon equals hf.

The surface, if it absorbs any light energy, must accept whole photons, not pieces of a photon. Actually, it is the electrons within the surface that accept the photon energy. If a single photon gives the electron an amount of energy that is greater than its *binding energy* (the energy required to overcome all the forces holding the electron within the surface), then the electron can escape from the surface. If the photon does not give the electron sufficient energy, it cannot escape from the surface; it will simply "rattle around" inside the solid and dissipate the energy acquired by absorbing the photon. Usually it is not possible for the electron to store up the energy from successive photon absorptions; either it escapes or it dissipates the energy before it has a chance to absorb another photon. If the photon energy absorbed is greater than the binding energy, then the excess energy will appear as kinetic energy (energy of motion) of the photoelectron.

Einstein's theory predicted that if an experiment were carried out to measure the maximum kinetic energy of the photoelectrons as they emerged from the surface, then a graph of the maximum kinetic energy versus frequency of the light would be a straight line starting at some threshold frequency, which would be characteristic of the photocathode material and surface conditions. Moreover, it was predicted that the slope (or steepness) of the line, in the proper units, would be exactly equal to Planck's constant. Such a graph is shown in Fig. 7.6. Experimental confirmation of some of Einstein's predictions was first reported in 1912, and finally an American, Robert A. Millikan, in 1914 published the results of a comprehensive set of measurements verifying Einstein's "heuristic" theory and showing that the photoelectric effect could be used to measure Planck's constant *h* independently of the blackbody radiation problem.

Einstein received the Nobel Prize for his predictions of the laws of the photoelectric effect (not for relativity theory!), and Millikan also received a Nobel Prize for his experimental studies of the effect. Interestingly enough, Millikan (like most of his contemporaries) did not like Einstein's theory. He felt Einstein had obtained the right answer, but that the effect was still not well understood!

4. Implications of the Photon Hypothesis

The photon hypothesis carried with it some implications about the nature of light that would not be expected from the electromagnetic theory. For one thing, it was necessary to consider that the photons were geometrically compact; that is, the photon travels like a bullet, not like a wave. Otherwise the energy of a single photon might be spread over a wave front several feet—even several yards—in diameter. Then when the photon was absorbed, all of its energy would have to be "slurped up" instantaneously from all parts of the wave front and concentrated at the location of the photoelectron—something not permitted by the theory of relativity, because the energy would have to travel at a speed greater than the speed of light. Einstein insisted that because the energy of the photon was absorbed at one particular point, the photon itself had to be a very concentrated bundle of energy.

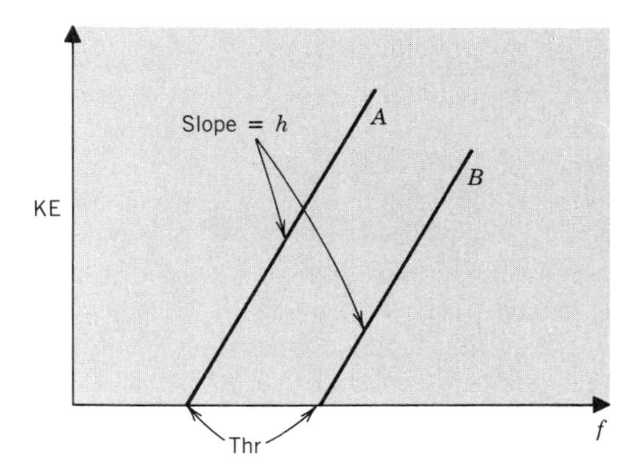

Figure 7.6. Einstein's prediction of relationship between maximum photoelectron energy and frequency. KE, maximum kinetic energy of photoelectrons; *f*, frequency of incident light; *A, B,* data for two different materials; Thr, threshold frequency.

Actually, there is other evidence that the absorption of light requires the photon hypothesis. One very commonplace piece of such evidence is the granularity of underexposed photographs. If one takes a photographic negative and exposes it for the proper length of time, a positive print is obtained. If, however, the negative is exposed to only a small amount of light, the print will simply show an unrecognizable pattern of a few dots. If the exposure time in successive prints is increased by small amounts, the dots begin statistically to group themselves into the bare outlines of a recognizable image. With increasing exposure times, a sufficient number of dots is accumulated that the statistics become overwhelming, and the final print is obtained. This is shown in Fig. 7.7. Each dot represents the absorption of a photon at a specific point, confirming that the photon energy is geometrically concentrated.

A more dramatic demonstration of this "bullet" nature of light is shown by the Compton effect. In this effect, a beam of X rays (known to be very short wave-length electromagnetic radiation) is allowed to "bounce" off some electrons. It is found that the electrons recoil under bombardment by the X rays, and that the X rays also recoil. In fact, the wavelength of the X rays is increased (and hence the frequency is decreased) as a result of the recoil. This means that the energy of the photons decreases. (It is as if a photon had elasticly collided with an electron, just like a cue ball with a billiard ball.) It is even possible to calculate the momentum of a photon (according to a formula to be discussed later) and show that in this collision, as in any elastic collision, both the total momentum

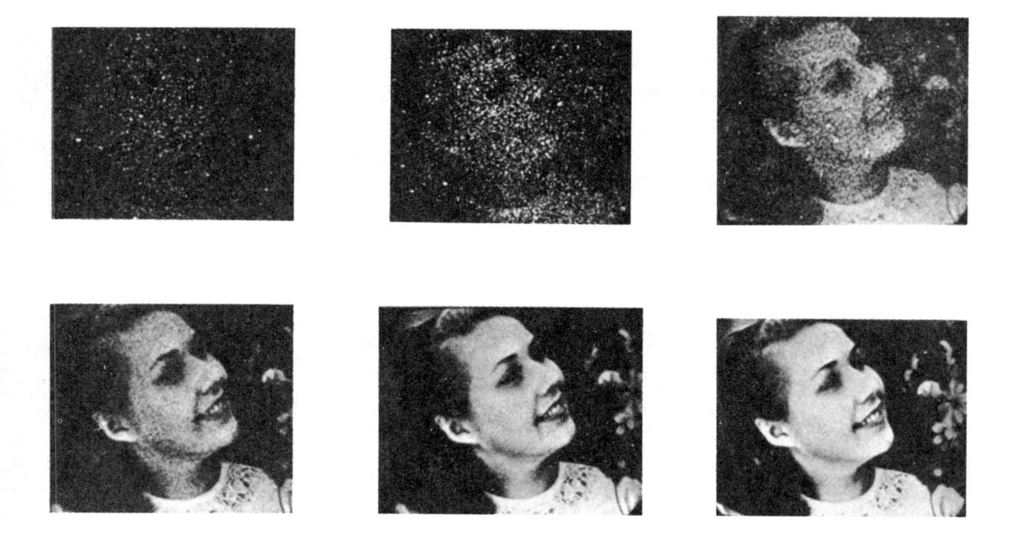

Figure 7.7. Granularity and statistics of blackening of photographic images. The number of photons involved in forming increasingly detailed reproductions of the same image. (*a*) 3000 photons. (*b*) 12,000 photons. (*c*) 93,000 photons. (*d*) 760,000 photons. (*e*) 3,600,000 photons. (*f*) 28,000,000 photons. (Courtesy Dr. Albert Rose.)

and the total energy of the colliding particles (the photon and the electron) are conserved.

Still, the evidence for the wave nature of light is overwhelming: Interference effects (as demonstrated in the colors of soap bubbles), diffraction effects (apparent in the absence of sharp shadows of objects and in the behavior of diffraction gratings), polarization effects (which prove that light waves vibrate perpendicular to the direction of propagation), the fact that ordinary light travels slower in glass than in vacuum, the astounding success and unifying power of Maxwell's electromagnetic theory of light, were all phenomena that were well understood and precisely explained on the basis of a wave theory of light.

OPTIONAL SECTION 7.1 *Conflicting Theories of the Nature of Light*

The idea that a beam of light is like a stream of particles, called by the generic name *corpuscles,* was not original with Einstein—it was tentatively discussed by Isaac Newton in the 1670s. In fact for the following 150 years, there had been a substantial controversy similar to that between the geocentric and heliocentric theories of the universe, but with much less theological and philosophical intensity. (A detailed technical discussion of this controversy is given in the book by Whittaker, listed in the bibliography for Chapter 6.) Newton hesitantly suggested a corpuscular theory of light as an alternative to a theory proposed a short time earlier by Robert Hooke, a rival of his, that light represents a wavelike disturbance of the ether, similar to waves in a fluid, in the same way that sound represents a disturbance of the air.

As is clear from Chapter 6, by this time the ether was considered to be ubiquitous—filling all space, including any space between atoms or molecules in ordinary matter, whether solid, liquid, or gas. The properties of the ether depended somewhat on whether it was in free space or within ordinary matter. In Hooke's theory, colors of light are not present originally in a beam of white light, but rather are induced as a result of the oblique passage of the beam across the interfaces between two media, say air and glass, as in a prism (see Fig. 7.8). This implied further that the colors can be changed by repeated passages through a prism. Newton showed experimentally, however, that white light consists of a mixture of colors that can be revealed by the use of a prism, but that cannot be altered any further by repeated uses of the prism. Newton objected further that sound waves and water waves give effects behind corners and obstacles, whereas light travels in straight lines and by and large cannot be detected in the geometrical shadow of obstacles, unless a mirror or change of medium is introduced. (It was already known at that time that shadows of the edges of objects were not perfectly sharp, i.e., that light could be found very slightly within the geometrical shadow of an obstacle. The corpuscular theory explained this in terms of a variation of the density of the ether at the very edge of the object.)

Within a few years the wave theory of light was reformulated by Christian Huygens in such a way as to conform with Newton's experimental results on the production of colors by prisms. Although Huygens was unable to overcome

Newton's point about the straight line motion of light, he also felt that the corpuscular theory required that there should be some sort of effects if two beams of light crossed each other. They were not observed, but about this time the phenomenon of double refraction in certain materials was discovered. This phenomenon requires that a beam of light splits into two beams traveling at different speeds in different directions in the same material. Newton insisted that this could not be reasonably explained by a light wave having the same general characteristics as a sound wave.

As a result of these objections by Newton and others, for over 100 years, well into the nineteenth century, the general consensus among scientists was that light was corpuscular in nature, although there were some facts that were not easily explained by a corpuscular theory. For example, when a beam of light shines on a smooth glass surface, part of the beam is transmitted through the surface and part is reflected. How does a corpuscle in the beam decide whether it is to be reflected or transmitted? There were no obvious reasons for such an occurrence. With a wave theory, however, particularly as formulated by Huygens, it was entirely natural that part of the wave disturbance would continue on into the glass medium and part be reflected from the glass surface.

At the beginning of the nineteenth century, Thomas Young in England became convinced of the value of the wave theory of light. He recognized that waves could exhibit the phenomenon of interference (discussed in Section E3 below) and performed experiments to demonstrate this phenomenon. He showed that under the proper experimental conditions light could be observed well into the geometrical shadow region and that interference effects could also be found there. Young also suggested that light waves should be different from sound waves or water waves—that they should be transverse waves rather than longitudinal waves. (The distinction between transverse waves and longitudinal waves is discussed in Section B4 of Chapter 6.) Young felt that transverse waves would overcome Newton's assertion that a wave theory could not account for double refraction. This suggestion was seized upon by Augustin Fresnel, a brilliant French civil engineer, who then used it to give a complete explanation of double refraction and the polarization of light, which was far superior to the explanation that the advocates of the corpuscular theory had proposed.

One of the most decisive ways to determine which of two competing theories is correct is to find experiments or observations for which they predict opposite results. This was the case for the competing geocentric and heliocentric theories of the universe—the geocentric theory predicted that the size of the planet Venus as measured during its various phases would be essentially unchanged, while the heliocentric theory predicted that full Venus would be considerably smaller than crescent Venus. Galileo's telescope observations confirmed the heliocentric theory prediction. The corpuscular theory of the eighteenth century assumed that light travels faster in a medium such as glass or water than in the air or the ether. Huygen's wave theory required that light travel slower in glass or water than in the ether. In 1850 the speed of light in water was measured and was found to be only ¾ of the speed of light in air, thereby favoring the wave theory.

As a result of all these developments, by the middle of the nineteenth century the consensus of the scientific world was very much in favor of the wave theory of light. To cap it all off Maxwell's electromagnetic theory appeared in 1864 and gave a very thorough and sound theoretical basis for the wave theory. None of this seemed to faze Einstein, however. In 1905 he was already proposing that the ether—the very medium in which light waves were a disturbance—should be discarded, and now he was proposing to revive a corpuscular theory of light!

Thus early twentieth-century physicists were faced with a dilemma. In certain experiments, primarily those dealing with the emission and absorption of light, the photon hypothesis worked very well. In other experiments, primarily those dealing with the propagation of light (i.e., the way in which light gets from one place to another), the wave hypothesis worked very well. One wit remarked that on three alternate days of the week the experiments supported the photon theory, whereas on three other days the evidence supported the wave theory; thus it was necessary to use the seventh day of the week to pray for divine guidance!

Ultimately it was suggested that light has a dual nature, that the properties of light can be linked to wave properties or to particle (bullet) properties depending on the details of the particular experiment and their interpretation. The wave and particle properties are different aspects of the "true" nature of light. In fact, the wave and particle properties are intimately connected with each other. It is necessary to use the frequency f, a wave property, to calculate the energy E ($= hf$) of the photon. Similarly, as will be seen later, it is necessary to use the wavelength λ to calculate the momentum, a particle property.

D. The Nuclear Atom and Atomic Spectra

1. Early Concepts of Atomism

The word **atom** literally means indivisible. In Western thought, the concepts of atomism can be traced back some 2500 years to the Greeks Leucippus and Democritus, who considered that a bit of matter could be divided into small bits, which could be divided into even smaller bits, which could be divided further, and so on, until ultimately there would be reached the smallest bits of matter that could not be divided any further and hence were impenetrable and indivisible. This idea was expounded further by the Roman poet Lucretius some 500 years later in a long, poetical discourse on Epicurean philosophy. In these earliest views, there were only four types of atoms, associated with the four prime substances of Aristotle. Starting with the British chemist John Dalton in 1808, it was recognized that there were a number of elemental substances, each of which had its own kind of atom, perhaps distinguished by having its own peculiar shape, and perhaps having attached to it a small number of "hooks and eyes" by which it could be joined to other atoms to form a **molecule.** A molecule is thus defined as the smallest possible bit of a substance (not necessarily an elemental substance). For example, a molecule of water is composed of two hydrogen atoms and one oxygen atom. Those substances that are not elemental are now called *compounds.* Although compounds can be decomposed into *elements,* and hence molecules can be separated into constituent atoms, the elements and their atoms were initially considered as not being further decomposable or divisible, respectively.

The concept of atoms and molecules in motion led to the kinetic-molecular theory of matter and the recognition that chemistry is essentially a branch of physics. During the nineteenth century it became possible to estimate the size of atoms and molecules. Atoms are of the order of a few angstroms in diameter. (The angstrom is a very small unit of length; there are 254 million angstroms in one inch. In scientific notation, 1 angstrom $= 10^{-10}$ meters.) Many substances contain only one atom or only a few atoms per molecule; biological molecules, on the other hand, may contain several hundred or even several thousand atoms. The atoms in a molecule or a solid exert attractive forces on each other that overcome their random thermal motions and bond them together. It is possible to make a crude but useful analogy between the bonding forces and "elastic springs" connecting the various atoms with each other (the springs are used instead of the hooks and eyes mentioned earlier), but obviously the actual manner of bonding must be different.

2. Electrical Nature of Matter; Rutherford's Planetary Atom

Not too long after Dalton's work became known it was realized that matter has electrical characteristics. The final verification of the existence of the electron in 1897 confirmed that atoms themselves have an electrical structure. In fact, some 150 years earlier Boscovich, a Serbian scientist, had argued on essentially metaphysical grounds that atoms could not be the hard, impenetrable objects of the original conception, but had to have some spatial structure.

Of particular significance to the realization that atoms themselves have an internal structure was the study of atomic spectra, which was developed to a high art during the nineteenth century. Under proper conditions all substances can be made to emit light of various colors. By the use of suitable devices such as prisms, this emitted light can be decomposed into its various constituent colors or spectrum, as shown schematically in Fig. 7.8. Well before the end of the nineteenth century it was known that each different chemical element has its own characteristic spectrum. In fact, it is possible by the study of the spectrum emitted by any substance to determine the chemical composition of that substance.

There are various ways of causing a substance to emit its spectrum: It can be thermally excited by the application of heat, say by burning it in a flame; it can be electrically excited by passing high-voltage electricity through it when it is in the gaseous state; or it can be illuminated with other light, causing it to *fluoresce.* When the spectra (plural for spectrum) of even the simplest substances—the

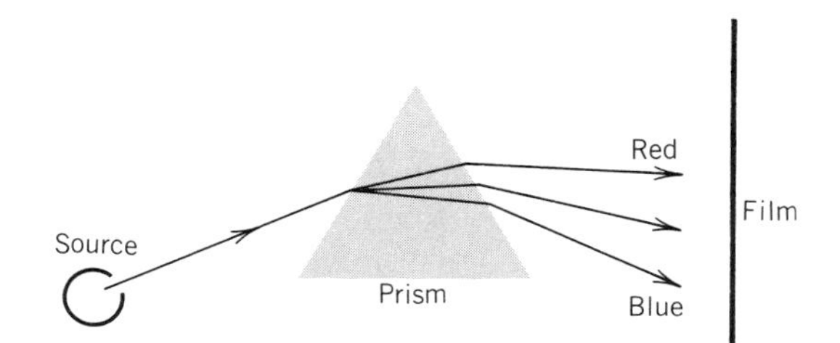

Figure 7.8. Dispersion of light into a spectrum. Light from the source is dispersed by the prism into a spectrum of colors (frequencies), which is recorded on a photographic film.

elements—are studied, it is seen that they are quite complex, although in principle they can be explained in terms of the electromagnetic theory of light.

As pointed out in Section B1 above, as a useful explanation, each atom may be considered as containing a number of electrical oscillators capable of vibrating at various characteristic frequencies. If these are set into motion by various means involving the transformation of heat or electrical energy into mechanical energy, then they will radiate electromagnetic waves having a frequency corresponding to the characteristic frequency of the oscillators. The number of different colors in the spectrum of a substance would correspond to the different types of oscillators present and "excited" in the atom, and the relative brightness of the colors would depend on the number of oscillators of a given type and the effectiveness with which they are excited. A study of the simplest spectrum known, that of the hydrogen atom, led to the conclusion that the structure of even the hydrogen atom must be as "complex as that of a grand piano."

At the end of the nineteenth century and the beginning of the twentieth century, a number of experiments began to reveal some information about the structure of the atom. Atoms were shown to contain both positive and negative electrical charges. As already noted in connection with the photoelectric effect, the negative charges are electrons that can be "split off" from the atom. The discoverer of the electron, J. J. Thomson, suggested that the atom could be considered as like a blob of raisin or plum pudding, with the electrons being the raisins imbedded in the massive, positively charged "blob" of pudding. However, a New Zealander, Ernest Rutherford, working in England, then demonstrated that the atom seemed to be mostly empty space, with most of its mass concentrated at its center. Nevertheless, the chemical and kinetic-molecular properties of the atoms show that this apparently empty space is part of the size of the atom.

Rutherford then proposed the following model of the atom, which is the basis of the description in most current popular discussions of atomic structure. Almost all of the mass of the atom is concentrated in a very tiny volume, called the **nucleus,** at the center of the atom. This nucleus carries with it a positive electrical charge. Outside of the nucleus are a number of tiny bits of matter, the **electrons,** each of which carries a negative electrical charge. There are enough electrons so that their total negative charge is exactly equal to the positive charge on the

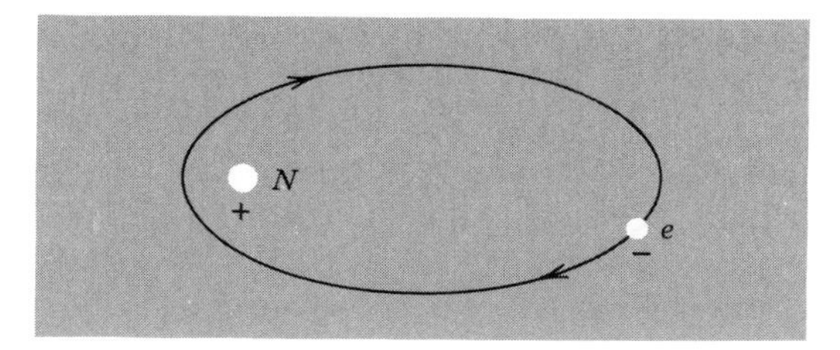

Figure 7.9. Rutherford–Bohr–Sommerfeld model of a hydrogen atom. *N*, nucleus; *e*, orbiting electron. Note that although the orbit is shown as highly elliptical, the nucleus is still at the center of the atom (actually at one focus point), because as shown by Sommerfeld and others, there are other effects that cause the entire orbit to precess (i.e., rotate about the nucleus as a center).

nucleus. The electrons are in motion about the nucleus, traveling in elliptical orbits in the same way as the planets travel in elliptical orbits about the Sun, except that the attractive force causing the elliptical orbits is electrical rather than gravitational (see Chapter 6, Section B1). In other words, the atom is like a miniature solar system. The only difference between various types of atoms is in the amount of mass and charge of the nucleus and the number of orbiting electrons. In the case of the hydrogen atom, there is one electron orbiting the nucleus; for helium there are two electrons; for lithium, three electrons, and so on. Figure 7.9 shows the model for hydrogen.

There is, however, a serious shortcoming for Rutherford's model as proposed: it is unstable. An orbiting electron is necessarily an accelerating electron because it constantly changes its direction of motion. But according to classical electromagnetic theory, an accelerating electron (or any accelerating electric charge) must radiate energy away.[3] This means that the electron must eventually spiral into the nucleus. Calculations showed that this ''eventuality'' would take place in about 0.01 millionths of a second, accompanied by a flash of light. But this does not happen. Moreover, the calculated spectrum of the light flash, even if it were to occur, is nothing like the spectrum of light emitted by excited hydrogen atoms. These were serious flaws.

3. Bohr's Improved Planetary Atom

Despite the problems associated with Rutherford's model, his experimental evidence that almost all of the total mass of the atom is concentrated in the nucleus was incontrovertible. Niels Bohr, a young Danish physicist who was working with Rutherford in 1912, and who was greatly influenced by the quantum ideas

[3]In Chapter 6, electromagnetic waves were described as being caused by a vibrating or oscillating electric charge. The essential feature of the vibrating charge is that its velocity is always *changing;* that is, it is always undergoing acceleration.

of Planck and Einstein, undertook to modify Rutherford's model in the light of these ideas. He published his results in 1913. He felt that the atom must be stable, and hence there must be certain electron orbits for which no radiation is emitted, despite the requirements of electromagnetic theory. Because electrons in different orbits have different energy values, only certain particular orbits could be allowed, just as the oscillators in the blackbody radiation problem could have only certain allowed values of energy. Bohr then supposed that an atom could change its energy only by having one of its electrons change from one orbit to another orbit of different energy. Thus an atom could emit a quantity of energy—a photon, for example—by having an electron "jump" from one orbit to an orbit of lower energy. Similarly, it would only absorb light of a given frequency if the photons of that frequency were of just the right energy to match the energy difference between the orbit the electron was in and some other allowed orbit, so that the electron could "jump" into the other orbit.

By a study of the experimentally measured spectrum of the hydrogen atom, and an ingenious use of the idea that when the orbital dimensions get sufficiently large, quantum ideas should give the same result as classical ideas, Bohr inferred the following postulates for the structure of an atom:

I. **Most of the mass of the atom is concentrated in the positively charged nucleus.** Under the influence of the attractive electrical force exerted by the nucleus, the negatively charged electrons orbit the nucleus in certain allowed stable, nonradiating, orbits. For simplicity, these orbits are taken to be circular.

II. **There is a definite rule for determining which particular orbits are allowed.** The angular momentum (a quantity of angular motion that is analogous to linear momentum for straight line motion) of the electron, which is constant for a given orbit, must be related to Planck's constant h. Specifically, the angular momentum must be equal to a whole number n, multiplied by h, and divided by 2π. (In the case of a circular orbit, the angular momentum is the mass m of the electron multiplied by its speed v and the radius r of the orbit, that is, $mvr = nh/2\pi$, where n is a whole number, 1 or 2 or 3 or 4, etc.) The different allowed orbits are characterized by the number n, which is called a *quantum number,* and $h/2\pi$ is a quantum of angular momentum.

III. **An atom can absorb or emit radiation only in the form of photons corresponding to the energy difference between allowed orbits.** The frequency f of the light absorbed or emitted must be related to the energy difference, E, of the orbits by Einstein's relation $E = hf$.

Using these postulates, and the already measured values of the charge and mass of the electron and of Planck's constant h, Bohr was able to calculate the allowed energies of the hydrogen atom, and thus the frequencies of the light emitted in the hydrogen spectrum.[4] His calculated values agreed with the measured values to within about 0.01 percent. Moreover, he predicted that there should be some particular wavelengths in the ultraviolet portion of the spectrum that had not yet been measured. These were searched for and found, just as he had predicted. He also calculated from his theory the size of the hydrogen atom in its unexcited state to be about one angstrom, in agreement with experimental measurements.

Just as in the case of Newton's theory of planetary motion, further elaborations and refinements of Bohr's theory led to even better agreement between theory and experimental measurements. A German physicist, Arnold Sommerfeld, introduced elliptical orbits and corrections (required by relativity theory) to the mass of the electrons because of high speed. Two Dutch physicists, Samuel A. Goudsmit and George E. Uhlenbeck, considered that the electron itself must be spinning on an axis, which should contribute magnetic effects to the calculation of the energy of the electron. The concept of a spinning electron also made it possible to explain the various magnetic properties of matter. All of these ideas led to the introduction of additional quantum numbers besides the quantum number n introduced by Bohr. These quantum numbers were related to the shape of the elliptical orbit, the orientation of the plane of the orbit in space, and the angular momentum of spin of the electron. Thus, in addition to only certain energy orbits being allowed, of all the possible different shape elliptical orbits having the same energy, only certain shapes of the ellipses were allowed. Additionally, the electron spin was quantized to have only one allowed value of its spin angular momentum, as distinguished from its orbital angular momentum, and only two possible orientations of its axis of spin.

For all of its phenomenal successes, and the seminal nature of its concepts, the Bohr theory of atomic structure had several serious shortcomings. On the one hand, Bohr's fundamental postulates were quite arbitrary and not derivable from existing theory. Yet, on the other hand, he made free use of existing classical theories. There was no justification of the assumption, for example, that the electromagnetic theory of radiation, which had been proved to work very well for orbits the size of an inch, and which Bohr used in inferring his postulates, should not work for orbits the size of an angstrom. After all, radio communication, which is based on classical electromagnetic theory, is quite successful. Moreover, new quantum numbers were just introduced as needed, again without fundamental justification, for the sole purpose of explaining experimental results.

How many different kinds of quantum numbers were going to be necessary,

[4]Actually Bohr used a much more complicated line of reasoning, but these postulates can be inferred from his work and are usually used in presenting the Bohr model. Moreover, Bohr, like most other physicists at that time, did not accept Einstein's photon concept until the discovery of the Compton effect in 1923, ten years later (see Section C4 above). A further point worth noting is that although the energies deduced from Bohr's model are different from those assumed by Planck in his analysis of the blackbody radiation problem, Planck's analysis is still valid, because the radiation emitted by the blackbody is independent of the details of the model. All that is required is that the energies of the constituent atoms or oscillators change by discrete amounts. Einstein later carried out a different detailed analysis of the blackbody problem in which he introduced the concept of stimulated emission of radiation. This concept is the basis on which a laser works.

and what rhyme or reason would be related to their appearance? In fact, the quantum number for electron spin is not even a whole number, but has the value ½. In addition, detailed experimental measurements of certain fine features of the spectrum led to the conclusion that, in certain circumstances, other quantum numbers might also not be whole numbers, in disagreement with the implied idea that various quantized quantities had to come in complete bundles.

Furthermore, the attempt to extend the quantitative calculations to atoms having more than one electron encountered serious snags. Calculations for the helium atom, which has two electrons, failed totally. It seemed that it would be necessary to introduce new special hypotheses and assumptions for each different kind of atom.

There were many questions to which the Bohr theory gave no answers or even inklings of answers. For example, although it could account for the specific frequencies in the spectrum of a particular atom, it said nothing about how bright or intense the light emitted at these frequencies should be. In fact, there are even a number of frequencies that should be in the spectrum but that are never observed. It turned out to be possible to deduce certain rules, called *selection rules,* to predict which frequencies are observed and which are not; but the Bohr theory gave no clue, good or bad, as to the existence of these rules. All in all, it was recognized that the Bohr theory was an incomplete or "stopgap" theory.

Nevertheless, the unifying power of Bohr's concepts was so great that they could be applied, if only in a qualitative or semiquantitative way, to many different areas and fields of physics and chemistry. The Bohr–Sommerfeld–Rutherford model of the atom is still discussed in all introductory high school and college courses and popular discussions dealing with atomic structure. It is the best simple "picture" available, despite all its shortcomings.

OPTIONAL SECTION 7.2 *Energy Level Diagrams*

Bohr's theory was particularly useful because it introduced the twin concepts of *energy levels* and *stable energy states* into the study of atomic and nuclear physics. These two concepts make it possible to analyze a very wide variety of experimental measurements even when there is no real understanding of what the measurements mean. While the Bohr theory has been discarded, the concepts of energy levels and stable energy states are important features of the currently accepted theories.

An energy level diagram is a graph along a vertical scale of the energies that an object such as an atom or an electron may have. Figure 7.O.1 shows the possible energies of an electron in the hydrogen atom. The energies of the first few allowed orbits are shown as horizontal lines. The energies are given in *electron volts* (abbreviated as **eV**) and are also negative. An electron volt is a very small unit of energy equal in size to 1.6×10^{-19} joules (for comparison, a 100 watt light bulb converts 100 joules of electrical energy into heat in one second). The fact that the energies are negative is a matter of convenience; in all cases it is differences of energy between two energy levels that are important. The zero value of the energy scale corresponds to an electron being an infinite distance away from the nucleus of the atom.

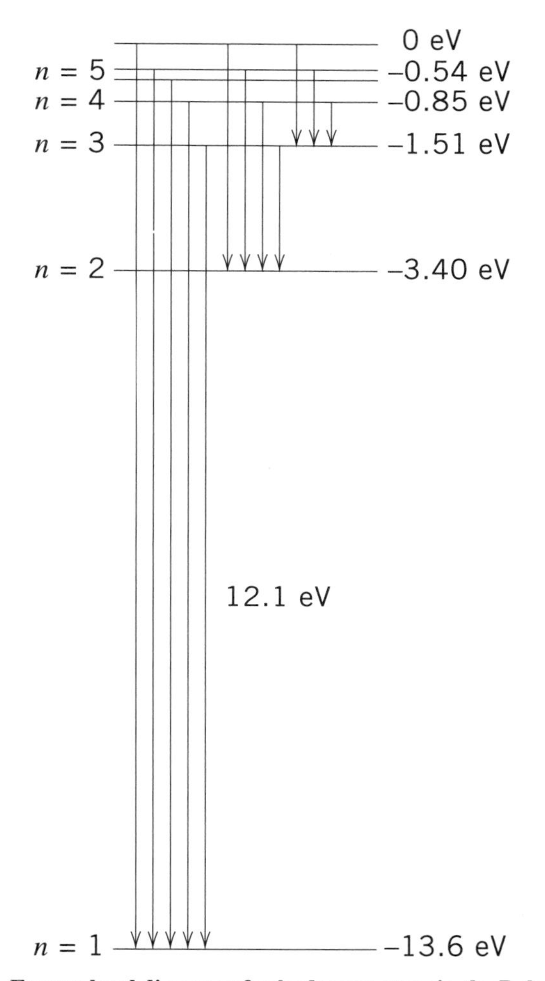

Figure 7.O.1. Energy level diagrams for hydrogen atom in the Bohr–Sommerfeld theory. Energy levels considering only circular orbits. $E = -13.6/n^2$ eV, for different values of n, the principal quantum number. Vertical arrows indicate possible transitions between energy levels that result in the emission of a photon with energy equal to the difference in energy of the two levels.

Associated with each energy level are one or more states of motion of the object (in this case, the electron) that can be maintained for some period of time, that is, for some period of time the energy of the object is constant (conserved). It may be for only a short period of time that the object is in a particular energy state, but it is definitely not zero.

In Bohr's original analysis of the hydrogen atom, the energies of the allowed orbits are equal to $-13.6/n^2$ eV, where n is the quantum number for a particular orbit. The radius of the orbit is equal to 0.5 n^2 angstroms (there are 254 million angstroms in one inch). In Fig. 7.O.1 the energy levels are labeled with the value of n. Also shown in the figure are vertical arrows representing the energy change when an electron moves from one orbit to another. For example, when an electron moves from the orbit having $n = 3$ to the orbit having $n = 1$, the energy change is -13.6 eV minus -1.51 eV or -12.1 eV. This means that a photon of energy 12.1 eV is emitted. Using Einstein's relationship $E = hf$, the frequency of the photon can be calculated. The wavelength λ of the photon is calculated from $f\lambda = c$, where c is the speed of light. The result is the following relationship between the wavelength in angstroms and the energy in eV: $\lambda = 12400/E = 12400/12.1 = 1025$ angstroms.

Figure 7.O.2 shows the energy level diagram for the hydrogen atom after taking into account the possibilities of elliptical orbits and of electron spin. These possibilities mean that the individual stable energy states allowed by the simple Bohr theory become subdivided into additional states, which are characterized by additional quantum numbers. Now the diagram is expanded to a two-dimensional diagram, in that the allowed energy levels are arranged in horizontal columns according to the value of the quantum number k, which was introduced by Sommerfeld to allow for the ellipticity of the orbit. Orbits having the same value of n, but different values of k, are considered to represent different energy states. In this case the energies of the different states are slightly different from each other because of effects originating in Einstein's special theory of relativity or in the magnetic effects introduced by a ''spinning'' electron.

Some of the photons that might be expected from transitions of an electron from one energy level (state) to another are not observed. For example, the transition between the level having $n = 1$ and $k = 1$ and the level having $n = 3$, $k = 3$ is not observed. It was therefore concluded that there were certain ''selection rules'' that were in force. These rules dictated that in order for light to be emitted as a result of the transition from one state to another the values of the quantum number k in the two states had to differ by exactly 1. It was not at all clear why these rules existed.

As things developed it was often the case that theoretical calculations gave values for the energies of photons that were significantly different from actual measured values. Using the basic concepts of the Bohr scheme, it was possible to carry out the inverse process, that is, use the experimental measurements to determine what the energy levels actually are, and even to deduce the values of quantum numbers or recognize the existence of additional quantum numbers. This made it possible to evaluate some of the assumptions put into theories. It was these evaluations that led to the realization that the Bohr theory, while pointing out the general nature of results to be expected from a complete theory, was itself quite inadequate because it could not give any reason for the various selection rules or any basis for determining what new quantum numbers might be valid in any particular situation.

OPTIONAL SECTION 7.3 *Bohr Theory and the Periodic Table*

Even though the Rutherford–Bohr–Sommerfeld model was incomplete and ultimately unable to account quantitatively for atomic spectra, in its most refined form it did lead to an understanding of X-ray spectra. This in turn led

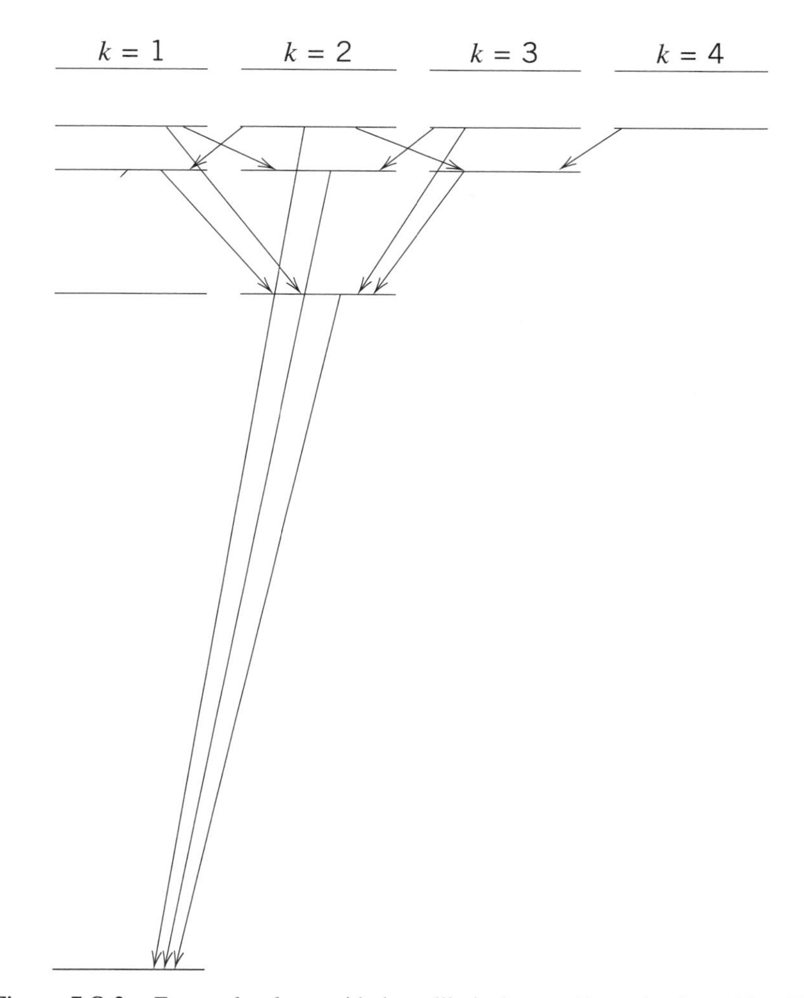

Figure 7.O.2. Energy levels considering elliptical as well as circular orbits. The energy levels for a given value of n are only very slightly changed (by roughly .01 percent, and by different amounts for different values of k, the "orbital quantum number"). For a given value of n, k has integer values from 1 up to n. When $k = n$, then the orbit is circular; when k is less than n, the orbit is elliptical, becoming more so as k becomes smaller. The (almost) vertical arrows indicate those transitions between levels that result in the emission of a photon, as allowed by "selection" rules. The quantum numbers n and k are peculiar to the Bohr–Sommerfeld theory. Somewhat different quantum numbers are used in the later theories.

to fundamental insight into the **periodic table of the elements,** which had been introduced in 1869 by Dmitri Mendeleev, a Russian chemist, in an effort to classify the large number of chemical elements that were already known at that time. Considering the chemical elements in the order of their atomic masses, Mendeleev found that each successive element could be placed into one of eight classes in order in almost the same way as a dealer distributes cards in a card game. There were some notable exceptions to this scheme, but it did reveal a repeating pattern for many of the chemical and physical properties of the basic chemical elements as their atomic mass increases. There was strong evidence that the same sort of repeating pattern was true of the atomic spectra of the elements. All of this implied that the atoms of the different elements are constructed from even more basic ''building blocks.'' Each different atom is composed of different numbers of these basic building blocks ''arranged'' in some way that results in the repeating pattern of properties shown in the periodic table.

The electron is one of these basic building blocks. The nuclear model of the atom, coupled with the analysis of X-ray spectra as discussed below, suggested that the nucleus itself is an assembly of the other building blocks, and that these building blocks had positive electric charge as well as mass.

X rays were discovered in 1895. By 1912 it was recognized that they are very short wavelength (hence very high frequency) electromagnetic waves (see Table 6.1). This means that X-ray photons are very large bundles of energy. By 1914 it was known that the chemical elements can be caused to emit characteristic X-ray spectra and that the Bohr theory could be used to explain the X-ray spectra of the elements. According to the Bohr theory, the X-ray spectra of atoms are generated when the electrons in orbits closest to the nucleus change from one orbit to another.

When the X-ray spectra of the elements are included in the periodic table, they do not show the same repeating pattern as the chemical properties, but rather the energy of their photons increases steadily as the atomic mass increases. This implies that as additional building blocks are added to the nucleus there will be an increase of positive electric charge as well as an increase of mass for the following reason. In the Bohr theory the energy of the allowed orbits depends on the square of the positive electric charge ''seen'' by the negatively charged electron. Those electrons closest to the nucleus, whose orbital energies are involved in the emission of X rays, are affected by the full charge of the nucleus. Therefore the energy of their photons increases as the amount of charge on the nucleus increases. Hence the nuclei of elements having greater atomic mass must have more positive electric charge, and also more orbiting electrons.

On the other hand those electrons furthest from the nucleus are much less affected by the positive charge on the nucleus because the other electrons partially cancel out the effects of the positive charge of the nucleus. Since the addition of more positive charge to the nucleus requires that there be more electrons, the additional electrons will mostly cancel out any effects on the outermost electrons. As a result the photons of the outermost electrons of all atoms are of similar energies, and the spectra resulting from the outer electrons will tend to follow the repeat pattern of Mendeleev's periodic table.

The study of X-ray spectra, while helpful in contributing to an understanding of the periodic table, raised some additional questions. Why are some electrons in inner orbits and some not? The same question could be raised about the orbits of the planets—why are some planets close to the Sun and some very far from the Sun? Kepler had an answer for that question in terms of Pythagorean solids (see Chapter 2), but such an answer is not acceptable in the twentieth century. The Bohr theory gives no answer to this question. In fact the answer must be given as a postulate called the **Pauli Exclusion Principle,** which in its simplest form states that no two electrons in the same atom can have the same set of four quantum numbers. Since each allowed electron orbit in an atom can be specified in terms of four quantum numbers, the Pauli Exclusion Principle compels the electrons to be distributed among the possible orbits that can be derived from the Bohr theory, or rather from better theories.

The Pauli Exclusion Principle when coupled with the correct theory of the atom makes it possible to explain completely all the features of Mendeleev's periodic table of the elements. In the modern version of the periodic table it is necessary to arrange the elements according to their **atomic number** rather than their atomic mass. The atomic number is equal to the number of protons in the nucleus, which is also equal to the number of orbiting electrons. The **proton** is a nuclear building block that has the same mass and electric charge as the nucleus of the hydrogen atom. The nucleus has another building block called a **neutron,** which has almost the same mass as the proton, but no electric charge. Thus even before the correct theory of the atom was developed it was the "patched up" version of the Rutherford–Bohr–Sommerfeld model of the atom, with quantum numbers introduced almost arbitrarily, that led to the understanding of the periodic table and the way in which different atoms are constructed.

4. De Broglie's Wave Hypothesis

Within a decade of Bohr's modification of the nuclear model of the atom, a young French physicist, Louis de Broglie, as part of his doctoral dissertation, suggested a basis for justifying Bohr's postulates. Recognizing the significance of the equivalence of energy and mass proposed in Einstein's relativity theory, de Broglie suggested that because both mass and light are forms of energy, they should be describable in the same terms. Thus because Einstein had shown in his analysis of the photoelectric effect that light exhibited both wave and particle properties, so, too, matter must exhibit both wave and particle properties. Moreover, concomitant with the symmetry and intimate relationship between space and time demanded by relativity theory, there is also a symmetry and intimate relationship between momentum and energy. The argument proceeds roughly as follows: According to Einstein (see Chapter 6) $E = mc^2$ for any particle, and the momentum of any particle, $p = mv$ where v is the velocity of the particle, which for light must equal c. Therefore $p = mc$, and this implies $p = E/c$. But according to Einstein, the energy of a particle of light, a photon, equals hf. Therefore the momentum of a photon is hf/c. However, for light waves there is a relationship between frequency f and wavelength λ, namely, $f = c/\lambda$. Therefore by simple algebra the momentum p of the photon must equal Planck's constant h divided by the wavelength, that is, $p = h/\lambda$. This is the formula mentioned in the discussion of the photoelectric effect (Section C4). There is some "sleight-of-hand" involved in this reasoning. It says that a photon has mass. Fortunately, in the final formula mass no longer appears, so we need not worry about the mass of a photon.

De Broglie suggested that this relationship must hold for matter as well as for light, and in particular it must be true for the electrons in an atom. He believed that the wave nature of the electron determines how electrons get from one place to another, just as the wave nature of light determines how photons get from one place to another. Inverting the relationship between wavelength and momentum, he could use the velocity of the electron (because momentum is mass times velocity at low speeds) to calculate its wavelength.

Turning his attention to the question of which orbits are allowed in an atom, de Broglie pointed out that if a whole number of wavelengths could fit exactly around the circumference of an orbit, then as the electron circles around the orbit under the "guidance" of its wave nature, the wave would reinforce and sustain itself in a pattern called a *standing wave pattern.* If, however, a whole number of wavelengths did not exactly fit the orbit, then on successive circuits around the orbit the waves from the different circuits would be out of step with waves from previous orbits and so would ultimately cancel out the wave pattern entirely. Thus such an orbit could not be sustained. These ideas are illustrated in Fig. 7.10.

For any circular orbit it is relatively easy to calculate the velocity of the electron in terms of the radius r of the orbit. Since the circumference of an orbit is $2\pi r$, the standing wave condition simply means that $n\lambda = 2\pi r$, where n is a whole number. Using his relationship between wavelength and momentum, de Broglie was then able to derive by simple algebra Bohr's angular momentum rule for the allowed orbits (Bohr's second postulate). In other words, the reason that only certain orbits are allowed in an atom is that the wave nature of the electron can establish a stable wave pattern only for certain orbits. For all other orbits the wave pattern cannot be stabilized.

Of course, the existence of a clever analogy alone does not establish a scientific principle. Fortunately, within a few years independent experimental evidence of the wave nature of the electron was found by investigators working in England, Germany, and America. (In England, the investigators included G. P. Thomson, the son of J. J. Thomson, who had discovered the electron a generation earlier.) An essential property of waves is that of diffraction, and the experimentalists showed that electrons could be diffracted in the same way as X rays. In fact, by

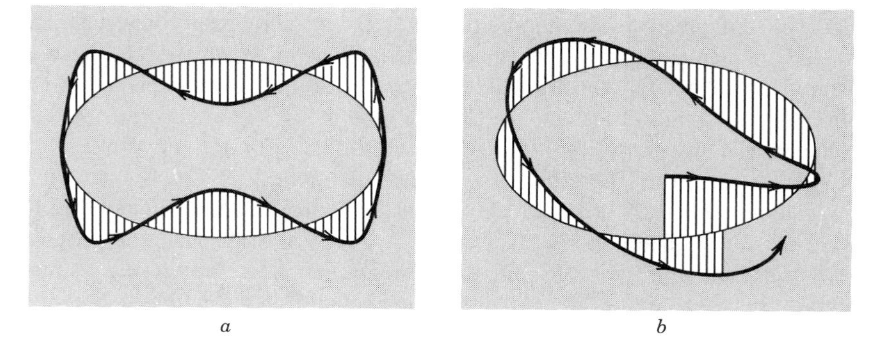

Figure 7.10. De Broglie waves for a circular orbit. (a) Reinforced waves. (b) Nonreinforcing or canceling waves.

diffraction experiments with crystals it is possible to measure the wavelength of X rays and electrons. Thus it became possible to measure the wavelength of a beam of electrons and show that it was exactly as would be calculated from the velocity of the beam, the mass of the electron, and de Broglie's wave hypothesis.

As has so often been demonstrated in the nineteenth and twentieth centuries, advances in one area of science often have unpredictable effects on other areas of science and human endeavor. The ability to perform diffraction experiments with electrons (and with neutrons, subatomic particles about 1800 times more massive than electrons) has become a standard part of the repertoire of pure and applied science laboratories throughout the world. The electron microscope, which is found in many medical and biological research centers and is designed according to principles of wave optics, permits studies of objects that are too small to be seen with ordinary light microscopes.

De Broglie's wave hypothesis, in its simple form, was of rather limited applicability to the detailed understanding of atomic structure; but it was quickly expanded and elaborated to give a complete, comprehensive, and powerful new theory of the nature of matter, capable of solving problems and answering questions with which Bohr's theory could not cope. Moreover, the ramifications and implications of this new theory have extended to almost every area of science and those areas of philosophy dealing with material knowledge. Some of these ramifications will be discussed further below.

E. Quantum Theory, Uncertainty, and Probability

1. Dual Nature of Matter and Light

De Broglie's significant insight was that matter, which is a form of energy just as light is a form of energy, should be describable in terms of propagating waves as well as in terms of particles moving under the influence of various forces. In the same manner as Einstein viewed light waves as determining how photons get from one place to another, matter waves should determine how particles, in particular, electrons, get from one place to another. Instead of using Newton's laws of motion (or other principles based on Newton's laws) to calculate the motion of particles, it is necessary to use some other laws or equations to determine how waves are propagated from one place to another.

2. Schroedinger Wave Theory

In 1926 an Austrian mathematical physicist, Erwin Schroedinger, published a general theory for the propagation of matter waves. (This theory was an outgrowth of a seminar he had been asked to give on de Broglie's wave hypothesis.) Schroedinger's theory dealt with the propagation of waves in three dimensions, whereas de Broglie's theory was essentially a one-dimensional theory (it considered only waves traveling around the circumference of an orbit, not radially or perpendicularly to the plane of the orbit). One-dimensional waves are easily visualized:

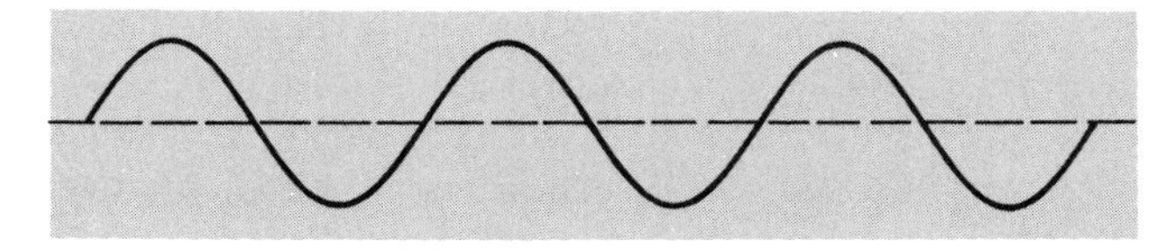

Figure 7.11. One-dimensional waves.

The waves set up along the strings of a violin or guitar or piano travel along the length of the string (Fig. 7.11); the sound waves set up in an organ pipe or horn travel along the length of the pipe (one does not consider waves running across the pipe or horn). On the other hand, the waves set up on the surface of a drum are two-dimensional (also transverse[5]), as shown in Fig. 7.12. Similarly, waves created by dropping pebbles in water are two-dimensional (as are the waves shown in Fig. 6.5). Of course, the sound waves we hear at a concert are three-dimensional.

Matter waves are tuned by "Mother Nature" in the same way as the waves in musical instruments or auditoria are tuned by their designers. The length of the string, the diameter of the drum, the length of the organ pipe, the position of various ports on wind instruments and frets on string instruments, the dimensions of the auditorium, all determine certain fundamental resonant wavelengths for each system. These dimensions are all involved in *boundary conditions,* which can be specified mathematically. In addition, the *propagation velocity* (i.e., the speed with which the wave travels) determines the frequencies that will be resonant for the specific boundary conditions. It is the violinist or the piano tuner who controls the speed by adjusting the tension of the string, thereby determining the resonant frequency. (The violinist also varies the boundary conditions by "fingering" the strings.)

In the case of matter waves (say for an electron) the boundary conditions are controlled by the environment in which the electron finds itself. For example, the environment for the electron inside an atom is different from the environment for an electron traveling down the length of a television picture tube. In the atom, the boundary conditions are three-dimensional, whereas in the picture tube they are one-dimensional. Moreover, the speed of the matter wave for an electron is determined by the potential energy situation in which it finds itself: Within an atom the potential energy and hence the speed of the electron is controlled by the electrical attraction of the nucleus and the electrical repulsion of all the other electrons in the atom. In a television picture tube the potential energy is determined by the electrical voltage applied to the tube. Therefore the matter waves

[5]As noted in Chapter 6, waves are described as transverse or longitudinal, depending on whether the direction of the vibratory disturbance is perpendicular (transverse) to the direction of propagation of the wave or parallel (longitudinal) to the direction of propagation of the wave. Waves within the bulk of liquids or gases are longitudinal; waves in solids can be either transverse or longitudinal or a combination of both. Waves on a drumhead are transverse, as are the waves on a string. Light waves are transverse; sound waves in air are longitudinal. See Chapter 6, Section B4, for a brief discussion of light waves.

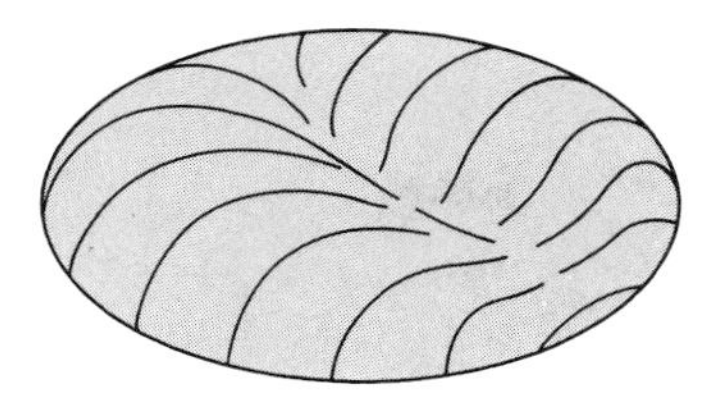

Figure 7.12. Two-dimensional waves.

that describe the electron will be different in the atom from what they are in the television picture tube. (This corresponds to the fact that in the Bohr model, an electron in an orbit travels a very different path than it would travel in a television picture tube—different dispositions of forces are acting on it.)

For each particular kind of wave phenomenon (i.e., vibrating strings, vibrating drumhead, sound waves, water waves, light waves, matter waves, etc.), despite all the possible variations in boundary conditions and propagation velocity, one particular governing equation permits the calculation of the wave disturbance at any point and time. This equation is called a **wave equation.** The wave equation has a somewhat different mathematical form for the different types of wave phenomena; that is, the mathematical form of the equation for waves on a string is different from the mathematical form of the equation for sound waves in a three-dimensional medium, which is in turn different in form from the equation for electromagnetic light waves, and so on.

A particular form of wave equation for matter waves was developed by Schroedinger; this equation is called **Schroedinger's wave equation.** Schroedinger's wave equation does for matter waves what Newton's laws of motion do for particles in motion. Unfortunately, it is not possible to draw a few simple graphs to show the essential nature of Schroedinger's equation. For one thing, Schroedinger's equation and the solutions to it usually involve the use of *complex numbers.*[6] Thus the solutions that describe the matter waves are not readily shown or represented by the kinds of graphs or pictures that are used for other waves. The solutions, of course, are described by appropriate mathematical formulae.

The exact form of the solutions to Schroedinger's wave equation are determined, as already indicated, by various boundary conditions and the potential energy relationships for the particular situation. These solutions, then, are three-dimensional waves. In the case of electrons in atoms these three-dimensional waves define the location of the electrons within the atom, and thus they take the place of the "orbits" of the Bohr theory. In common usage the wave solutions are even called orbits, even though they are conceptually quite different from the orbits of Bohr theory. Another term that is used rather than orbits is **shells**—we say that the electrons are found in shells around the nucleus. Even these shells are not geometrically distinct from each other; they overlap and penetrate each

[6]Complex numbers usually contain the square root of -1, and have some of the characteristics of two-dimensional vectors. A discussion of complex numbers and their significance would be too large a digression from the principal goals of this introduction to quantum physics. Suffice it to say that they are extremely useful for proper mathematical description of many physical phenomena.

other somewhat. A more representative description is to say the electrons in the atom are found in "clouds" around the nucleus.

3. The Meaning and Interpretation of Schroedinger Waves

Before discussing the results of the application of the Schroedinger theory to atomic physics, it is necessary to define some basic concepts involved in wave motion. For particle motion, the basic concepts are position, velocity, acceleration, mass, energy, and so on. For wave motion, however, some of the basic concepts are **amplitude, phase, propagation velocity,** and **interference** (see Fig. 7.13). If we think of a wave as a propagating or spreading disturbance (pulsation) in a medium, or as the disturbance of an electromagnetic field, the amplitude of the wave is the maximum disturbance (at any point in space) from the equilibrium value as the wave proceeds. At that point in space, the disturbance increases from zero to the maximum value (amplitude) in one direction, decreases to zero again, reverses direction and increases to the maximum value, decreases to zero again, reverses back to the original direction, increases to the maximum value, and so on; and if it is a sustained steady wave, the cycle of disturbance at a point waxes and wanes repeatedly. At any instant, the disturbance may be in a maximum phase or a minimum phase or some phase in between.

The phase of the wave refers to what part of the disturbance is occurring at the particular point in space at a particular instant of time. At other points in space the disturbance will also go through a cycle, but not necessarily at the same time as at the first point. Actually all the various points in space affected by the wave act like the simple harmonic oscillators discussed at the end of Section B3 above, but as one looks at different successive points along the wave in the direction of propagation at *one instant* of time, the phases of the oscillators vary from zero disturbance to maximum in one direction, back to zero, and then to maximum in the opposite direction, back to zero again, and so on. As time goes on, the phases change in a progressive manner in the direction of propagation, and the speed with which the changes progress is called the phase propagation velocity.

Sometimes more than one wave of disturbance is present in a medium at the same time. If the circumstances are correct, then the disturbances caused by the individual waves alone are added to each other when they are all present. If these

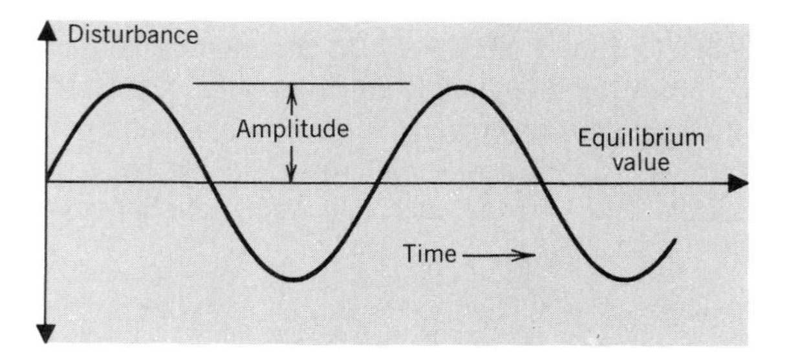

Figure 7.13. Characteristics of waves.

disturbances have the same wavelength and frequency, then an interesting phenomenon called interference takes place. The resulting disturbance will also be a wave of the same wavelength and frequency, whose amplitude and phase will depend on the amplitudes and relative phases of the individual waves (Fig. 7.14). If all the individual waves have the same phase at a particular point at a given instant of time, then the resulting wave will have an amplitude equal to the sum of the amplitudes of the individual waves, and hence will be quite large. This is called constructive interference.

On the other hand, it is also possible for the individual waves to have such phase relations that at any given point, when one wave disturbance is at its maximum amplitude in a given direction, another wave may be at its maximum amplitude in the opposite direction, with the result that the two waves cancel each other

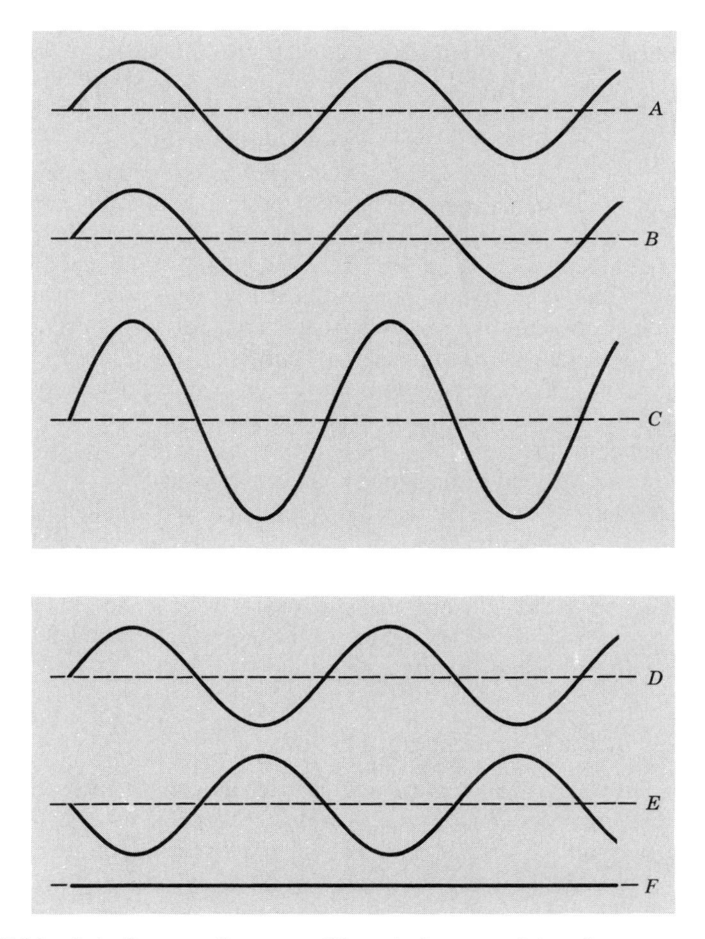

Figure 7.14. Interference of waves. Wave *A* plus wave *B* interfere constructively to give wave *C*. Wave *D* plus wave *E* interfere destructively to give no wave at all as represented by the straight line *F*.

completely and there is no net disturbance at all, just as if there were no wave present whatsoever. This is called destructive interference. These are the two extreme cases; there will be various degrees of constructive and destructive interference between the two extremes.

Interference effects are used to explain the appearance of colors in soap bubbles and oil films on water on one hand, and the operation of the Michelson interferometer on the other hand. (Figures 7.13 and 7.14 show these quantities and concepts for waves involving real numbers; the ideas are similar for the complex numbers involved in matter waves, but not as easy to depict.)

The energy transmitted by a wave disturbance is related to the amplitude of the wave. The rate of transport of energy by the wave is not directly proportional to the amplitude of the wave, however, but rather to the square of the amplitude. This is true for light waves and sound waves. Returning now to Schroedinger waves (i.e., matter waves), the logical question arises: does the wave transport energy (according to relativity theory, matter is a form of energy)? What is the medium that carries the wave or alternatively, what is waving? What kind of wave is the Schroedinger wave? The answer is (and this is a postulate of the theory) that the Schroedinger wave is a **probability wave**! The wave "carries" the probability of finding the electron (for example) at a particular point in space.[7] By analogy with electromagnetic waves, it is postulated that this probability is proportional to the "square of the modulus" of the amplitude of the probability wave. (It is necessary to say "square of the modulus" of the amplitude rather than just square of the amplitude because the Schroedinger wave involves complex numbers; the modulus is a particular characteristic of a complex number that is analogous to the magnitude of a vector quantity.)

One significant aspect of a probability wave is that we cannot talk about finding an electron at a particular point in space because we cannot localize a wave at one particular point in space. A wave is necessarily spread out, and thus the probability of finding an electron at one point is also spread out. We say that the electron is "smeared out." This idea has some interesting implications. Before going on to these implications, it is worthwhile discussing what Schroedinger theory says about atomic structure, as contrasted to Bohr theory. The probability calculated from the Schroedinger wave amplitude is a "real" (ordinary) number, and therefore it is possible to draw graphs of the probabilities calculated from the Schroedinger waves. These are discussed below.

4. Success of the Schroedinger Theory

Applying his wave equation to the hydrogen atom, with the appropriate potential energy and boundary conditions, Schroedinger was able to show that acceptable solutions of his wave equation are possible only for certain discrete values of energy. This is similar to de Broglie's idea that only certain waves would fit into allowed orbits. The discrete values of energy are exactly the same values as

[7]This interpretation of the Schroedinger wave was first put forward by Max Born, a German physicist. Schroedinger himself felt that the electron in an atom was somehow spread out over the wave.

calculated previously by Niels Bohr. The probability waves can be specifically calculated, thus giving the actual probabilities for finding the electron when it has a given value of energy. The result is that for a given energy, the associated electron orbit is not a sharply defined circle or ellipse but rather a "cloud" of "probability density" extending over a fair-size region of space (Fig. 7.15). Where the cloud is most dense, the electron is most likely to be found, and where it is least dense, the electron is least likely to be found. As it turns out, the cloud is always fairly dense at a distance from the nucleus equal to the radius calculated from the Bohr theory. This should be expected because the Bohr theory worked fairly well for the hydrogen atom.

What is quite different from the Bohr theory is the extent of the cloud. For example, there is a small probability that the electron could be found right at the nucleus. Even more surprising, there is a very small probability that the electron could be found a mile away from the nucleus. It is a very small probability—much less than one chance in a billion years—but it is, in principle, mathematically not zero, even though for all practical purposes it is essentially zero. Moreover, as shown in Fig. 7.15, the clouds associated with the various possible energy levels often have more than one region where the probability is greater than in the immediately adjacent regions.

The Schroedinger picture also makes it possible to solve the problem of the stability of the orbits, which Bohr had to handle by the first of his special postulates (Section D3 above). Using the probability waves, it is possible to calculate the probability of finding an oscillating electric charge in the atom when an electron is at a given energy level. The result of the calculation is that there is zero net oscillating electric charge, and therefore zero radiation of electromagnetic energy. Thus the orbit is energetically stable.

It is also possible to understand how radiation may be emitted when an electron changes from a higher energy orbit to a lower energy orbit. During the time that the electron is changing from the higher energy state to the lower energy state, it can be said to be in both energy states simultaneously; that is, the probability waves associated with both states are active, and they interfere with each other. When the probability of finding an oscillating electric charge is calculated under these conditions, it is usually found to have a value greater than zero, but the value oscillates with a frequency corresponding to the difference of the frequencies of the two simultaneous waves. This frequency is exactly the same frequency as would be calculated from Bohr's third special postulate (Section D3 above). The calculation simply expresses the idea that the frequencies of each of the waves corresponding to the two energy states are "beating" against each other.[8] The

[8]The phenomenon of "beats" between waves of slightly different frequencies is most easily demonstrated by simultaneously sounding two tuning forks of very slightly different frequencies. The resulting sound will oscillate in loudness with a beat frequency equal to the difference between the frequencies of the two tuning forks. A similar effect is noted in poorly tuned radios that happen to be receiving two stations of slightly different broadcast frequencies simultaneously. A very loud annoying whistle is heard, and the frequency of the whistle is 10 khz, which is the difference between the frequencies of the two stations. The same principle is used in tuners for most radios and television sets, where it is called *heterodyning*.

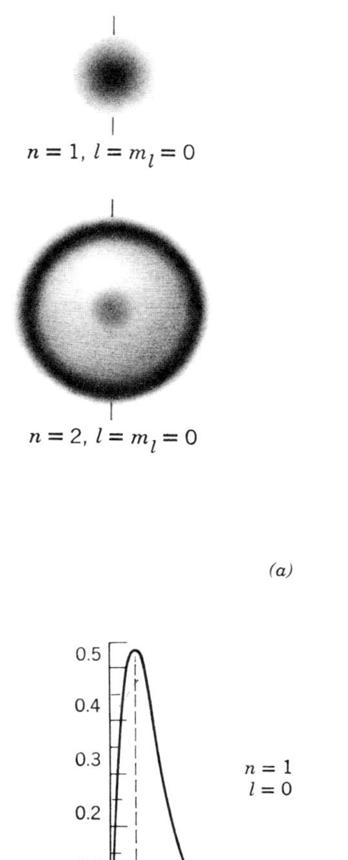

(a)

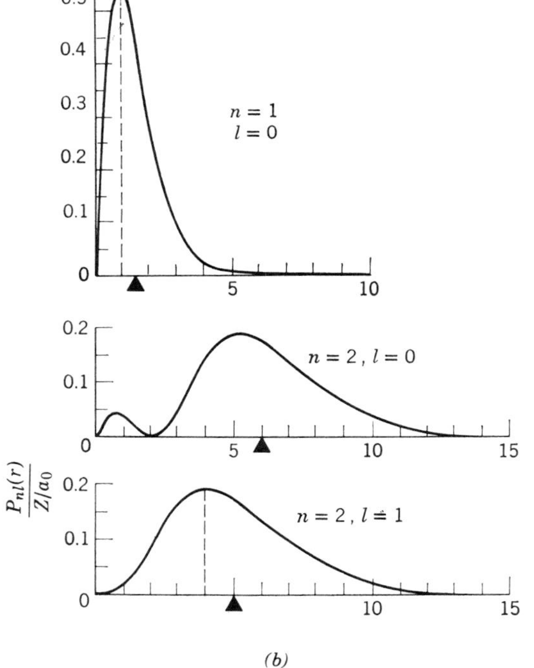

(b)

Figure 7.15. Probability density for a hydrogen atom. (a) Three-dimensional. Each figure represents a different ''orbit,'' having different values of the various quantum numbers. (b) Graph of total probability density at a given radius for the same ''orbits'' as above. The dotted lines show the radii of the corresponding Bohr orbits. (Reproduced by permission from Robert Eisberg and Robert Resnick, *Quantum Physics,* 2nd ed., John Wiley & Sons, New York, 1985.)

mathematics also permits calculating the strength or intensity (actually, the amplitude) of the electrical oscillations. Thus the theory is able to determine the brightness of the spectral lines emitted. Moreover, it explains the various selection rules that have been found to govern whether transitions between certain energy levels even occur and contribute their corresponding frequencies to the observed spectrum. This is simply beyond anything the Bohr theory can do.

Moreover, the various quantum numbers, which were introduced on a rather arbitrary basis in the old theory, arise naturally from the various requirements imposed by the boundary conditions and the inherent geometrical symmetry of the problem. They are interpreted somewhat differently than in the old theory. In the simple Schroedinger theory, there are three quantum numbers associated with the three dimensions of space. (A fourth quantum number, usually associated with electron spin, is not inherent in the theory, as will be discussed below.) The quantum numbers represent certain quantities that are conserved in a particular problem and are quantized; that is, they are constant but can have only discrete values and not a continuous range of values. Furthermore, the quantum numbers, although integers, do not represent the number of bundles or units of the conserved quantity, but are used to calculate the discrete values of the conserved quantity. In the case of the simple hydrogen atom, these quantities are the total energy of the atom, the magnitude of the orbital angular momentum of the electron, and one component of the vector which represents the angular momentum.[9] It is even possible to show that for the waves that have high quantum numbers, the regions of high probability begin to look increasingly like orbits in the classical sense. This illustrates the idea that quantum physics gives essentially the same results as classical physics in the domain where classical physics has been proven valid, as discussed in Section A2 above.

Although they are not inherent in the Schroedinger wave mechanical theory as originally formulated, it is also possible to graft onto the theory corrections for magnetic effects on the energy due to the spinning electron, and thus make the refined calculations that were not possible for the Bohr theory and its elaborations by Sommerfeld and others. Moreover, mathematical techniques have been developed to apply the Schroedinger theory to all atoms, and to achieve excellent quantitative agreement with experiment. This was an area where calculations based on the Bohr model had failed miserably. (Recently it has been shown how to make the Bohr theory apply to some of these cases; however, this was done only after the Schroedinger theory showed what the correct answer should be.)

The Schroedinger theory, although powerful and useful, does have certain limitations. Unlike the Bohr theory, however, it is possible to work around these limitations. The chief limitation, from a fundamental and conceptual viewpoint, is that Schroedinger's wave equation does not satisfy the requirement imposed by the theory of relativity that the form of the equation be the same for all observ-

[9]Angular momentum, like ordinary linear momentum, is a vector quantity. As may be inferred from Fig. 3.13, and the discussion in Chapter 3, Section C5, any three-dimensional vector can be considered to be the sum of three mutually perpendicular vectors, which are called components of the original vector.

ers. Attempts to modify the Schroedinger theory to meet this requirement have met with limited success. In fact, it is for this reason that the Schroedinger theory does not itself satisfactorily treat the problem of the spinning electron and the various magnetic effects associated with the spinning electron. There is no basis in the Schroedinger theory for determining the electron spin and its associated quantum number.

Within a few years of Schroedinger's work, P. A. M. Dirac, an English mathematical physicist, published a new theory, called a relativistic quantum theory, which incorporated the demands of relativity theory from the very beginning. In this theory, there are no waves involved; indeed there are no quantities involved that would permit drawing simple pictures or models of an atom. In the theory, the idea of electron spin and the magnetic effects that are pictorially ascribed to electron spin turn out to be consequences of the requirements of relativity, just as magnetic effects in general can be attributed to relative motion of electric charges. Dirac's theory yields all the quantum numbers of Schroedinger's theory and additionally it yields the half-integer quantum number associated with electron spin. It becomes quite clear why there are four quantum numbers associated with an electron: Boundary conditions and mathematical symmetries must be associated with the four dimensions of space–time. The Dirac theory also predicts that in addition to the existence of the negatively charged electron, there should also exist a *positively* charged particle having the same mass as the electron. This particle, now called a **positron,** was discovered some four years after Dirac predicted its existence.

The Dirac theory is not always convenient to apply, and for most purposes the Schroedinger theory can be readily modified to allow for the effects of electron spin. Thus the Schroedinger theory is very useful for solving problems throughout the domain of atomic, molecular, and solid-state physics.

5. Heisenberg Uncertainty Principle

The introduction of probability into the discussion of a physical situation means that there is some uncertainty. Even if something is very highly probable, it is necessarily slightly uncertain. Often when something is said to be probable or uncertain, the statement is made only because there is not enough time or a good enough measuring instrument available to make a determination with certainty. In quantum physics, however, it is believed that the **uncertainty is inherent in the nature of things** and cannot be removed regardless of how well the measurements may be made. Simply stated, it means that if predictions are made based on past events and a complete, well-understood theory, the results may not be certain. Thus the direct and rigid connection between cause and effect is destroyed, because it is not certain that the cause will lead to the exact calculated effect.

To see how this occurs, it is necessary to discuss what is involved in making or specifying a precision measurement. Suppose that for some reason it were necessary to measure the diameter of a round object such as a basketball. Suppose further that the surface of the ball is assumed to be perfectly smooth and perfectly round, so that there need be no concern about superficial irregularities. The ball might be set up against a suitable measuring stick, and the person making the measurement would try to look perpendicular to the measuring stick to see exactly where the projection of the extreme edges of the ball would fall on the scale. If this person has a fairly good eye for such things, the measurement might come out to be 12½ inches. The question might then be asked: Is it exactly 12.5 inches, or is it possibly 12.499 inches or maybe 12.501 inches?

If the question is sufficiently important, then the person making the measurement might get a large machinist's micrometer caliper, adjust it with the right "touch" so that the jaws of the caliper just make contact with the ball (but do not squeeze it), and determine that the diameter of the ball is 12.500 inches. Again the question might arise: Is it possibly 12.4999 or 12.5001? The person making the measurements might then determine that the last question is not answerable with his instrument. The error of his measurement is possibly as much as .0001 inch. It might be only .00001 inch, but he cannot prove it with his instrument, and with his sense of touch. In adjusting the instrument to fit snugly against the ball, he might inadvertently squeeze it out of shape by .0001 inch, particularly if it is a soft ball.

Two conclusions may be drawn from the foregoing discussion. One is that, at least in the physical world, it is not possible to know something such as the diameter of a ball unless it can be measured. In other words, knowledge must be based on experimental observation or measurement. This statement is at the core of the philosophical school called logical positivism. The other conclusion is that the attempt to measure an object or a situation may disturb or distort the thing that is to be measured. (In the case of the soft basketball, the distortion was only about .0001 inch; but this set a limit to the precision to which the diameter of the ball could be measured, and hence a limit to the precision to which the diameter could even be known.)

An attempt to measure something that distorts what is being measured is not uncommon. It is fairly common knowledge among those who study the utility of various medications in combating illness that the test patients who take these medications often report they feel better whether or not the medicine is effective. The fact that they are being observed changes their medical response. It becomes necessary to give some of the patients a placebo (a pill that has no effect at all), without advising them about whether they are receiving the placebo or the true medication. The idea is to observe patients without their knowledge. If the observer is clever enough, a way can be devised so that the patient's subjective reactions will not distort or interfere with the observations.

In considering observations on matter, then, the question arises as to whether it is always possible to find a way to make a measurement to any precision desired, without disturbing the experimental situation. It may be very difficult to do, and may involve more effort than the knowledge is worth, but it is a matter of principle that needs to be understood. Suppose, for example, that it was desired to carry out an experiment to observe an electron and to verify the probability waves described in the Schroedinger theory. (This would be a very significant experiment.) Schroedinger felt that the solutions to his equation did not merely give a probability of finding the electron at some point in space, but actually meant that the electron was itself smeared out throughout the region of the electron cloud. Others felt that it should be possible to see the electron, albeit with only a probability, at specific points in space. Maybe it even would be possible to see the electron actually spinning on its axis.

How does one see an object as small as an electron, or even an atom? One, in effect, looks at it with light. But a wavelength of the light with which we actually see is about 5000 times larger than the diameter of an atom. In the same sense that an ocean wave sweeping over a small pebble is not disturbed by the presence of the pebble, a light wave sweeping over an atom would not be noticeably affected by the presence of the atom, and we cannot see any effect of the presence of the atom on the light wave. The wavelength is too long. One should use ''light'' with a wavelength that is as small as the atom or electron; in fact, the wavelength should be much smaller. Very short wavelength electromagnetic radiation—namely, X rays or even gamma rays—should be used to ''see'' the electron.

But now the dual nature of electromagnetic radiation comes into play. The X ray or gamma ray photons that are guided by their waves have a considerable amount of energy and momentum, according to the ideas of Einstein and de Broglie (Sections C4 and D4 above). In fact, their energy and momentum are perhaps a thousand times greater than that of the electron being observed. As soon as one of the photons interacts with an electron, the collision will be so violent as to knock the electron completely out of its orbit (Compton effect); and thus the attempt to measure the position of the electron with great precision totally disturbs the situation. It becomes necessary to make the measurement with longer wavelength photons, so as to avoid disturbing the electron unduly, and therefore the precision of the measurement is not as good as desired.

It is necessary to review what is involved in making measurements on the microscopic level. According to the quantum mechanical theory of measurement, the act of observation of the electron will force it into a particular quantum state that it would not otherwise have entered. Before the measurement it was not known in which orbit the electron could be found; the measurement actually ''put'' it into some orbit.

These ideas were first expressed by Werner Heisenberg, a German physicist, in the form of a principle called the **Heisenberg uncertainty principle** or the **Heisenberg principle of indeterminacy.** In words, the principle states that *it is impossible to measure, predict, or know both the position and momentum simultaneously of a particle, with unlimited precision in both quantities.* In fact, the error (or uncertainty) in the position multiplied by the error in the momentum must always be greater than Planck's constant h divided by 2π. For example, if the error in one of the quantities is numerically smaller than about $0.01h$, then the error in the other quantity must be greater than $100/2\pi$, or about 16. It should be noted that Heisenberg's principle applies to quantities in specific related pairs (e.g., the position in a given direction and the momentum in that direction). It also applies to the energy of the particle and the time at which it has that energy. It does not apply to a simultaneous measurement of the momentum and energy of the particle. Generally speaking, in most measurement situations, the uncertainty required by Heisenberg's principle is quite small and beyond the reach of the available measuring instruments; however, there are certain situations in which the principle plays a useful role. The attempt to measure the position of an electron in an atomic orbit, as discussed above, is one such situation.

The principle has had a profound effect on discussions of metaphysics and fundamental concepts of knowledge. It destroys the idea that the universe is completely determined by its past history. This idea was strongly suggested by Laplace almost 200 years ago as a result of the success of Newtonian mechanics. (See the discussion in Section E of Chapter 3.) Laplace stated that if the positions and velocities of all the bits of matter in the universe were known at one time, and if all the various force laws were known, then the positions and velocities of all these bits of matter could be calculated and predicted for any future time. All future effects would be the result of earlier causes. Even if the task of measuring all these positions and velocities were humanly impossible, and even if the discovery of all the appropriate laws were impossible, nevertheless the positions and velocities *did* exist at a previous time and the laws *do* exist; therefore the future is predetermined.

But Heisenberg's uncertainty principle says this is not so. It is, in principle, impossible to make the measurements with sufficient precision or even to calculate from them the future positions and velocities, because we cannot know the future positions and velocities. (According to the positivist mode of thought, if we cannot measure, we cannot know or predict, nor can "Nature" know or predict.) Therefore, there are limitations on causality, but this does not mean that the future is completely unknown. We can calculate probabilities that things will occur in the future, and statistically these calculations will be borne out. It is only for the individual electron that we cannot predict.[10]

Heisenberg's uncertainty principle does more than set limits on our knowledge and inspire metaphysical debates. It tells us, for example, that all motion does not cease at the absolute zero of temperature. If this were true, then when a particular substance in a container is brought to absolute zero temperature, the momentum of its individual molecules would be zero with zero uncertainty. Heisenberg's principle would then require that the position of the molecules would be so uncertain that they could very well not even be in the container! On the other hand, if the molecules are indeed somewhere within the container, then the uncertainty principle requires that there be some slight uncertainty in their momentum. Therefore their momentum could very well not be zero, but in fact

[10]Not all physicists accepted the uncertainty principle and its consequences. Einstein, for one, did not like it, and he had a long series of arguments with Bohr and others about its validity. He made numerous attempts to refute the principle or to find examples where it would lead to obvious error or paradox. Ultimately Einstein conceded that predictions based on it were valid, but insisted that there must be some more satisfying principle that would account for the results of quantum theory but would still preserve causality completely. His primary motivation was a philosophical intuition that "God does not play dice with the universe." Nevertheless the prevailing mode of thought among most physicists today is that the uncertainty principle is valid and useful, and that there are limitations on causality.

greater than zero. This means that they are in motion, and have energy, even at zero kelvins.

The same logic can be used to provide one reason that electrons cannot remain within the atomic nucleus. If an electron were within the nucleus, then its position would be known with a maximum uncertainty roughly equal to the diameter of the nucleus. This then determines a minimum uncertainty in the momentum of the electron. The momentum must be at least as large as this minimum uncertainty. This means that the energy of the electron would have some minimum magnitude that turns out to be large enough to overcome the electromagnetic force of attraction of the positive charges inside the nucleus, so that the electron can escape from the nucleus.

Another application is to the theory of nuclear forces. To see this, the uncertainty principle should be expressed in terms of the uncertainty in the energy of a particular particle and the uncertainty in the length of time the particle is at that energy. The theories of nuclear forces (discussed in Chapter 8) require that special particles called **virtual particles** exist for very short periods of time. These particles have mass and thus energy is needed to create them. Where does this energy come from? If the duration of their time of existence is taken as the uncertainty in the time at which they existed, then the uncertainty principle says that there must be an uncertainty in their energy. The smaller the time duration the greater the energy uncertainty. If the energy uncertainty is greater than the energy needed to create the particle, then it can exist during the necessary time duration. In terms of the energy-money analogy used previously, the creation of the virtual particle is equivalent to over-drawing a bank account for a brief period of time, but making up the shortage of money before the bank knows for sure that the account is overdrawn. In this case, the smaller the time period during which the shortage exists, the greater the amount that can be overdrawn! Therefore the energy uncertainty of the virtual particle is large enough to equal the amount of energy needed to create it!

Other applications are to the existence of the very precisely defined energy levels necessary for the operation of lasers and for certain types of gamma ray emitters.

F. THE USE OF MODELS IN DESCRIBING NATURE

1. Models as Analogs or Synthesizing Constructs

Almost all of the preceding discussion of the nature of matter and energy has been based on the use of models. The models supposedly represent the "true reality" in the sense of Plato's Allegory of the Cave. This has been characteristic of the historical development of physics. The caloric model of heat described it as a colorless and weightless fluid. Bohr's model described the atom as a miniature solar system. In discussing the nature of an electron, it was thought for a while to be like a very tiny spherical ball that could spin on its axis, and then later it was considered to be like a wave. Light is considered as exhibiting a dual nature.

It may well be asked why these models should be made or even why old or flawed models should be discussed when it is known that they are not correct. Perhaps we should not try to find the ''true reality''—whatever we observe or whatever exists is real, and perhaps there is nothing else underlying existence. There is, however, a very practical reason for making models: They are convenient and make it possible to sum up in a relatively few coherent words an intricate collection of physical phenomena. They make it possible to assimilate and integrate into the human mind new facts and knowledge, and relate them to previous knowledge. When an atom is said to be like a miniature solar system, then a picture immediately springs to mind that shows that an atom is mostly empty space, that it should be possible to remove one or more electrons, and so on. Thus the model serves not only as an analogy, but also makes it possible to synthesize (in our minds) some other possible properties of atoms that we might otherwise not have guessed.

2. The Reality of Models

But is the model real? After all, if a hobbyist builds a model airplane to be set on a shelf in a room, it is definitely not the real thing. It is only a representation of the real thing, and we must keep this in mind in our examination of the model. For example, we must not think that the engines are made of plastic; or if it is a flying model, we must not think that the engines used in the model are like the real engines or even provide the model with the same flight characteristics as the real object. There are many pitfalls that have led physicists astray (to mix metaphors), when models are used. These have led to a number of paradoxes and to attempts to construct rival models for the same set of phenomena.

Although recognizing the claimed utility of models for practical purposes as discussed above, many physicists have argued that the sooner we can abandon the use of pictorial models, the sooner we can obtain a profound understanding of physical phenomena. After all, a model is not useful if it cannot be treated mathematically to see if it will lead to a detailed *quantitative* account of the experimental data. The model must be expressible in terms of equations. That being the case, why not abandon the model entirely and just write down the basic mathematical assumptions that lead to the equations? All that is necessary is that the assumptions not be inconsistent with each other and that there not be too many of them. Then, we need not be concerned about whether these assumptions make sense in terms of some simple-minded model.

Thus the model now becomes a set of equations, and there should be no question as to whether the model is real or not. An atom is not a set of equations.[11] In this vein, various economists construct ''models'' of the world economic sys-

[11]Interestingly enough, 100 years ago there was a significant school of thought among physicists that argued that atoms were not real, they were only mental constructs—they could not be seen with any instruments, and their existence was only inferred—and thus one should not talk about the kinetic-molecular model of matter.

tem and make predictions—usually dire—about what will happen if certain trends continue.

3. More Subtle Models for Quantum Mechanics

At about the same time that Schroedinger was developing his wave mechanical model, Heisenberg and others were developing strictly mathematical models for atomic phenomena. Heisenberg felt that the Rutherford–Bohr–Sommerfeld model was too full of inconsistencies, and that a model of the atom should make direct use of the experimental observations, taken primarily (but not entirely) from studies of the spectra emitted by the atoms under various excitation conditions. He made use of the properties of mathematical quantities called matrices (plural of matrix), and his theory was called **matrix mechanics.** He used different assumptions than Schroedinger—he did not like Schroedinger's theory. Schroedinger, on the other hand, found Heisenberg's assumptions "repellent." Interestingly enough, very quickly Schroedinger (and others as well) was able to show that both his theory and Heisenberg's theory were mathematically equivalent! Schroedinger started out with the idea that physical phenomena were by nature continuous, and that under certain circumstances (with proper boundary conditions) quantization and quanta appeared. Heisenberg, however, started out with the assumption of quantization of phenomena at the fundamental level. He also incorporated a formulation of his uncertainty principle into his assumptions. Quite often in detailed calculations both Schroedinger's and Heisenberg's approaches are used together, because Schroedinger's theory can be used to calculate certain quantities needed in Heisenberg's theory.

An even more abstract formulation was used by Dirac when he developed his relativistic quantum mechanics. This formulation is consistent with the use of both wave mechanics and matrix mechanics.

OPTIONAL SECTION 7.4 *Quantum Electrodynamics*

One of the most persistent questions in physics deals with the inherent nature of forces, whether electromagnetic, gravitational, or nuclear. Since the dominant force between and within atoms is the electromagnetic force, this question has led to the development of **quantum electrodynamics,** or for short, **QED.** QED, which is largely based on Dirac's relativistic quantum theory, also necessarily deals with light, its emission, its propagation, and its absorption. QED has served as a model for the development of **quantum chromodynamics** (**QCD**), the theory of the strong nuclear force or color force (discussed in Chapter 8), and presumably will serve as a model for the gravitational force.

One of the architects of QED, the late Richard Feynman, has given a popularized version of this theory in a book entitled *QED: The Strange Theory of Light and Matter,* cited in the references at the end of this chapter. The book states that QED avoids the apparent paradoxes and difficulties embodied in the dual wave–particle model of matter and radiation and in the Heisenberg Uncertainty Principle. A pictorial presentation is used that lends itself nicely not only to simplified presentations of the type used here but also to the elaborate mathematical calculations involved in QED. The following material shows something of the way in which this pictorial presentation is used to describe light.

Feynman asserts unequivocally that light is fundamentally corpuscular in nature, but these corpuscles must not be thought of as material particles subject to Newton's laws of motion, but rather as concentrated packages of energy subject to some special rules. These rules are adopted as basic postulates. This is entirely reasonable—Newton's laws of motion are not proved, but rather adopted as basic postulates. Additional postulates are that the emission (creation) and subsequent absorption (destruction) of a photon must be looked upon as space–time events (see Chapter 6) subject to laws of probability.

These postulates can be applied to the case of an individual photon that is emitted at a specific place and time (the first event) and that is absorbed at a second specific place and time (the second event). The probabilities of concern are the probability of being emitted, the probability of getting from the first place to the second place, and the probability of being absorbed. The emission and absorption probabilities are calculated in terms of the interaction of radiation with matter, whereas the probability of getting from the first place to the second place deals with how light propagates through space. In determining the probability that light will actually travel in space from the emission event to the absorption event, it is necessary to consider the possible world lines (Chapter 6, Section D6, Fig. 6.15) connecting the two events. This means that *all* possible unobstructed paths between the two points, whether they be straight line paths or not, whether the light must be reflected from a mirror or not, whether the light must travel through different media at different speeds, and so on, must be considered. Figure 7.O.3 shows schematically the two events, both in space–time and in space.

If only one world line, corresponding to one specific path in space, is considered and all other paths are blocked, then there will be a specific small probability that the photon will travel from the first event to the second event. This probability is calculated in the following way: Imagine a pointer that could rotate while the photon travels along the world line. The rate of rotation is f revolutions per second, corresponding to the frequency of the photon. When the photon gets to the second event, the pointer stops rotating. It will be pointing in some direction, but the probability will be calculated as the square of the length of the pointer. If that one world line is blocked, and some other world line, with quite a different path in space, is unblocked, the probability of travel from the first event to the second event will be calculated similarly. If the length of this path is different, the pointer will be pointing in a different direction because the length of time it takes the photon to reach the second event will be different. However, the length of the pointer will still be the same and so the probability will be the same—the square of the length of the pointer.

When both world lines are open, the probability of travel between the two events is determined by combining the probabilities of both world lines in the following way: add the two pointers vectorially (see Chapter 3, Section C5, Fig. 3.13), and calculate the square of the length of the vector sum. This result-

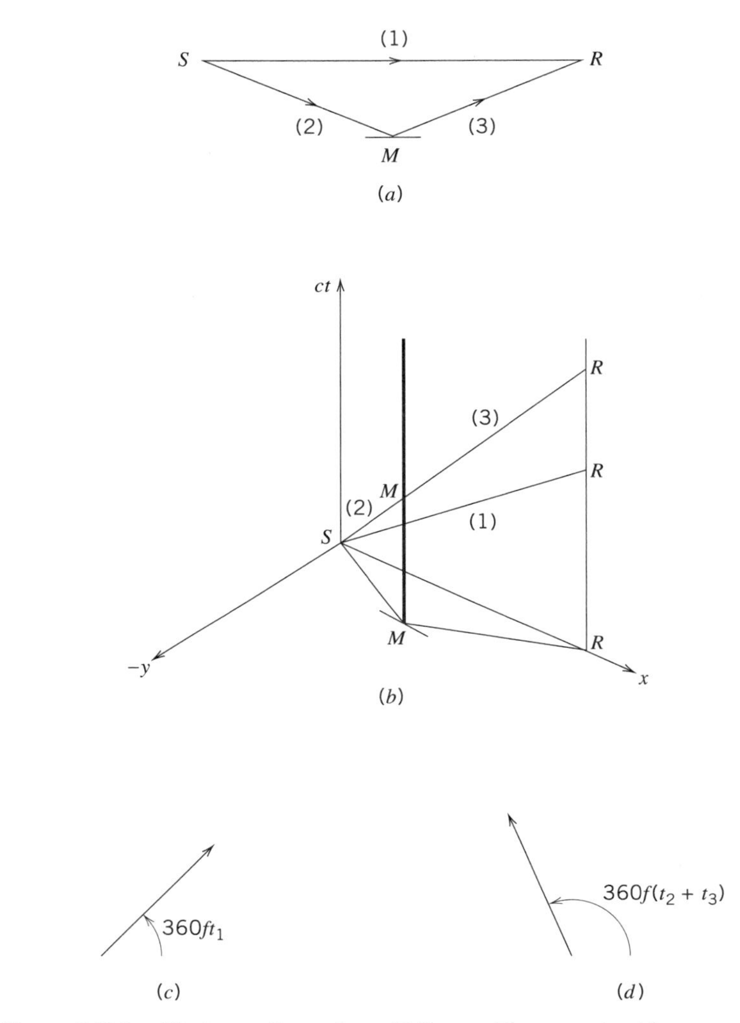

Figure 7.O.3. Photon paths and world lines. Photon emitted by source *S* travels to receiver *R* by either of two paths: directly from *S* to *R* by means of path (1) or from *S* to mirror *M* by path (2) and from *M* to *R* by path (3). (a) The two paths are chosen to lie in a common plane. (b) World lines for the two paths. The *x*-*y* plane shows the projection of the world lines of the photon onto the spatial plane of the two paths. (c) and (d) Position of the rotating probability "pointer" for the two paths. *t*1, *t*2, and *t*3 are the length of time needed to traverse the correspondingly numbered paths or path segments.

ing length and probability will depend very sensitively on the directions in which the two pointers are pointing (see Fig. 7.O.4). If they happen to be pointing in the same direction, then the vector sum of the pointers will be twice as long as each pointer, and the probability will be four times as much ($2^2 = 4$) as for each world line alone. If the pointers happen to be pointing in opposite

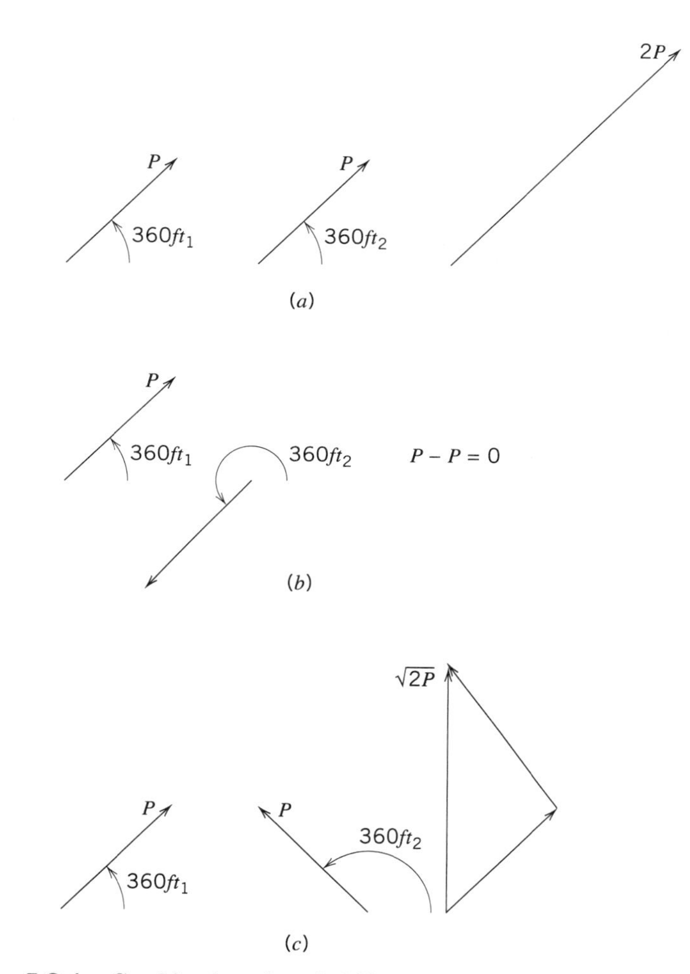

Figure 7.O.4. Combination of probability "pointers." The probability pointers for two hypothetical photon paths are shown. The pointers for each path have the same length, but different orientations dependent on the path and the time needed to traverse the path. The angle of orientation is given in degrees of arc and is calculated from $360 \times f \times t$, where f equals the energy of the photon divided by Planck's constant ($f = E/h$) and t is the time to traverse the path. If the calculated angle is greater than 360 degrees, then 360 is subtracted repeatedly to obtain an angle less than 360 degrees. (a) The probability pointers for the two paths have the same orientation, and their combined probability "amplitude" is $2P$, where P is the probability amplitude for any one world line. This corresponds mathematically to constructive interference. (b) The probability pointers are in opposite directions; the resulting probability amplitude is 0. This is destructive interference. (c) The probability pointers are at 90 degrees with respect to each other, and the resulting probability is $\sqrt{2}\,P$.

directions, then their vector sum will be zero and the probability will be zero. These two cases are examples of constructive and destructive interference discussed in Section E3 of this chapter, where they were explained in terms of waves. In this discussion waves are not mentioned, so we need not wonder what is the medium in which the wave travels or what is waving. We do not need waves; we only need to postulate the rules for determining probability and for calculating combinations of probabilities.

In fact it is not possible to restrict the considerations to just one or two world lines, but rather all possible world lines must be considered, and the probability of propagation by means of a bundle of world lines must be calculated. The calculations are more intricate than when using a wave model of light, but they are manageable. It is often necessary to make approximations to simplify the calculations, but the effects of such approximations can be evaluated.

At the beginning of this chapter it was stated that it is often conceded that there might be yet a further development of an overarching theory that would be more general than relativistic quantum mechanics. It is sometimes speculated, or even hoped, that this new theory will contain within it new underlying concepts or new variables of which we are currently not aware. These are sometimes called *hidden variables.* Those who are uncomfortable with the uncertainty principle, even though they are currently compelled to concede its apparent validity, believe that these hidden variables will once again restore complete causality to physics. There are indeed new theoretical developments that seem to hold the possibility of uniting not only relativity theory and quantum theory, but also gravitational theory and the theory of various forces active within the nucleus of the atom as well. Some of these are discussed in the next chapter. However, the desired hidden variables are not included in these developments. Moreover, there have been recent experiments that seem to show that most hidden variable theories are inconsistent with the experimentally confirmed results of quantum theory and therefore should be rejected.

G. THE IMPACT OF QUANTUM THEORY ON PHILOSOPHY AND LITERATURE

Quantum theory apparently sets limits on the rigorous causality associated with classical Newtonian physics. The Heisenberg principle is viewed by some writers as making possible, on the most fundamental level of the material universe, the concept of free will, because free-will decisions are necessarily unpredictable. It is not that an individual electron will "choose" to behave in an unpredictable manner but rather that nonphysical influences (for example, the human will or divine intervention) may affect the behavior of the material universe. In this respect free-will decisions are indistinguishable from random decisions. However, it must be emphasized that while the specific behavior of individual atoms and molecules may not be predictable, the overall statistics resulting from the astronomical number of atoms and molecules that make up a simple biological cell is so overwhelming that the average behavior of the very large number of atoms in a biological cell is highly predictable; and significant deviations will occur only very rarely. Thus it is not at all clear that we should believe that a universe governed according to modern physics allows any more freedom of choice than one governed according to classical physics.

Some writers have interpreted the uncertainty principle as demonstrating that there are limitations to what can be known in a material sense, saying that we can claim knowledge only about things that can be measured. Things that in principle cannot be measured are therefore unknowable and therefore nonexistent in a material sense.

As can be readily imagined, many thinkers take umbrage at such assertions as lack of causality or limitations on knowledge. Thus there has been a continuing lively debate, almost from the very inception of quantum theory, among physicists, philosophers, and philosophers of science on these matters, which are at the very root of basic concepts about knowledge. Philosophers tend to regard physicists as rather naive about such matters, and physicists tend to regard philosophers as somewhat out of touch with physical reality. However, as already mentioned, there is some disagreement among physicists themselves as to how quantum theory should be interpreted. A detailed annotated bibliography and guide to some of the numerous articles and books on the ''proper'' interpretation of quantum theory and its metaphysical–philosophical foundations and implications is given in the article by DeWitt and Graham listed in the references for this chapter.

The prevailing view among physicists as to the significance and interpretation of quantum theory was developed by a group of physicists called the Copenhagen School, among the most famous of whose members were Neils Bohr and Werner Heisenberg. An example of their mode of thought is the *complementarity principle,* which was introduced by Bohr in 1928 in order to present the Heisenberg uncertainty principle in more general terms. The complementarity principle states that on the scale of atomic dimensions and smaller, it is simply not possible to describe phenomena with the completeness expected from classical physics. Some of the measurements or knowledge required for a complete description from the classical viewpoint, such as position and momentum, are contradictory or mutually exclusive, if the definitions of these quantities are properly understood in terms of how they are to be measured. Whenever one of the quantities is measured beyond a certain level of precision, the other is distorted to such an extent that the disturbing effect of the measurement on it cannot be determined without interfering with the original measurement. This result is inherent in the definitions of the quantities being measured. The principle goes on to state that there is cause and effect, but what quantities can be used in describing a cause-and-effect relationship must be understood. Such things as position and momentum of an electron in an atom are not the proper quantities to measure but rather the state function (an alternative and better name for the wave function of the electron in the atom).

Werner Heisenberg and Niels Bohr. (Photo by Paul Ehrenfest, Jr., American Institute of Physics Niels Bohr Library.)

On another level, the idea that there are fundamental limits on the measurement of certain quantities in physics has suggested to workers in other fields that there may be analogous fundamental limits in their own disciplines on quantities to be measured or defined. The attempt to measure these quantities beyond a certain level of precision inherently distorts certain other complementary quantities. Thus the idea of complementarity may be significant for other branches of knowledge.

There is very often a strong temptation to apply major principles of physics in an overly simplistic manner to the description of the universe at all levels. Thus, following Newton's work, some argued that the universe was predetermined in all aspects and behaved like a large elaborate clockwork. This influenced the early development of the social sciences. Following the development of the entropy concept and the understanding of the second law of thermodynamics, still others asserted that the universe, taken as a whole, must be running down. Some social critics have suggested that the degradation associated with the second law is evident in modern art and even in modern society. Quantum mechanics, on the other hand, has been hailed as making possible free will and an end to all considerations of predeterminacy. Obviously, such conclusions are overdrawn.

How then should the application of physics to fields outside the physical sciences and technology be viewed? Aside from enhancing our appreciation of the beauty and grandeur of the physical universe, physics also offers new ideas and outlooks that must be justified primarily by the needs of the other areas of endeavor. Thus physics can furnish useful analogies on which to base new ways of analysis and insightful figures of speech.

Such uses of physics are sometimes apparent in literature. Indeed, concepts of modern physics, both from relativity and quantum theory, have affected many works of modern fiction. Such writers as Joseph Conrad, Lawrence Durrell, Thomas Pynchon, William Gaddis, Robert Coover, and Robert Pirsig have employed various of these concepts in their writing, with varying degrees of success. This is discussed in the book by Friedman and Donley on Einstein and in the article by Alan J. Friedman on contemporary American fiction, cited in the references for this chapter.

Study Questions

1. What is meant by the range of validity of a theory?
2. Over what range is Einstein's special theory of relativity valid?
3. What is the difference between a blackbody radiator and a cavity radiator?
4. Give some examples from "real life" of situations where the conditions approach those of a cavity radiator.
5. Give some examples of the use in "real life" of good or poor absorbers/emitters.
6. What is meant by a partition function?
7. What is the ultraviolet catastrophe?
8. What is a quantum? A photon?
9. How did Planck's quantum theory violate common sense?
10. What was paradoxical about the results obtained in experimental studies of the photoelectric effect?
11. What is meant by the dual nature of light?
12. What is the definition of an element? A compound? A molecule? An atom?
13. What was wrong with Rutherford's original model of the atom?
14. What were Bohr's postulates for the planetary model of the atom?
15. What does the theory of relativity have to do with various models of the atom?
16. What are quantum numbers? To what are they related?
17. What is electron spin?
18. What was the reason that de Broglie introduced his wave hypothesis?
19. What is the medium for Schroedinger waves?

20. What role does the Schroedinger wave equation play in quantum mechanics?
21. What is the difference between an electron shell and an electron cloud?
22. Do Schroedinger waves actually exist?
23. What is constructive interference? Destructive interference?
24. What does the orbit of an electron in an atom look like?
25. If the Schroedinger theory is so much better than the Bohr theory, why is the Bohr theory still being taught?
26. What is the Heisenberg uncertainty principle?
27. Does something like the Heisenberg uncertainty principle apply in the social sciences? Justify your answer and discuss the consequences of your answer.
28. Does anyone know what an atom "really looks like"? Discuss your answer.

PROBLEMS

1. The energies involved in the kinetic-molecular model of matter are really very small by comparison with the energies encountered in everyday large-scale life. For example, the energy used to heat a cup of coffee is calculated in Problem 4 of Chapter 6, and is equal to 7200 joules. According to Optional Section 7.2 the energy of an electron in the first Bohr orbit is -13.6 electron volts. Calculate this energy in joules.
2. In many climates solar energy is used to heat water. Assume that the average effective wavelength of the sunlight incident on a solar heater is 5000 angstroms. (a) Calculate the wavelength in meters. (b) Calculate the frequency of this light. (c) Calculate the energy of a photon of this light. (d) Calculate how many photons would be needed to heat the cup of coffee of the preceding problem.
3. (a) Calculate the wavelength of the emitted light resulting from a "jump" of an electron from the fourth to the third orbit of hydrogen. (b) The normal range of wavelengths of electromagnetic radiation detectible by the human eye is from about 4500 angstroms to about 6800 angstroms. Can the wavelength calculated in part (a) be seen by the human eye?
4. It can be shown that the total energy of an electron in a circular Bohr orbit is equal to $-$(K.E), where $+$(K.E.) is the kinetic energy of the electron in orbit. The potential energy is negative, that is, $-$(P.E.) $= -2$ (K.E.). (a) Calculate the energy of an electron in the orbit having $n = 4$. (b) The mass of the electron is 9.11×10^{-31} kg. What is the speed of the electron in the orbit having $n = 4$? (c) Calculate the momentum of the electron of part (b). (d) Calculate the de Broglie wavelength of the electron in that orbit. (e) Calculate the radius of the orbit in this case.

REFERENCES

Bryce S. DeWitt and R. Neill Graham, "Resource Letter IQM-1 on the Interpretation of Quantum Mechanics," *American Journal of Physics,* 39, July 1971.

Richard P. Feynman, *QED: The Strange Theory of Light and Matter,* Princeton University Press, Princeton, N.J., 1985.
A nonmathematical discussion of quantum electrodynamics, the theory that "explains" all ordinary matter and that has been verified to a greater accuracy than any other theory known to date. Feynman, who shared a Nobel Prize for his contributions to the theory, was famous also for his ability to explain very abstruse subjects clearly and simply.

Richard P. Feynman, Robert B. Leighton and Matthew Sands, *The Feynman Lectures on Physics,* Vol. III, Addison-Wesley, Reading, Mass., 1965.
See comment in references for Chapter 4.

Kenneth W. Ford, *Basic Physics,* John Wiley & Sons, New York, 1968, Chapters 23 and 24.
See comment in references for Chapter 6.

A. J. Friedman and Carol Donley, *Einstein as Myth and Muse,* Cambridge University Press, Cambridge, 1985.

Alan J. Friedman, "Contemporary American Physics Fiction," *American Journal of Physics,* Vol. 47, May 1979, pp. 392–395.

G. W. Gamow, *Mr. Tompkins in Paperback,* Cambridge University Press, London, 1972, Chapters 7, 8, 10, and 10½.
See footnote 11 in Chapter 6, Section D.

Werner Heisenberg, *Physics and Philosophy: The Revolution in Modern Science,* Harper, New York, 1958.
Presents the philosophical viewpoint of one of the great architects of the quantum theory.

Banesh Hoffman, *The Strange Story of the Quantum,* Dover Publications, New York, 1959.
An introductory, essentially nonmathematical discussion of quantum mechanics, written for the lay public.

Gerald Holton and Stephen G. Brush, *Introduction to Concepts and Theories in Physical Science,* 2nd ed., Addison-Wesley, Reading, Mass., 1965, Chapters 26, 28, and 29.
See comment in references for Chapter 1.

Stanley L. Jaki, *The Relevance of Physics,* University of Chicago Press, Chicago, 1966.
See comment in references for Chapter 1.

Robert Bruce Lindsay and Henry Margenau, *Foundations of Physics,* Wiley, New York, 1936.
See comment in references for Chapter 6.

Edward M. MacKinnon, *Scientific Explanation and Atomic Physics,* University of Chicago Press, Chicago, 1982.
A scholarly history of the development of various atomic theories written from the viewpoint of a philosopher. One of his concerns is the interplay between the development of quantum physics and the developments of atomic theories.

Edmund Whittaker, *A History of the Theories of Aether and Electricity, Vol I. The Classical Theories; Vol. II. The Modern Theories.* Philosophical Library, New York, 1941; Tomash Publishers, Los Angeles and American Institute of Physics, New York, 1987.
See comment in references for Chapter 6.

Murray Gell-Mann. (American Institute of Physics, Meggers Gallery of Nobel Laureates.)

Chapter 8

CONSERVATION PRINCIPLES AND SYMMETRIES

Fundamentally, things never change

A. Introduction

Since at least as early as the times of the ancient Greeks, humanity has considered that there are a few fundamental building blocks, or particles, of which everything is made. Leucippus and Democritus (circa 400 B.C.) believed that matter can be divided only so far, at which point the smallest possible particle will be reached. Democritus called this most fundamental particle an **atom** (which means ''indivisible''). Aristotle realized that there must be more than one kind of fundamental particle in order to account for the different characteristics of different materials. Aristotle considered everything on Earth to be comprised of various amounts of earth, water, air, and fire, his ''fundamental'' elements.

With the development of the science of chemistry, it was recognized that there are about 100 different **elements** and that all matter consists of various combinations of these elements. These elements range from hydrogen and helium, the lightest ones, to uranium, the heaviest naturally occurring element. In 1803 John Dalton, an English chemist, proposed that each element has a characteristic atom that cannot be destroyed, split, or created. He proposed that atoms of the same element are alike in size, weight, and other properties, but are different from the atoms of other elements. The name *atom* was chosen because of the clear relationship with the original concept of Democritus.

We now know that Dalton's atoms are divisible and are made up of more truly ''fundamental'' particles. By the 1930s the atom was known to consist of a nucleus containing neutrons and protons with electrons circulating around the nucleus (see Chapter 7). For a time, these three particles were considered to be the basic building blocks of matter. However, numerous experiments performed since the 1940s indicate that neutrons and protons are made up of even more basic particles. In fact, these experiments performed at various high-energy particle accelerators (so-called atom smashers) revealed hundreds of different kinds of particles that can be produced in collisions between atomic nuclei. Throughout the 1950s and 1960s physicists became increasingly concerned about this proliferation of known subnuclear particles and wondered if they could all be ''fundamental'' building blocks.

Even as the number of subnuclear particles discovered experimentally was continuing to increase, many physicists were struck by the fact that many reactions expected among the various subnuclear particles were not observed. The only explanation for these missing reactions seemed to be that there were new kinds of conservation laws that would apparently be violated if these reactions took place. Slowly, these new conservation laws were discovered. From previous experience, physicists knew that each conservation law implied a certain structure or symmetry in nature.

In 1961 Murray Gell-Mann at the California Institute of Technology and Yuval Ne'eman, an Israeli physicist, discovered an important new classification scheme of the subnuclear particles based on the symmetries implied by the new conservation laws. This new classification scheme, in turn, led Gell-Mann to suggest that the large number of subnuclear particles were all made up of only a few particles, which he named **quarks.** Subsequent research has led physicists to accept the quark model as correct, and physicists hope that the quarks are now indeed the long-sought fundamental building blocks of nature.

This chapter will begin by discussing the structure of the nucleus, starting with our understanding about the time of the discovery of the neutron in 1932. We will follow the subsequent evolution of the understanding of the nature of the nucleus and the amazing force that holds it together and discuss the discovery of some of the new subnuclear particles. Following a brief summary of our knowledge of subnuclear particles as of about 1960, we will see how the careful study of conservation laws governing the interactions among the subnuclear particles led to the discovery of important new symmetries in the physical laws governing characteristics of subnuclear particles. We will see how these new symmetries led to the quark model and study its problems and successes. Finally, we will try to summarize our present knowledge of the "fundamental" building blocks and indicate possible future developments. The search for the fundamental building blocks of nature has a long scientific history. One of the main goals of this chapter will be to see how the study of conservation laws and their related symmetries has led to one of the most significant developments in the history of physics.

B. The Nuclear Force and Nuclear Structure

As discussed in the preceding chapter, we know that the atom is a nuclear atom; that is, it has a nucleus. The nucleus contains more than 99.9 percent of the mass of an atom. Furthermore, this mass is confined to a very small volume. Various experiments indicate that the diameter of a typical nucleus (e.g., a carbon atom nucleus) is less than 10^{-14} meters (one hundred-thousandth of a billionth of a meter), which is only about 0.01 percent of the diameter of the atom. Because nearly all the mass of an atom is concentrated into such a very small volume, the density of matter in a nucleus is very high. The density of matter in a nucleus is about 10^{17} kg/m^3 or a hundred million billion kilograms per cubic meter![1] We should not be surprised to find that there are new physical phenomena involved with the structure of the nucleus that are not seen at all on a larger, more familiar, scale.

For more than 60 years it has been known that the nucleus contains protons and neutrons. The neutron mass is slightly greater than the proton mass and both are about 1.7×10^{-27} kilograms. The diameter of either a neutron or proton is known to be slightly more than 10^{-15} meters. A proton has a positive charge exactly equal in magnitude to the negative charge of an electron; a neutron has no net charge. The number of protons in a nucleus determines what element the

[1]Numbers such as 10^{-14} or 10^{17} are referred to as numbers in "scientific notation," where the exponent indicates how many decimal places to move the decimal point. Thus 10^{17} is the number 1 followed by 17 zeros and the decimal point, and 10^{-14} is the number 1 preceded by 13 zeros and the decimal point. See discussion in Chap. 1.

neutral atom will be (a neutral atom has a number of electrons surrounding the nucleus exactly equal to the number of protons in the nucleus). Nuclei generally have about equal numbers of neutrons and protons, except for the heavy elements such as lead, bismuth, and uranium, which have significantly more neutrons than protons. Nuclei with the same number of protons but different numbers of neutrons are said to be **isotopes** of the same element. They will have the same numbers of surrounding electrons and thus will behave identically chemically.

It is important to note that if we consider only the two basic forces we have discussed so far in this text, we must conclude that any nucleus would be completely unstable. A nucleus comprised of protons and neutrons is a nucleus containing many "like" charges and no "unlike" charges. Because like charges repel, we know that the electromagnetic force is trying to break all nuclei apart (except for hydrogen which has only one proton in its nucleus). What then holds nuclei together?

The only other force discussed so far is the gravitational force. If we try to see whether the gravitational attraction between the neutrons and protons in a nucleus can hold the nucleus together, we find that gravity is much too weak. If we assume that the gravitational force between neutrons and protons in a nucleus is given by the same formula that correctly yields the gravitational attraction between the Moon and the Earth, or the Earth and the Sun, or any two large objects, we find that gravity is weaker than the electromagnetic repulsion by a factor of about 10^{39}!

Clearly, another kind of force exists in order to explain why nuclei are stable. This new force is called the nuclear force and has been studied for over 60 years. The nuclear force is known to be very strong and attractive. It acts between neutrons and protons in a nucleus but has no effect on the surrounding electrons. This force, known as the **strong nuclear force,** is about 100 times stronger than the electromagnetic repulsion inside a nucleus. Thus nuclei with more than about 100 protons are unstable because the **electromagnetic repulsion,** which continues to add up, is finally greater than the attractive nuclear force, which does not continue to add up (physicists say it saturates).

The strong nuclear force is very different from either the electromagnetic or gravitational force. Both of the latter are infinite-range forces (i.e., their forces act at all distances), although they become weaker at larger distances (the inverse square law as discussed earlier in this text). In contrast, the strong nuclear force is *finite* ranged. Two nucleons (either neutrons or protons) must almost, but not quite, touch each other before the strongly attractive nuclear force "turns on." This finite-ranged characteristic of the nuclear force has resulted in it sometimes being referred to as the nuclear "glue," because glue works only between two objects brought into contact with each other. The strong nuclear force is different from the electromagnetic and gravitational forces also in that there is no known simple mathematical expression that permits one to calculate the strength of the force for a given situation. The nuclear force is known to depend on several different variables in rather complicated ways. Although nuclear physicists have studied the nuclear force for many years, they have been unable to find any simple expression to describe the strength of the force. Later in this chapter we will see why this situation exists.

In addition to the electromagnetic, gravitational, and strong nuclear forces, one more basic force is known to exist. This force, known as the **weak nuclear force,** is responsible for a certain kind of radioactive decay of nuclei known as beta decay. Radioactive decay refers to the spontaneous emission by a nucleus of one or more particles. The decaying nucleus transmutes itself into another kind of nucleus. In the simplest form of beta decay, a neutron inside a nucleus changes into a proton and an electron, and the electron is emitted from the nucleus. The weak nuclear force is a factor of about 10^{11} times weaker than the strong nuclear force, and we will not need to consider it further for our simple discussion of nuclear structure. We should note that there are several other kinds of radioactive decay besides beta decay, in which other kinds of particles may be emitted.

Historically, studies of naturally radioactive substances provided our first clues regarding nuclear structure. The most commonly emitted particles from radioactive elements were given the names of the first three letters of the Greek alphabet—they were named alpha, beta, and gamma particles. We now know that an alpha particle is the nucleus of a helium atom, which consists of two protons plus two neutrons; that a beta particle is an electron; and that a gamma particle (more often called a gamma ray) is a high-energy quantum of electromagnetic energy, also known as a photon. Although nuclear physicists still study natural radioactive decay, more often they study nuclei by bombarding stationary target nuclei with high-velocity beams of various particles, such as protons, alpha particles, or electrons. Such collisions provide a way to add particles and energy (from the kinetic energy of the incident particle) to nuclei in a precisely determined way, in order to study how nuclei transmute into other nuclei or break up into constituent particles. Such experiments are called nuclear reactions because they result in the separation or production or combination of the basic particles, or groups of particles, from the nucleus. Nearly all of the information presented here regarding nuclear structure and the kinds of particles found inside nuclei has been obtained from careful studies, performed at various accelerator laboratories around the world, of particle collisions with nuclei.

OPTIONAL SECTION 8.1 *Nuclear Reactions and Energy Levels*

In many ways nuclear reactions are like chemical reactions. For example, the chemical reaction of hydrogen and oxygen to form water is described by the following chemical "equation":

$$2H_2 + 1O_2 \rightarrow 2H_2O$$

This equation states that the reaction of two hydrogen molecules and one oxygen molecule will result in the formation of two water molecules. The equation is said to be *balanced* because the number of hydrogen atoms, 4, and the number of oxygen atoms, 2, are the same on both sides of the equation. The nuclear reaction resulting from the bombardment of boron nuclei with "slow" neutrons to obtain lithium nuclei and helium nuclei is represented by the following "equation":

$$_0n^1 + {_5}B^{10} \rightarrow {_3}Li^7 + {_2}He^4$$

where n, B, Li, and He are symbols for the neutron, the boron nucleus, the lithium nucleus, and the helium nucleus, respectively. The subscript before the symbol indicates the amount of electric charge on the particle (in units of $+e$, the magnitude of the charge on an electron) and the superscript after the symbol indicates the total number of nucleons (neutrons plus protons). This equation is also balanced because the amount of electric charge, $5e$, and the total number of nucleons, 11, are the same on both sides of the equation. There are other quantities that need to be balanced, that is, conserved, and this is discussed in Section C below.

Note that in the chemical equation the symbols for the reaction components are for the chemical elements, and the same symbols, that is, hydrogen and oxygen, appear on both sides of the equation. This reflects the fact that for chemical reactions the nuclei of atoms are indeed indivisible. On the other hand, for nuclear reactions the nuclei of atoms are divisible, and so the same chemical element symbols quite often do not appear on both sides of the equation.

While energy release or absorption is present in both chemical and nuclear reactions, in nuclear reactions it is often necessary to call specific attention not only to energy changes, but also to the *form* of the energy changes, as shown in the following two-step reaction:

$$_0n^1 + {_{47}}Ag^{109} \rightarrow {_{47}}Ag^{110*} \rightarrow {_{47}}Ag^{110} + \gamma$$

where Ag is the symbol for silver, γ is the symbol for a gamma ray photon, and the asterisk indicates an "excited" nucleus. Another example of a nuclear reaction involving significant amounts and forms of energy changes is

$$_6C^{11} \rightarrow {_5}B^{11} + {_1}\beta + \nu$$

where C is a carbon nucleus, B is a boron nucleus, β (beta) is an electron or positron (identical with an electron except for having positive, rather than negative, electric charge) and ν is a neutrino, a particle having zero electric charge and as yet immeasurably small, possibly zero, rest mass. The total of the relativistic masses of the positron and the neutrino may be as large as 1 or 2 MeV (1 MeV $= 1 \times 10^6$ eV $= 1.6 \times 10^{-13}$ joules, see Chapter 7, Optional Section 7.2).

Just as the detailed study of chemical reactions and of atomic spectra has been the key to fathoming the structure of atoms, so too the detailed study of nuclear reactions and the energies involved (nuclear spectra) have been the keys to understanding the structure of the nucleus. In particular, there are nuclear energy level diagrams just as there are atomic energy level diagrams (Chapter 7, Optional Section 7.2) characterized by quantum numbers related to the properties of the nuclear building blocks. Moreover, just as in the case of atomic spectra, there are various selection rules that govern whether transitions between the energy levels can take place. The possibilities are much "richer" (complex) because the energy differences between levels can be emitted in the form of mass of newly created particles as well as photons.

Moreover, in addition to revealing the structure of the nucleus, the study of nuclear reactions and energy level diagrams has led to inferences about the fundamental forces and conservation principles discussed in Sections C and D below. These nuclear reactions can come either from spontaneous radioactive decay (see Optional Section 8.2 below) or from reactions induced using particle beams from accelerators.

Optional Section 8.2 *Radioactivity and Half-Life*

Much of the public interest in nuclear physics centers around radioactive materials and their half-lives. The **activity** of any amount of radioactive material is defined as the number of nuclei per second that will decay, that is, emit alpha rays or beta rays or gamma rays, and so on. The activity at any instant of time depends, among other things, on the number of radioactive atoms that have not yet decayed. It also depends on the **decay constant** of the particular species (i.e., isotope) of radioactive material in question. The **half-life** is the length of time that must pass before the activity (or equivalently, the number of radioactive atoms) decreases by a factor of two. The decay constant and the half-life are simply related to each other: Their product is equal to 0.693, which is the natural logarithm of 2. Measured values of half-lives for different radioactive isotopes vary from approximately 10^{-5} seconds to approximately 10^{9} years.

The activity of radioactive material is essentially the same kind of quantity as the brightness or intensity of the emission lines in atomic spectra and can be calculated in accordance with the principles of quantum mechanics as discussed in Chapter 7, Section C4, except that more sophisticated versions than Schroedinger's wave mechanics must be used. This means that only the probability that a given atom will decay can be calculated.

Heisenberg's uncertainty principle is also applicable to nuclear energy levels, in that the energy states are known to very-high-precision because the length of time they will have that energy is very uncertain (long half-life). This precision was used in one of the tests of Einstein's general theory of relativity mentioned in Chapter 6, Section E3. The "red-shift" of light in a gravitational field was measured using gamma rays as the light whose frequency was measured.

Shortly after the basic characteristics of the strong nuclear force were realized (namely, its strength and finite range), an extremely important step was taken toward understanding the basic nature of this force. In 1935 the Japanese physicist

Hideki Yukawa proposed a theory that explained the nuclear force as due to an exchange of finite-mass quanta (or particles). This kind of description of a force is known as a quantum-field theory, because the force field is suggested to consist of "virtual" quanta, which are exchanged between nucleons when they come within the range of the nuclear force. A "virtual" particle is one that may exist for only a very short time and cannot be observed experimentally except by adding a large amount of energy to make it "real." Virtual particles are allowed by the Heisenberg uncertainty principle discussed near the end of the previous chapter.

A quantum-field theory was known to exist for the electromagnetic force and was first suggested for the strong nuclear force by Werner Heisenberg in 1932. In the quantum-field theory of the electromagnetic force, the quanta that are believed to be exchanged between charged particles are the quanta of light, or photons. Each electrical charge is surrounded by a cloud of "virtual" photons, which are being constantly emitted and absorbed by the charged particle. When two charged particles are brought near each other they exchange virtual photons, so that one charged particle emits a photon and the other charged particle absorbs it, and vice versa. This constant interchange between the two results in the electrical force between them. In this quantum-mechanical theory of force fields, it is the zero mass of the photons that makes the electromagnetic force infinite ranged.

According to Yukawa, because of the short range of nuclear forces, a nucleon is surrounded by a cloud of virtual quanta *with mass,* which the nucleon is constantly emitting and absorbing. If another nucleon is brought close to the first, a particle emitted by one may be absorbed by the other, and vice versa. This exchange of particles results in a force being exerted between the two nucleons. Yukawa was able (using Heisenberg's uncertainty principle) to estimate the mass of the exchange particles. The result was that, if his quantum-field theory was correct, there should exist a new particle with mass about one-seventh the neutron or proton mass. Note that this mass is still about 270 times the mass of an electron, so that Yukawa's new particle clearly did not correspond to any known particle at that time. In contrast to the quantum-field theory for the electromagnetic force, Yukawa's proposal required that there be a new undiscovered particle. Clearly, the experimental discovery of Yukawa's particle would provide strong support for the quantum-field description of forces.

In 1947 the predicted particle was discovered to be emitted in nuclear reactions induced by high-energy cosmic rays (mostly protons) from space. The particle was named the pi-meson, or for short, the pion. In 1949 Yukawa was awarded the Nobel Prize in physics for his work on this subject.

Subsequent theoretical and experimental investigations showed that the nuclear force is not explained simply by the exchange *only* of pi-mesons. It is now known that to account for various characteristics of the nuclear force within the framework of the quantum-field theory, one must include the simultaneous exchange of more than one pi-meson and even the exchange of other, more massive, field quanta. Nevertheless, the basic correctness of Yukawa's theory is almost universally accepted.

In fact, it is now believed that all basic forces are properly described by quantum-field type theories. The gravitational force, because it is infinite ranged (like the electromagnetic force), must have a zero mass exchange quantum. The necessary particle has been named the *graviton* and is generally believed to exist, even though it has not been observed experimentally. Because the gravitational force is so weak, each graviton is expected to carry a very minute amount of energy. Such low-energy quanta are expected to be difficult to detect, and most physicists are not surprised that the graviton has escaped experimental verification. Several groups of scientists around the world are performing elaborate (and difficult) experiments to try to detect gravitons.

On the other hand, the exchange particles for the weak nuclear force, called the *intermediate-vector bosons,* are so massive that only a few very high-energy particle accelerators ("atom-smashers") can deliver enough kinetic energy to "make" these particles in the laboratory. The weak nuclear force is not only very much weaker than the strong nuclear force, but also very much shorter ranged. Thus by the same kind of argument that enabled Yukawa to estimate the mass of the pi-mesons, nuclear physicists knew that the masses of the intermediate-vector bosons must be very large. In 1982 and 1983 physicists at the European Center for Nuclear Research (CERN) used a new experimental technique involving colliding beams of high-energy protons, which enabled them to achieve much higher reaction energies than would be possible with stationary targets. They were able to identify the intermediate-vector bosons in the reaction products. These particles are named W-particles and Z-particles. The masses of these particles were found to be close to those previously estimated and are about 100 times the mass of a proton. Although these Ws and Zs have now been observed in other laboratories as well, they can still only be made by using high-energy colliding beams.

It is interesting to note that the concept of a quantum-field theory, which "explains" a force as the exchange of virtual particles, is actually just the latest attempt by scientists to understand the "action-at-a-distance" problem discussed earlier in this text. Just how two objects, not directly in contact with each other, can have effects on each other has long been a serious, fundamental question. (This question was raised even in Newton's time.) The hypothetical ether, primarily postulated to explain how light travels through empty space, was also believed to help explain the action-at-a-distance problem. Now, the quantum-field theory, generally accepted by most physicists, may finally have provided the explanation to this long-considered problem. Even though Yukawa's original idea for the description of the strong nuclear force as due to the exchange of *only* pi-mesons is known to be incorrect, his prediction of a new particle based on his theory, and its experimental discovery, led to the general acceptance of the quantum-field description of the basic forces. Later in this chapter we will see that the reason why more than one kind of exchange particle is required to describe the strong nuclear force may be that the strong nuclear force is not actually one of the basic forces; rather, it is more correctly described as a residual interaction left over from a much stronger, more basic force that exists *inside* nucleons.

Yukawa's theory had another important effect on the development of nuclear physics. His prediction of yet another particle besides neutrons and protons that might exist inside nuclei provided much of the original motivation to search for new kinds of "fundamental" particles. In fact, a different particle, now called the *muon,* was discovered shortly before the pion, and was, for a short time, mistakenly thought to be Yukawa's predicted exchange particle. We now know that the muon is not at all like a pion, but is basically like an electron, except that it is about 200 times heavier. The discoveries of the muon and the pion were just the beginning. With the aid of new, more powerful accelerators, more and more kinds of particles were discovered. By about 1957, 18 so-called elementary particles were known, including ones with names such as sigma, lambda, xi, and K-meson (see Table 8.1). During the late 1950s and 1960s, the list grew until only numbers could be assigned for names of the new particles and well over 100 "elementary" particles were known to exist. Such a proliferation of particles removed the apparently simple description of the basic building blocks of matter known in the 1930s. Physicists generally were dismayed by the apparently complicated collection of fundamental particles. The Greek idea that simple is beautiful continues to dominate scientific thinking.

Even before the number of known subnuclear particles began to grow so rapidly, nuclear physicists recognized that the different particles fell naturally into a few different groups or "families" according to their characteristics. For each particle, physicists tried to determine its mass, electric charge, spin (a quantum number describing its intrinsic angular momentum as discussed in the previous chapter for electrons), and some other properties to be discussed below. The particles were then assumed to be in families with other particles possessing similar properties. As an example, the 18 "elementary" particles known in 1957 are listed in Table 8.1 separated into the four known families.

The **baryons** are particles with rest (i.e., proper) masses equal to or greater than the rest mass of the proton. (*Baryon* means "heavy particle.") These particles are all known to interact with each other via the strong nuclear force (as well as the electromagnetic force if they are charged). Note that they all have half-integer spin as one of their inherent properties. Except for the proton, all the baryons are known to be unstable and usually will spontaneously decay (i.e., break up into other particles) in a small fraction of a second. A short "lifetime" means that the particle will exist for only a short time after it is created in a nuclear reaction. When it decays, it turns into one or more longer-lived particles that will in turn decay until only stable particles (like protons and neutrons) are left. (Even a neutron, when isolated, will spontaneously decay with a mean lifetime of about 15 minutes, and physicists are currently trying to determine if a proton actually is completely stable—that is, has an infinitely long lifetime.)

The **mesons** are particles with rest masses less than the baryons but greater than the leptons, the next family. These particles all have zero or integer intrinsic spin values. Besides a neutral and a charged pi-meson (Yukawa's predicted particle), there exists a neutral and a charged K-meson, or kaon for short. Note that the mesons are all unstable, with a characteristic lifetime of a small fraction of a second.

The **leptons** have rest masses less than those of the mesons and have half-

Table 8.1. Fundamental Particles (circa 1957)

Family	Particle	Symbol	Charge	Spin	Mass (MeV)	Lifetime (sec)
Photon	photon	γ	0	1	0	Stable
Lepton	electron	e^-	−1	½	0.511	Stable
	muon	μ^-	−1	½	105.7	2.2×10^{-6}
	electron neutrino	ν_e	0	½	0?	Stable
	muon neutrino	ν_μ	0	½	0?	Stable
Meson	charged pion	π^+	1	0	139.6	2.6×10^{-8}
	neutral pion	π°	0	0	135.0	0.8×10^{-16}
	charged Kaon	K^+	1	0	493.8	1.2×10^{-8}
	neutral Kaon	K°	0	0	497.7	0.9×10^{-10} or 5.2×10^{-8}
	eta	η	0	0	548.7	$\sim 2 \times 10^{-19}$
Baryon	proton	p	1	½	938.3	Stable
	neutron	n	0	½	939.6	917
	lambda	Λ	0	½	1115.4	2.6×10^{-10}
	sigma plus	Σ^+	1	½	1189.4	7.9×10^{-11}
	neutral sigma	Σ°	0	½	1192.3	5.8×10^{-20}
	sigma minus	Σ^-	−1	½	1197.2	1.5×10^{-10}
	neutral xi	Ξ°	0	½	1314.3	2.9×10^{-10}
	xi minus	Ξ^-	−1	½	1320.8	1.6×10^{-10}

integer spin. (Lepton means "light particle" and meson means "medium particle.") This family consists of the electron and the muon, which acts like just a heavy electron. The neutrinos are each associated with either an electron or a muon and either have zero rest mass or at most a very small rest mass close to zero.

The **photon** is the quantum of electromagnetic radiation and is in a family all by itself. It has spin equal to one and (probably) zero rest mass.

Besides the particles listed in Table 8.1, for each particle there was known (or believed) to exist a corresponding **antiparticle.** An antiparticle has the same rest mass, intrinsic spin, and lifetime as its associated particle. It has, however, an exactly opposite sign of electric charge and certain other characteristics not yet discussed. The antiparticle of an electron, for example, is called a positron; it was discovered in 1932. The antiparticle of a proton is called simply an antiproton and was first observed experimentally in 1955. Some particles such as the neutral pion (π°) and the photon are believed to be their own antiparticles. When a particle and its antiparticle collide, they will annihilate each other and produce new particles and/or electromagnetic energy (photons). If we count antiparticles separately, then the list of subnuclear particles in Table 8.1 actually indicates 32 particles (15 particles and their antiparticles, plus the π° and photon).

The list of particles in Table 8.1 might not be considered too long and complicated, especially when organized into families as indicated. Unfortunately, the list began to grow rapidly in the late 1950s and throughout the 1960s. The proliferation of new subnuclear particles arose in the family identified as the baryons. More and more particles were discovered with larger rest masses and shorter decay times. Their lifetimes became so short that these new ''particles'' began to be called simply *resonances,* which directly implied that physicists were no longer sure that these new energy quanta were actually particles. Elementary-particle physicists gave up trying to find names for all these new baryons and labeled them simply by their rest mass energies. The somewhat complicated situation of Table 8.1 became much more complex and distinctly bewildering. The remainder of this chapter will be devoted to a discussion of how this proliferation of new particles was finally understood in terms of a simpler underlying structure by the study of conservation laws and their related symmetries.

C. Conservation Laws and Invariants

Physicists have recognized the importance of conservation laws for about 300 years. Newton presented his laws of conservation of mass and momentum in his *Principia* in 1687. The development of the law of conservation of energy was one of the most important nineteenth-century advances in physics (see Chapter 4). Nuclear physicists studying nuclear reactions in carefully controlled experiments soon realized that several different quantities were always conserved. Some of these were the already known conserved quantities such as energy, momentum, electric charge, and angular momentum. But eventually new quantities were discovered to be conserved, and physicists suspected that the careful study of these quantities would lead to a better understanding of the mysterious nuclear force.

Recall that conservation of a quantity simply means that it remains constant in total amount. Nuclear physicists can carefully prepare a nuclear reaction so that they know exactly how much total mass, energy, electric charge, momentum, and so on, is available before the reaction takes place. By the use of specialized instruments, they can also measure these same quantities after the reaction takes place. A careful comparison of the known amounts of each quantity before and after the nuclear reaction reveals which quantities are conserved (i.e., do not change). In such studies, it was discovered that nuclear reactions controlled by the strong nuclear force obey more conservation laws than are obeyed by gravitational and electromagnetic interactions.

In any interaction (whether it be between two billiard balls or between two subnuclear particles), the following seven quantities are always conserved:

1. Mass-energy
2. Momentum
3. Angular momentum
4. Electric charge
5. Electron family number
6. Muon family number
7. Baryon family number

In considering the last three quantities, one counts a particle as +1 and an antiparticle as −1. Thus these three laws say that the total excess of particles over antiparticles (or vice versa) within each family is always exactly maintained and further indicates that these "families" are somehow natural classifications. The first four conserved quantities have been known for about 100 years (some much longer) and have been discussed earlier in this text. These seven basic quantities are conserved by all four of the basic forces of nature, so far as we know. Let us now consider each of the basic forces separately with regard to known conservation laws obeyed by that force.

In any reaction in which the strong nuclear force dominates (and such is the case for most nuclear reactions), besides the seven basic quantities listed above, the quantities known as parity, isotopic spin, and strangeness are also conserved. The **parity** of a system involves its helicity or inherent right- or left-"handedness." For example, a right-handed screw would have right-handed helicity and a left-handed screw would have left-handed helicity. For nuclear particles, helicity deals with the orientation of the intrinsic spin vector of a particle with respect to its velocity vector (direction of travel). **Isotopic spin** is a quantum-mechanical quantity that describes the neutron or proton excess of a nuclear system. **Strangeness** is a quantum-mechanical quantity with no simple description, and will be discussed further below. Related to these conserved quantities is the fact that nuclear reactions are also observed to obey an "operation" known as **charge conjugation.** This last statement means simply that for any possible nuclear reaction, there is another possible reaction, which corresponds to changing all positive charges to negative ones and all negative charges to positive ones (actually one is interchanging all particles with their antiparticles). Some of these new quantities, such as isotopic spin and strangeness, would not be noteworthy if they were not conserved in nuclear reactions. Because they are conserved we must regard them as clues to fundamental characteristics of subnuclear particles.

The electromagnetic force is second in strength to the strong nuclear force, and interactions dominated by the electromagnetic force apparently conserve all the same quantities as the strong nuclear force, except for isotopic spin. That isotopic spin is not conserved by the electromagnetic force is related to the fact that isotopic spin involves neutron or proton excess—that is, the excess of particles with or without charge. Because the electromagnetic force depends only on whether or not an object is charged, it is not sensitive to the neutron or proton excess, but rather to the number of charged particles.

The weak nuclear force obeys only the seven basic conservation laws. When it was shown in 1957 that this force did not conserve parity, physicists were generally surprised, because such a failure indicated that the universe is fundamentally not ambidextrous. Such an inherent "handedness" seemed peculiar, and the weak nuclear force was considered to be somewhat enigmatic.

Finally, the gravitational force is so weak that there are no known reactions between individual subnuclear particles that are dominated by this force. Consequently it is not known which of the conservation laws are obeyed by the gravitational force at the microscopic level.

Before we end this brief discussion of known conservation laws, it is important to remember the implications of conservation laws. Once a law of nature is recognized, it means that we have discovered a *fact* of nature. A natural law (if it is correct) must be obeyed; there is no choice. A civil law that says that one must obey stop signs can be ignored (perhaps with some unpleasant consequences). A natural law is a *constraint* on how physical systems can develop. For example, when we know that baryon number is conserved, we have automatically ruled out a large number of nuclear reactions that would not conserve this quantity. By the time we add the constraints of conserving the other family numbers, charge, parity, and so on, we can often predict that only a few nuclear reactions are possible for a given set of starting conditions. If any of the other reactions are observed experimentally, it means one of our assumed conservation laws is wrong.

The history of the study of subnuclear particles actually proceeded in just the opposite way. Although more and more subnuclear particles were discovered, it was observed that most of the possible reactions expected between these particles did not occur. In order to explain why these reactions were forbidden, new conservation laws had to be invented. As it turned out, some of these newly discovered conserved quantities, such as isotopic spin and strangeness, eventually led to a new understanding of the fundamental building blocks of nature.

D. Conservation Laws and Symmetries

The new conservation laws led to a new understanding of the fundamental building blocks of nature because of **symmetries** that these conservation laws implied. Symmetries in nature are something with which we are all familiar. Many natural objects have a left side that is just like the right side, or a top that is identical to the bottom. Many flowers, crystals, snowflakes, and even human faces are either perfectly symmetric, or almost symmetric. By symmetric, in these instances, we mean there is some rotation that can be applied to the object, which

Figure 8.1. Snowflake with 60° symmetry. (Richard Holt/Photo Researchers.)

will result in it looking the same (or nearly the same) as when it started.[2] This is not the same as *mirror symmetry,* which interchanges the right and left sides. A snowflake, however, might be symmetric through several different rotations, such as 60 degrees, 120 degrees, 180 degrees, and so on. Figure 8.1 shows a highly symmetric snowflake that will look exactly the same when rotated through any angle that is a multiple of 60 degrees.

Physicists have generalized our common understanding of symmetry to say that a **symmetry operation** is any well-defined action that leaves the object looking the same (or like its mirror image) as when one started. Physical symmetry operations must be expressed as mathematical operations. For example, the symmetry operations discussed for a snowflake are rotations. The parity operation represents mirror symmetry. Of interest for our subject here is the realization that the basic laws of physics are left unchanged by many of these so-called symmetry operations. For example, none of the laws of physics is changed in a closed system if the entire system is rotated by any arbitrary angle. Similarly, the laws of physics

[2]This interest in symmetry was at the root of the ancient Greek search for perfection in the heavens, as discussed in Chapter 2, Section A2.

are unchanged if a closed system is simply moved in space (provided that *all* relevant objects are moved with it) or if the system is displaced in time. These are three examples of symmetry operations that can be performed.

The three symmetry operations mentioned above that do not change the laws of physics appear trivial, and to some extent, they are. We would not expect scientists in China to discover that their laws of physics are different from ours, or scientists 100 years from now to find different physical laws governing the universe. However, each of these trivial symmetry operations corresponds to one of the known conservation laws. Using simple mathematical analysis, it can be shown that the fact that the laws of physics are unchanged by a translation (change of location) in space corresponds to the law of conservation of momentum. *Corresponds* means that the one necessarily implies the other. The fact that the laws of physics are unchanged by a translation in time corresponds to the law of conservation of energy. That the laws are unchanged by a rotation corresponds to the law of conservation of angular momentum.

The relationships between symmetry operations and conservation laws do not end with just these three "trivial" symmetry operations. The combination of all three (space, time, and rotation) corresponds to Einstein's relativity principle, sometimes known as Lorentz invariance. The fact that the laws of physics are unchanged by a change of phase in the quantum-mechanical wave function (see the previous chapter) corresponds to the law of conservation of electric charge. The symmetry operations that correspond to the other known conservation laws are not easily described geometrically but are mainly operations in the mathematical formulation of quantum mechanics.

The correspondence between a symmetry operation and a conservation law works both ways; either one implies the other. This correspondence makes these relationships extremely important. Every new conservation law implies a new "quantum number" and a symmetry operation that will leave the laws of physics unchanged. From every newly discovered conserved quantity a new symmetry in the laws of physics may be inferred. It was precisely such a relationship that enabled Gell-Mann and Ne'eman to discover the new symmetry implied by the known conservation laws for nuclear reactions, including the newly discovered (at the time) conserved quantity called strangeness.

E. The Quark Model

Now that we have discussed the important relationship between conservation laws and underlying symmetries, we can consider how the discovery of new conserved quantities led to the development of a new model for the fundamental building blocks of nature. This new model started when, independently, Gell-Mann and Ne'eman recognized that if baryons and mesons (two of the "families" of particles discussed above) were organized by their strangeness and isotopic spin quantum numbers, simple patterns emerged. Recall that strangeness is a quantum number introduced to explain why many expected reactions among the known particles were observed relatively infrequently, and that isotopic spin describes the different charge states that strongly interacting particles can have.

These simple patterns can be seen most easily if we use a new quantity, Y, called hypercharge, defined as

$$Y = S + B$$

where S is the strangeness quantum number and B is the baryon number. The simple patterns obtained for baryons with intrinsic spin of ½ and for mesons with intrinsic spin of 0 are shown in Fig. 8.2. The location of each particle is determined by the values of its hypercharge quantum number Y and its isotopic spin quantum number I_3 (actually the so-called z-component of the isotopic spin quantum number). The location of each particle is uniquely determined. The symbols Ξ^-, Σ^+, and so on, refer to the particles listed in Table 8.1. Note that the baryon classification includes the well-known neutron and proton plus some of the newer, "stranger" particles.

The meson classification includes the pi-mesons originally predicted by Yukawa plus some recently discovered mesons. Both of these patterns are hexagons with a total of eight members each (two at the center). Gell-Mann referred to these classifications as the *eightfold way*. The important concept here is not what hypercharge and isotopic spin are, but that by using them simple patterns are obtained. It is equally important to remember that these quantum numbers were "invented" (or discovered) in order to obtain needed conservation laws that could explain why some nuclear reactions were allowed and others were forbidden.

In physics, it is often said that the real test of any new model or theory is whether it can correctly predict something new. Although Gell-Mann's and Ne'e-

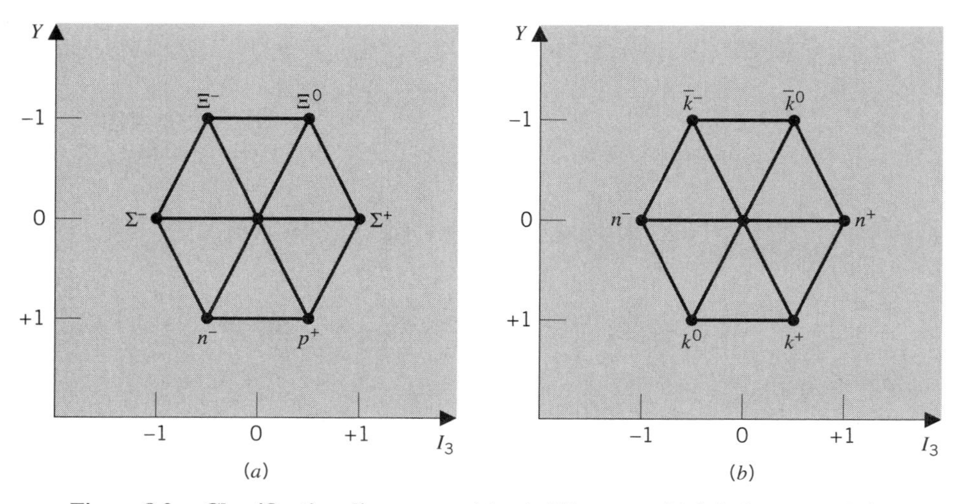

Figure 8.2. Classification diagrams. (a) spin ½ baryons. (b) Spin 0 mesons. Y, hypercharge; I_3, z-component of isotopic spin.

man's new way of classifying particles clearly leads to some simple patterns, unless their classifications lead to some new knowledge not previously known, it would all be merely an interesting scheme. Remarkably, the new classification scheme immediately provided a prediction of a new particle that had never been observed experimentally. Figure 8.3 shows the plot for baryons with intrinsic spin of $\frac{3}{2}$ similar to the plots of Fig. 8.2. All the particles represented on the plot were known except for the particle called Ω^- (omega minus) shown at the very bottom of the figure. The pattern (or symmetry) of the figure is obvious; it is an upside down triangle. Simply by its location in the figure, Gell-Mann could predict the hypercharge (and therefore the strangeness and the isotopic spin of the new particle). Further considerations allowed him to determine the charge and to estimate the mass of the Ω^-, as well.

Given the predicted properties of this new particle, experimental physicists at the Brookhaven National Laboratory on Long Island, New York, immediately set out to discover if this new particle, the Ω^-, actually existed. The general plan was to try to pick out for study at least one reaction that should lead to the production of the Ω^-, while still obeying all known conservation laws. If the Ω^- was created, then the experimenters would be able to recognize it by its mass, charge, spin, and other quantum characteristics. A likely reaction was selected and, in a dramatic verification of the fundamental importance of the new classification scheme, the Ω^- particle was observed experimentally in November, 1964, only seven months after the prediction of its existence.

The experimental discovery of the Ω^- particle clearly demonstrated that there was a basic significance to the new classification scheme. In 1963 Gell-Mann showed that the groupings of the particles according to their patterns in the plots of hypercharge versus isotopic spin were accurately described by a branch of

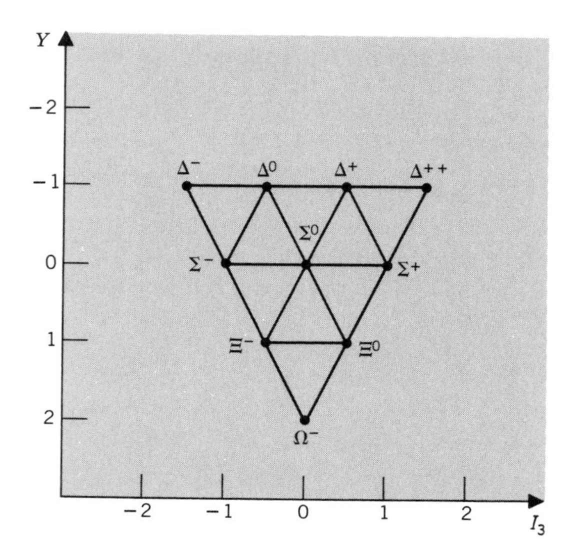

Figure 8.3. Baryon decuplet classification diagram. Y, hypercharge; I_3, z-component of isotopic spin.

mathematics called *group theory* (not surprising because the particles were separated into groups). In particular, these different groups were described by a mathematical representation known as *SU(3)*. It is not important that we understand the mathematics of *SU(3)* here, except to know that it describes groups that comprised combinations of only three fundamentally different objects. (This is the same kind of reasoning used when scientists inferred from Mendeleev's periodic table, discussed in Chapter 7, that atoms were made up of more basic objects!)

Thus Gell-Mann (together with another physicist, George Zweig) hypothesized that the known baryons and mesons (the families described by the new classification) were comprised of three even more fundamental particles. Gell-Mann named these new particles **quarks** (from a simple rhyme in James Joyce's novel, *Finnegans Wake*). Because the great proliferation of new subnuclear particles was occurring in the baryon and meson families, Gell-Mann's hypothesis, if successful, would result in a great simplification in the list of fundamental particles.

Gell-Mann then deduced the characteristics such as mass, electric charge, spin, and so on, that each of the three quarks must have in order for combinations of them to describe the known characteristics of the baryons and mesons. Some of the characteristics of the quarks seemed peculiar, especially the need for them to have charges that are either one-third or two-thirds the charge on an electron or proton. It had been thought for many years that the charge on an electron was the smallest unit of electrical charge. Gell-Mann found that baryons must each be made up of three quarks, and mesons of two quarks. Corresponding to each of the three quarks is an antiquark, which has the same relationship to its quark as other antiparticles to their corresponding particles (such as opposite charge, same mass, etc., as discussed earlier). A meson is actually a quark and antiquark pair. All of the known baryon and meson groups (in the new classification scheme) were accounted for on the basis of the quark model. Even the existence of the Ω^- particle was correctly predicted by the quark model.

The apparent success of, and consequent simplification resulting from, the quark model was great enough that many experimenters immediately set out to try to detect the quarks experimentally. Many physicists were skeptical from the start as to the likelihood of success, because no particles with a fractional charge had ever been seen, and each of the three quarks is predicted to have a fractional charge (either one-third or two-thirds the electron charge). To date, no convincing experiments have been able to detect an individual quark, and no free quark has been observed experimentally. To further complicate the situation, recent discoveries of more massive subnuclear particles have shown that the original three-quark model must be extended to include a fourth, fifth, and sixth quark. Thus the quark model will not represent such a simplification if too many are required.

In spite of the failure to detect an individual quark and the recent increase in the number of needed quarks, most nuclear physicists today are convinced that the quark model is basically correct. In fact, the need to add new quarks has provided some of the strongest evidence for the validity of the model. When a new particle was discovered that required a new quark in order to have its characteristics properly described, one could immediately predict other new particles that would correspond to other combinations of the new quark with the ''old''

Table 8.2. Quark Characteristics

Name	Spin	Electric Charge	Baryon Number
Up	½	+⅔	⅓
Down	½	−⅓	⅓
Charmed	½	+⅔	⅓
Strange	½	−⅓	⅓
Top (truth)	½	+⅔	⅓
Bottom (beauty)	½	−⅓	⅓

quarks. These new particles have consistently been discovered experimentally, further verifying the existence of the new quark and the correctness of the entire quark model.

The properties of the six known quarks are listed in Table 8.2. Gell-Mann's original three quarks are usually named **up, down,** and **strange.** They can account for all of the baryons and mesons of the 32 ''original'' elementary particles of Table 8.1. A proton, for example, is believed to consist of two up quarks and one down quark; a neutron of one up and two down quarks; and Yukawa's positively charged pi-meson of an up quark and an antidown quark.

The discovery of a new, relatively long-lived particle known as the J or Ψ (psi) particle in 1974 required the introduction of a fourth quark usually called the **charmed** quark. The discovery of another relatively long-lived particle called the Υ (Upsilon) particle in 1977 required the fifth quark called the **bottom** (or **beauty**) quark. Finally, experiments performed using colliding proton beams at the Fermi National Accelerator Facility near Chicago, Illinois, in 1993–94 indicate the existence of a sixth quark, called the **top** (or **truth**) quark. Besides these six kinds (or flavors as they are often called) of quarks, each quark can come in three different **colors,** usually chosen as **red, green,** and **blue.** Also, for every quark there is an antiquark, as discussed above. Thus there are 36 different quarks, but only six basic kinds.[3]

If the quarks are actually the fundamental building blocks of the baryons and mesons, there must be a new force, not yet mentioned, that holds quarks together inside a baryon or meson. This force must be extremely powerful in order to explain why free quarks are never observed. It was largely in attempting to explain this new force that the concept of color was introduced for quarks. The very strong force between quarks is believed to exist between quarks of different colors (or between a colored quark and its antiquark). This **color force** is understood to be somewhat analogous to the electric force that exists as an attractive force between charges of opposite sign. An electric force field can be understood as originating on a positive charge and terminating on a negative charge (or vice versa) as illustrated in Fig. 8.4. The electrical force between them can bind them together. Similarly, the color force between quarks is believed to originate on a

[3]The use of such words as *strangeness, charm, color, up, down,* and so on, to describe physical quantities while appearing to be whimsical, is just as legitimate as coining new Greek words such as *entropy, enthalpy,* and so on.

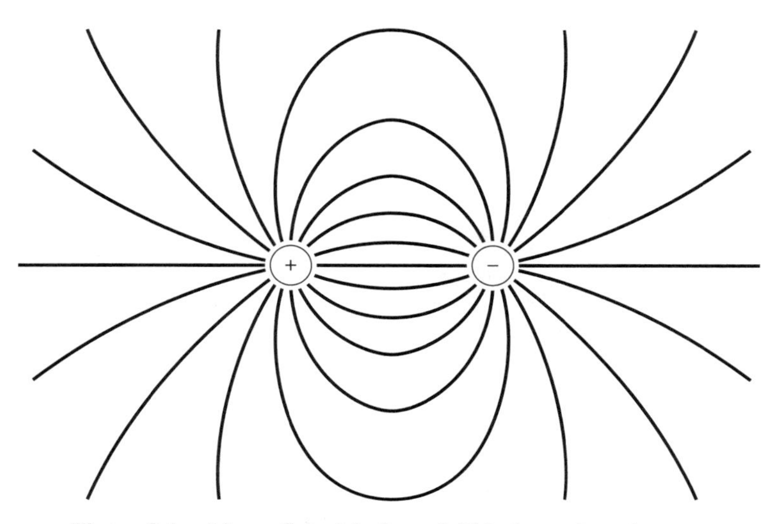

Figure 8.4. Lines of electric force field between two charges.

quark of some specific color and to terminate on *two quarks* of the other two colors or on *one antiquark* of the same color. One says that quarks feel the color force—just as charges feel the electric force. The color force fields for a baryon and a meson are illustrated in Fig. 8.5.

It is known that, for ordinary light, equal mixtures of red, green, and blue yield no observable color (i.e., white). In analogy, a baryon made up of one red, one green, and one blue quark is said to have no net color. A meson made up of a quark and antiquark of the same color is said also to have no net color (i.e., an antiquark has ''anti'' color). An amazing fact is that every particle discovered so far is understood to be made up of a combination of quarks that turns out to be a ''colorless'' combination. One always finds combinations of three quarks having one of each color, or a quark–antiquark pair. It appears that quarks always

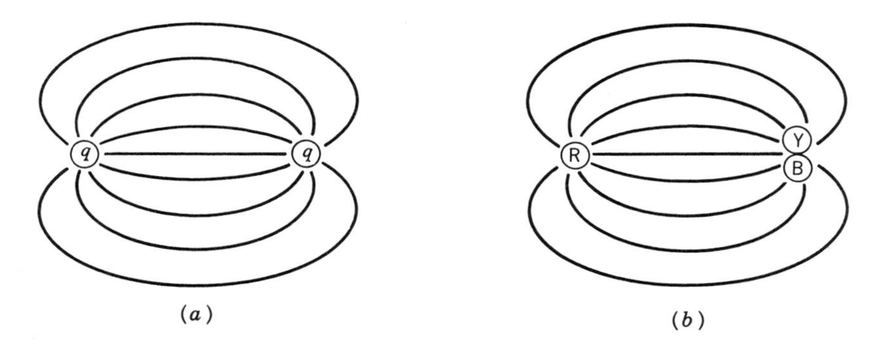

Figure 8.5. Lines of color force field. (a) Between a quark–antiquark pair in a meson. (b) Between three quarks in a baryon. q, quark; $\bar{q}$, antiquark; R, red quark; Y, yellow quark; B, blue quark.

come in threes or quark–antiquark pairs just so that there will be no net color. The color force requires this, so that the color force can always terminate within each particle.

The color force is very strong, because quarks are apparently so tightly bound inside particles that it has not yet been possible to find a ''free'' quark. In addition, it is believed that the color force between quarks *increases* rather than becoming weaker as quarks are separated. This is in contrast to the electric and gravitational forces between two particles, which become weaker as the particles are separated. The color force thus is a rather peculiar force when compared with the forces with which we are more familiar. The fact that the color force increases as quarks are separated may explain why quarks are so strongly bound inside particles and never seen individually. The color force is believed to also be described by a quantum-field theory where the exchange quanta are referred to as **gluons.**

Finally, before we end our discussion of quarks and the color force between them, we should reconsider the strong nuclear force between nucleons (neutrons and protons). It appears that the strong nuclear force is not a truly fundamental force. The force between nucleons is in some sense a leftover effect of the very strong color force between the quarks inside the nucleons themselves. Nevertheless, it is still important to study the nuclear force, because the nuclear force determines the structure of nuclei that exist inside all atoms of all materials and is responsible for energy production in stars. However, because the nuclear force is actually derived from the color force, only when we try to approach the study of the nuclear force from that perspective can we hope to be completely successful in understanding it. This possibly explains why nuclear physicists have been unable to obtain a simple description of the nuclear force, even after more than 60 years of study.

F. Summary of Present Knowledge

We can summarize the present understanding of the fundamental building blocks of nature by listing the known quarks, leptons, and the field quanta of the fundamental forces. This is done in Table 8.3. Note that the quarks and leptons have been divided into families that start with the more familiar, or first discovered, particles and proceed on to the most recently discovered ones. The quark members of the first family are the up and down quarks, which can describe almost all the most common subnuclear particles, such as the proton and neutron. The lepton members of the first family are the electron and its associated **neutrino.** (A neutrino is a very weakly interacting neutral particle, which is emitted in certain kinds of radioactive decay.) The quark members of the second family are the strange and charmed quarks, which are needed to describe the characteristics of some of the more peculiar particles in the host of subnuclear particles created in high-energy nuclear reactions. The lepton members of the second family are the muon and its associated neutrino. The muon acts very much like an electron, but it is approximately 200 times more massive. The third family has

Table 8.3. Fundamental Particles (circa 1995)

		Leptons		Quarks	
First Family	Particle	*e*(electron)	ν_e(electron neutrino)	u(up)	d(down)
	Charge[a]	−1	0	⅔	−⅓
	Mass[b]	$.5 \times 10^{-3}$	0?	$\sim.4 \times 10^{-2}$	$\sim.7 \times 10^{-2}$
	Discovered	1898	1954	1963	1963
Second Family	Particle	μ(mu)	ν_μ(muon neutrino)	c(charmed)	s(strange)
	Charge	−1	0	⅔	−⅓
	Mass	.11	0?	~1.5	~.15
	Discovered	1936	1962	1974	1974
Third Family	Particle	τ(tau)	ν_τ(tau neutrino)	t(top)	b(bottom)
	Charge	−1	0	⅔	−⅓
	Mass	1.78	0?	>41	~4.7
	Discovered	1975	?	1994	1977

Field Quanta

Force	Field Quantum	Mass (GeV)	Discovered
Electromagnetic	Photon	0	1905
Weak	Intermediate vector boson ($W^{\pm}$; Z°)	81;92	1983
Color (strong)	Gluon	?	—
Gravity	Graviton	0	—

[a]Charges are in units of the charge of the proton.

[b]Masses are in units of billions of electron volts (GeV).

the bottom and top (or beauty and truth) quarks, which are needed to explain the characteristics of some new particles discovered since 1977. The lepton members of the third family are the tau and its associated neutrino. The tau was discovered in 1975 and acts as an even heavier electron than the muon. Its associated neutrino has not been positively identified.

Finally, we list the field quanta of the fundamental forces. Following our discussion in Section E above, we replace the strong nuclear force with the strong color force, whose field quantum is called the gluon. Only the field quanta of the electromagnetic force and the weak nuclear force have actually been observed experimentally. The field quantum of the strong color force, the gluon, is believed to be too massive to be produced in present particle accelerators. On the other hand, because the graviton is believed to carry so little energy, it will be very difficult to observe. In spite of the fact that only two of the four field quanta of the fundamental forces have been observed, this field quanta description of the forces is almost universally accepted as correct by modern physicists. The forces are all due to the exchange between interacting particles of particular quanta. The theory that describes the interaction between quarks in terms of the exchange of gluons is known as Quantum Chromo-Dynamics, or QCD for short (''chromo'' means color). The basic structure of QCD is modeled after the structure of QED, the very successful theory of the electromagnetic force. (QED is discussed in Chapter 7.) QCD is a ''quantum field theory'' in which the strong color force is described as due to the exchange of color force field quanta, the gluons.

For some time, physicists considered the possibility that there might be additional families of quarks and leptons; however, experiments studying the decay strengths of certain particles appear to set a limit on the number of families at three. In addition, certain considerations of how the universe developed in the first few moments after the Big Bang (see Chapter 6) appear to also limit the number of families to three. Note that most of the ordinary particles of the universe are made up of only the first family of leptons and quarks, and only in very high-energy nuclear reactions do we produce particles made up of quarks from the higher families.

Although we must remember that the leptons and quarks listed in Table 8.3 all have associated antiparticles and that each quark comes in three colors, it must be concluded that if these *are* the fundamental building blocks of nature, then the situation is fairly simple—not a hopelessly long list of seemingly unrelated particles. There is one more consideration we must discuss in order to conclude that these are truly fundamental building blocks. The question is simply, ''Is there any evidence that the leptons or quarks are made up of something even smaller?'' Basically the question can be reduced to whether the particle has ''size'' or not. If a particle has spatial extent (i.e., size), then something must be in there that occupies that space. Whatever this ''something'' is, is then what the particle is ''made of.'' Only if the particle is a point (i.e., has no size) can it be a fundamental particle. (Note, however, that concentrating a finite amount of charge into a point would appear to require an infinite amount of energy, which has caused this question to remain somewhat controversial.)

The electron is known to be extremely small. It *appears* to be a pure ''point'' in space and has no size whatsoever. Although the limit on the size of a quark has not been set as small as for an electron, it also appears to be just a point particle. The uncertainty principle discussed in the preceding chapter can be used to help convince physicists that it is extremely unlikely that anything smaller exists inside an object known to be as small as an electron or quark. The present experimental evidence is consistent with a model of electrons and quarks as structureless particles with essentially geometric ''point'' properties down to dimensions as small as 10^{-18} meters. Thus physicists are becoming increasingly convinced that Table 8.3 is the correct list of the fundamental building blocks of nature (although there do exist speculations regarding structure inside quarks).

What then remains to be understood? First, much work needs to be completed to verify that the leptons and quarks are fundamental. We need to determine for sure that there are only three families. If the numbers of families is limited, why? Finally, physicists would like to gain a better understanding of the fundamental nature of the four basic forces. A single description can now be provided for the electromagnetic and weak nuclear force simultaneously, and some physicists feel that the strong color force will soon fall into this unified description as well. That would leave only the gravitational force to be combined into one overall unified field theory of all the forces (a goal toward which Einstein worked for over 40 years). There have been proposed various forms of such a theory and these are referred to as Grand Unified Theories (GUTs). It appears that we have made some tremendous steps toward a basic understanding of nature in the last few decades, but there remains much more to do.

STUDY QUESTIONS

1. As of about 1935, what did physicists think were the fundamental building blocks of matter?
2. Where is nearly all of the mass of an atom concentrated?
3. How many times more massive is a proton than an electron?
4. What force tends to break nuclei apart?
5. What force holds nuclei together?
6. Why do the largest nuclei have about 100 protons?
7. In what important ways is the nuclear force different from the electromagnetic force?
8. Why did Yukawa propose the existence of a new subatomic particle?
9. What is a virtual particle?
10. According to quantum field theory, what is the exchange particle of the electromagnetic force?
11. According to quantum field theory, why is the nuclear force finite ranged?
12. How does quantum field theory explain the "action-at-a-distance" problem?
13. Name the four main "families" of subatomic particles and their distinguishing characteristics.
14. What is an antiparticle?
15. What happens when an antiparticle and a normal particle collide?
16. What does the conservation of baryon family number mean for a nuclear reaction?
17. Which force obeys the largest number of conservation laws?
18. What did the lack of certain expected nuclear reactions imply?
19. What is a symmetry operation?
20. What is the relationship between conservation laws and symmetry operations?

21. What was the new particle predicted by the classification scheme of Gell-Mann and Ne'eman?
22. What kind of mathematics did Gell-Mann use to deduce that baryons might be made up of quarks?
23. Have individual quarks been "observed"?
24. How many basic kinds of quarks are now required to explain the structure of subatomic particles?
25. What is believed to be the force that holds quarks together?
26. What is the name of the exchange particle for the force between quarks?
27. How many fundamental particles are there now believed to be? Into how many families are these divided?
28. What is required for a particle to be a true "fundamental" particle?
29. What is the goal of a unified field theory?

PROBLEMS

1. The artificially produced radioactive isotope of iodine, ${}_{53}I^{131}$, is often used for treating thyroid problems. This isotope has a half-life of 8 days. How much time must pass before the radioactivity level will fall to $\frac{1}{16}$ of the value of the initial dose?
2. The radioactive isotope of carbon, ${}_{6}C^{14}$, is used for dating of archeological plant specimens (usually wood). This isotope has a half-life of 5700 years. The ${}_{6}C^{14}$ is produced from ${}_{7}N^{14}$ in the atmosphere by cosmic-ray bombardment. The ${}_{6}C^{14}$ is then taken in by the plant during photosynthesis. Once the plant dies, it no longer takes in fresh carbon and the ${}_{6}C^{14}$ begins to decay away. The age of the plant can then be determined by comparing the ratio of ${}_{6}C^{14}$ to ${}_{6}C^{12}$ (the normal isotope of carbon) in the plant to that in the atmosphere. How old is a specimen of wood with a ${}_{6}C^{14}/{}_{6}C^{12}$ ratio $\frac{1}{4}$ of that in the atmosphere?

REFERENCES

B. L. Cohen, *Concepts of Nuclear Physics,* McGraw-Hill, New York, 1971.
An advanced undergraduate text on introductory nuclear physics.

R. Eisenberg and R. Resnick, *Quantum Physics,* 2nd ed. John Wiley & Sons, New York, 1985.
A less advanced undergraduate text with good introductions to the basic ideas of modern physics, including nuclear physics.

H. A. Enge, *Introduction to Nuclear Physics,* Addison-Wesley, Reading, Mass., 1966.
Another well-written, advanced undergraduate text.

G. W. Gamow, *Mr. Tompkins in Paperback,* Cambridge University Press, New York, 1971, Chapters 12–15.
See footnote 11 in Chapter 6.

Particles and Fields, Scientific American reprints, W. H. Freeman, San Francisco, 1980. A series of reprints from *Scientific American* on recent advances in the subject of elementary-particle physics, written for the interested lay public. See also various articles in *Scientific American* written since 1980.

Chapter 9

EPILOGUE

It is natural to wonder what might be some of the next ideas in physics that could "shake the universe." In some ways, where physics is now is similar to where it was one hundred years ago. At that time, many scientists thought that all the major problems of physics had been basically solved. The recent developments in understanding of heat and energy and in electricity and magnetism had been combined with Newtonian mechanics and apparently explained all physical phenomena. Even the recent work demonstrating the existence of molecules could be explained using these developments within the framework of the kinetic-molecular model (see Chapters 3 and 4). There were only a few "minor" problems that existed, and most physicists believed that these could eventually be explained with the known physical laws. As we have seen, these "minor" problems led to the development of relativity and quantum mechanics, and these new ideas completely revolutionized thinking about the physical universe.

Perhaps physicists today are not quite so confident that we now know everything as some were a hundred years ago. (Maybe we have learned from history!) But in many ways, it again seems as if the picture is nearly complete. If we consider both relativity and quantum mechanics, we seem to have the basic physical laws necessary to describe things in the universe from the very large (the whole universe) to the very small (the quarks and mesons inside nucleons). We know that we still need to combine relativity and quantum mechanics into *one* comprehensive theory, but this could be viewed as a ''minor'' extension of our present understanding.

But is all so nice? Do we have any ''minor'' problems lying around now that could lead to completely new ideas in physics? The answer, we think, is—quite possibly.

First of all, we have the general task of bringing together relativity and quantum mechanics. Although this *can* be done for a number of specific situations, it still cannot be done as one elegant (i.e., simple) theory. Physicists have been trying hard to do this now for several decades and still have no nice comprehensive theory. Part of the problem in this regard is that we still do not know exactly what the form is for general relativity. We know certain basic principles that must be in any theory of general relativity, but the exact form is still hotly debated. Because we do not know this form for sure, and because we do not know the effects of combining quantum mechanics with relativity, many conclusions concerning astrophysics are uncertain. These include the true nature of black holes, neutron stars, and the Big Bang.

Next, we have the problem of the fundamental building blocks and basic forces of the universe. Although the present situation is clearly simplified compared to thirty years ago, it is still not so simple. As presented in Chapter 8, we now have six quarks (which come in three ''flavors'' each), plus three leptons with their associated neutrinos, plus three fundamental forces with their various exchange quanta. Most of these particles also have ''antiparticles.'' This picture is referred to as the ''standard model.'' This is still a fairly complex model with a lot of particles with very different sorts of masses and charges, and so on. Many physicists have spent a great deal of time trying to ''understand'' the standard model in terms of a simpler underlying theory. For example, physicists wonder why there are three basic forces in nature. One would like to see that these are all really just different manifestations of one basic force. Much work, and some progress, has been made on this subject, but a simple explanation still eludes us. Similarly, one would like to understand why there are six quarks, and not some other number. Physicists have made models (i.e., theories) to explain the standard model. But there are problems with all these models; for example, most of these models predict that a proton should not be completely stable—it should eventually break up into other particles. Several large experiments have been performed in different parts of the world (the U.S.A., Europe, and Japan) to look for the decay of the proton. So far, none of these experiments has seen any evidence for such decay. These results are generally viewed as a problem for understanding the standard model.

There are other ''minor'' problems in physics today. One such problem is concerned with the number of neutrinos coming out of the Sun. Assuming the standard model and given the nuclear reactions we believe must be occurring at the center of the Sun, one can calculate the number of neutrinos that should be emitted. This result is interesting because nothing else from these reactions can come directly out from the center of the Sun, so that this is our only way to directly look at these reactions. The problem is that we observe only about one-third the number of neutrinos expected. This result means that either we do not really know what nuclear reactions are occurring in the Sun or that something peculiar happens with the neutrinos after they are emitted. Both possibilities are seriously considered. The first possibility is considered to be a problem for nuclear astrophysics. The second possibility would again question the standard model.

Of course there are other minor problems in physics, and the ones presented here may not include the one or ones that eventually turn out to require new ideas that will really shake physics. In addition, besides some of these ''minor'' problems, there are also new areas of physics being developed. The possible impact on physics of these new areas is presently uncertain.

One such new area is the subject of **chaos** in physics. Chaos refers to certain situations where a small change in one of the parameters describing the system can cause a very large change in the system as a whole. Many physical systems actually behave this way. One example is the path of comets. Many comets are known and have well-known elliptical orbits with definite periods; however, if the orbit is perturbed slightly, by passing near a planet for example, the orbit may change drastically. Often such a change will eventually cause the comet to fall into the Sun. Thus a small change in the orbit can produce a drastic change later. A similar situation exists for particle trajectories in large circular accelerators. A small perturbation in one spot on the path can eventually cause the entire beam to disperse and no longer stay within the accelerator beam tube. Other examples can include populations of atoms in excited states that decay. A small change in the decay rate can cause *all* the atoms to collapse to the lower energy state. Many such situations are known and many more surely exist. The mathematics used to describe these ''chaotic'' systems is not that of simple Newtonian mechanics (although the individual particles may each obey Newtonian mechanics). As we understand this mathematics better we will be able to ''design'' such systems to actually change in ways we want. This area of physics is in its infancy.

Another example of a new area of study in physics is so-called **string theory.** String theory is actually a mathematical theory which describes stringlike (i.e., long, narrow, flexible) objects, which exist in higher dimensional states than four-dimensional space–time. Certain rules exist that govern how these strings can combine and separate. This theory has been used to try to describe the standard model of elementary particles and other physical situations. This area appears very promising to some physicists, but so far it has been more descriptive than predictive.

In addition to new ideas, new phenomena are observed. Some of these new phenomena may not be understandable in terms of known physics. For example, new ''high-temperature'' superconducting materials have been found. These materials, in which there is no resistance to electrical current, offer the possibility of making electrical devices that can operate with very little power consumption. Superconducting materials have been known for sometime and were understood in terms of known physics. These ''traditional'' superconductors work only at very low temperatures, near absolute zero. The new materials are superconducting at much higher temperatures (although still quite cold, viz., at about 100 kelvins above absolute zero). It is clear that the superconducting state of the new materials cannot be explained with the same theory that explains the traditional superconductors. If we can understand the new superconductors, we may be able to make materials become superconducting at even higher temperatures, perhaps close to room temperature. Such a breakthrough would revolutionize certain industries.

For a while, it appeared that certain experiments revealed the existence of a so-called "fifth force." ("Fifth" here is obtained by counting gravity, electricity-magnetism, the weak nuclear force, and the strong color forces as the "known" four. We now know that the electric-magnetic force and the weak nuclear force can be combined into one force referred to as the electro-weak force. See Chapter 8.) This fifth force seemed to be of "intermediate" range, that is, it was not infinite range like gravity nor short range like the nuclear force. It appeared to have a range of about 10 to 100 meters. Such a force would not easily fit into the standard model. Although new experiments seem to indicate that this force does not actually exist, some physicists still think it may be possible. The certain discovery of such a force would definitely shake physics.

The conclusion regarding future developments in physics is clear. We do not know what surprises are still awaiting us. Will there be yet another idea that shakes physics? It seems likely—but we do not really know what it will be (or we would have included it in this text). We *do know* that many important concepts remain to be developed. We have described some of these here, including combining relativity and quantum mechanics, understanding the standard model of elementary particles, unifying the four fundamental forces, exploring chaos, understanding nuclear astrophysics (the Sun, stars, neutron stars, black holes, etc.), expanding string theory, understanding high-temperature superconductivity, and so on. Certainly the work done in some of these areas will lead to new capabilities and technologies.

Physics includes the study of the fundamental objects and forces in the universe. Our experience is that as we pursue this study, we continually raise new questions as we answer old ones. Most physicists agree that physics will never answer ultimate questions such as: "What created the universe?" or "What is the purpose of life?" Nevertheless, it is amazing just how much of the physical universe can be understood. The universe does appear to be ordered and to follow certain rules. Clearly, we do not yet understand this orderliness completely and have not yet determined the final form of the rules. It is still a very exciting time for physics.

INDEX

BELOIT COLLEGE LIBRARY
0 20 22 0117424 2